The Analysis of Nuclear Materials and Their Environments

Claude André Degueldre

The Analysis of Nuclear Materials and Their Environments

Springer

Claude André Degueldre
Engineering Department
Lancaster University
Lancaster
UK

and

Department of Analytical and Mineral
Chemistry
University of Geneva
Geneva
Switzerland

and

Nuclear Energy Division
Paul Scherrer Institute
Aargau
Switzerland

ISBN 978-3-319-58004-3 ISBN 978-3-319-58006-7 (eBook)
DOI 10.1007/978-3-319-58006-7

Library of Congress Control Number: 2017939092

Printed on acid-free paper

This Springer imprint is published by Springer Nature
The registered company is Springer International Publishing AG
The registered company address is: Gewerbestrasse 11, 6330 Cham, Switzerland

Foreword

I am very pleased to see this new book on *The Analysis of Nuclear Materials and Their Environments* by Claude Degueldre. The challenge of analyzing nuclear materials is that normal procedures are often difficult to use because of the high-radiation field surrounding the material, say with spent nuclear fuel. At the other end of the spectrum, the challenge is to analyze environmental materials in which concentrations of radionuclide are extremely low. In both cases, it is important to determine the solid-state chemistry of the nuclear material and the speciation of the radionuclide in the environment. This is exactly the type of data that are the foundation of safety analyses required by regulators for nuclear facilities.

Professor Degueldre has written a book that systematically presents and updates the full range of analytical techniques that can be applied to nuclear materials and radionuclides in the environment. His treatment is comprehensive, dividing the techniques into two broad categories—those by which a material may be examined by passive techniques, such as particle scattering, and those that require an interaction of the energy of the probe with the sample, such as those that cause electron excitations. For both cases, the explication of each technique emphasizes an understanding of the interactions and detection methods for the probe energy and solid/liquid interactions that are the basis for the different techniques. By combining the discussion of the techniques into a single volume, the reader can also appreciate the different length scales at which each probe samples the properties of the nuclear material. This allows scientists to match the proper technique to the scientific question that is being asked. I appreciate the emphasis on the detection methods, as detection is often the major challenge of an analytical technique. As the author emphasizes, the way forward with these challenging analytical problems is to combine techniques, taking advantage of the range of particle/wave interactions with the samples. This book is a welcomed introduction to the panoply of approaches that are available to the modern analyst faced with the challenge of analyzing nuclear materials or radionuclides in the environment.

May 2017

Rodney C. Ewing
Frank Stanton Professor in Nuclear Security
Center for International Security and Cooperation
Stanford University

Preface

Nuclear materials and their environments require analyses before and during their utilization as well as after service during disposal not only from the current nuclear units but also from planned or foreseen nuclear installations or systems.

Prior analysis, sampling, and sample treatment must be carried out when the analytical technique is not applied in situ, in a non-invasive way, or in an in-line or on-line mode.

The analysis may be carried out in situ, for example using a remote system, or in an underground laboratory in the phase under consideration. The analysis may also be done ex situ with transfer of the sample and separation when needed. For all analyses, sample volume, mass or amount, the flux of reagent, the size of the analyzed part of the sample and the acquisition time or time of analysis are key parameters that may affect the detection limit.

The information required—such as the chemical or radioisotope activity, the mass or volume of the sampled and analyzed item, the concentration as fraction or molarity of dopants or contaminants and the type or size of structures in the studied phases—has to be determined in a multi-scale approach at the nuclear scale, atomic or molecular scale, at the microscopic or macroscopic structural scale, at the bulk scale, at the component or system scale, and/or at the environmental or geographical scale according to the requirements of the study. Identification concerns the actinides, fission products or activated products as isotopes or elements, but also their speciation that may not only be done at the molecular scale but also in a broader sense such as at the environmental level. The time scale ranges from the femtosecond, accessible during Free Electron Laser investigations to describe ultra-fast phenomena, through the nanosecond to the mega-second, then to the giga-second or penta-second the timescale of uranium-238 half-life or of the age of the fossil natural geo-reactors. The explored energy range along the analytical methods goes also from the nano-eV (Mössbauer or nuclear magnetic resonance spectroscopy) to the giga-eV (muon-tomography), for example.

Passive and active analytical methods have been revisited in this work, with examples of their utilization in transmission, injection, diffusion or reflection modes. The sampling area, beam size and reagent quantities are either macroscopic, microscopic or nanoscopic in size, while spatial-temporal conditions make excitation incidence versus detection directions possible through solid angles, with synchronous detection or with temporal delay.

In this work the investigated analytical techniques have been classified according to their interactions, if any, between incident waves, particles or injected reagents and the analyzed sample, and, for their detected or recorded signals. For passive techniques, excitations are absent and phonons, photons, leptons, neutrons or ions are detected or quantified for their energy, flux, activity, quantity or mass. For interactive techniques, irradiations or reagent additions are made with phonons, photons, leptons, neutrons or ions with a known energy, flux, activity or mass. The irradiation or injection is done locally while the reception may be carried out in a given space at a given angle from the stimuli direction or the incident beam, instantaneously or after a certain delay after irradiation.

The detection tools are spectroscopy, microscopy, radiography, and tomography. The reaction takes place within or without a specific field such as electrical, magnetic, flow or mechanical acceleration. The detected signal may be the same in nature as the incident one, with the same energy, elastic interaction, or a signal with lower energy and inelastic interaction, with particles being again phonons, photons, leptons, neutrons or ions. In addition to these analytical tools or techniques, neutral species such as atoms or molecules may also be used to interrogate the material. They are treated as ions from a mass and charge point of view. The techniques are classified according to increasing energy of reagents or incident particles or waves. The combination of all excitation or reagent addition and product detections makes the analytic potential very rich to perform the identification of molecules, elements or isotopes, their quantitative determination, and their spatial speciation.

There has been an optimization of techniques and the discovery of new analytical tools over the last century. Some of the techniques are found today to be obsolete, others re-emerge due to new interests; some may be completed by combining the potential of one technique with another. In addition, there has been a constant challenge in pushing the use of the analytical techniques toward lower detection limits, better lateral and depth resolutions, more extreme applications and more flexible uses.

As far as the nuclear materials are concerned, studies must reflect the demanding conditions of temperature, pressure and irradiation under which they are used. These materials act as barriers and their properties are investigated with emphasis on mechanical performances, durability, plasticity and stability when damaged or loaded by dopants or contaminants. These materials range from fuels for thermal or fast reactors, to structural materials. Fuels are analyzed prior to and after irradiation, after their reprocessing for recycling and later as waste forms. Macro-properties such as thermodynamic, thermophysical and mechanical as well as microstructural analysis of these materials have to be analyzed, for example comparing again properties prior to and after irradiation.

As far as the environments of nuclear materials are concerned, one has to think about the way the analyst and the environmental scientist would collaborate together to produce data that can be used by modelers or by authorities. The challenge is to understand the behavior of actinide elements, fission products and other contaminants in the environment. Biogeochemical pathways have to be described, quantified and understood. Transport of actinides, fission products and other contaminants in fluids such as air or water include particulate or colloidal phases. These analyses must be integrated in the analytical strategy as specific species for modeling their biogeochemical behavior.

Data are provided by the analysts for the scientists and the modelers. The problem is to understand the behavior of radionuclides in the systems or the material properties with regard to its integrity. In the environment, contaminant pathways have to be described. The contaminated systems interact with the local environment that may modify radionuclide speciation by physical-chemical processes. The analytical results must be integrated in the study for modeling their chemical and physical properties.

A challenge for future investigations will be to find and develop direct analytical probes for full nuclear material characterization at very low defect, dopant or contaminant concentrations to better characterize the damages, species or structures and to predict their behavior in homogeneous, heterogeneous or complex nuclear materials and in their environments.

Aargau, Switzerland Claude André Degueldre

Acknowledgements

The author would like to acknowledge several scientists who participated in discussions, providing material or notes. Specifically thanks are due to

- F. Aiouache, Lancaster University, UK
- R. Bellin, CEA Cadarache, France
- S. Caruso, Nagra, Switzerland
- M. Döbeli, ETHZ, Switzerland
- J. Griffin, Chemistry Department, Lancaster, UK
- M. Hertrich, Nagra, Switzerland
- M. Joyce, Lancaster University, UK
- A. Kerridge, Lancaster University, UK
- K. Patel, University of Cambridge, UK
- D. Shuh, BNL, USA
- S. Stefanovsky, All Bochvar Institut, Moscow, Russia.
- N. Toulhoat, CNRS/IN2P3, IPNL, France
- M. Tylka, ANL, USA
- E. Vance, ANSTO, Australia
- M. Veleva, PSI, Switzerland

The work has been the fruit of analytical practice over the last 45 years, mainly in Algeria, Belgium, France, Germany, Japan, Switzerland, the UK and USA. The work was however self-financed and its elaboration required intensive personal work. The data reported in this textbook are not specifically the data of the author of this work. They are taken from the open literature, and their use requires citation. The author of this work is consequently not responsible for their quality. His liability is restricted to the novel form of presentation of the analyses of nuclear materials and their environment.

Contents

Abbreviations

Analytical techniques ending with: S for spectroscopy, M for AFM microscopy

AAFM	AFM in acoustic mode
AAS	Atomic absorption spectroscopy
ACV	Alternating current voltammetry
AE	Acoustic emission
AES	Auger electron spectroscopy
AFM	Atomic force microscope
AM	Acoustic microscopy
AMS	Accelerator mass spectroscopy
APT	Atom probe tomography
BN	Backhausen noise
CE	Conductivity electrical
CM	Confocal microscopy
CoL	Colorimetry
CouL	Coulometry
DCP	Direct current polarography
DNAA	Delay neutron activation analysis
DPP	Differential pulsed polarography
DPV	Differential pulsed voltammetry
DRS	Diffuse reflection spectroscopy
EAFM	Electrochemical AFM
EC	Electrical conductivity
EDS	Energy dispersive X-ray spectroscopy
EELS	Electron energy loss spectroscopy
EIS	Electrochemical impedance spectroscopy
EMCD	Electron energy-loss magnetic chiral dichroism (under magnetic field)
EMPA	Electron microprobe analysis
EPR	Electron paramagnetic resonance (under electric field, polarization)

ERDA	Elastic recoil detection analysis
ERT	Electrical resistance tomography
ESCA	Electron spectroscopy for chemical analysis
ESIMS	Electrospray ionization mass spectroscopy
ESR	Electron spin resonance spectroscopy (or EPR)
ETOM	Electron tomography
ETV	Electrothermal vaporization
EXAFS	Extended X-ray absorption fine structure spectroscopy
ExEELFS	Extended electron energy loss fine structure
FEL	Free electron laser
FEM	Field electron microscopy
FIB	Focused ion beam (microscopy)
FTIRS	Fourier transform infra red spectroscopy
FTRAD	Fission Track Radiography
GC-MS	Gas chromatography-mass spectrometry
GDMS	Glow discharge mass spectrometry
GRAV	Gravimetry
HIM	Helium ion microscopy
HPLC	High-performance liquid chromatography
HR	Hight resolution
HV	Hight voltage
ICPMS	Inductively coupled plasma mass spectroscopy
ICPOES	Inductively coupled plasma optical emission spectroscopy
INAA	Instrumental neutron activation analysis
IRFT	Infrared Fourier transform spectroscopy
IRM/S	Infrared microscopy/spectroscopy
ISE	Ion selective electrode
LAICPMS	Laser ablation inductively coupled plasma mass spectrometry
LAICPOES	Laser ablation inductively coupled plasma optical emission spectrometry
LAMMA	Laser microprobe mass analysis
LEEM	Low-energy electron microscopy
LIBD	Laser-induced breakdown detection
LIPAS	Laser-induced photoacoustic spectroscopy
LIPDS	Laser-induced photothermal displacement spectroscopy
LSC	Liquid scintillation counting
MALDI	Matrix-assisted laser desorption ionization
MBAS	Mössbauer absorption spectroscopy (under magnetic field B)
MBES	Mössbauer emission spectroscopy (under magnetic field B)
MOAS	Mössbauer absorption spectroscopy (field-free)
MOES	Mössbauer emission spectroscopy (field-free)
MRI	Magnetic resonance imaging
MRTOM	Magnetic resonance tomography

MS	Mass spectroscopy
MuSR	Muon spin resonance spectroscopy
MuTom	Muon tomography
NAA	Neutron activation analysis
ND	Neutron diffraction
NERDA	Neutron elastic recoil detection analysis
NIRS	Near infrared spectroscopy
NMR	Nuclear magnetic resonance (under magnetic field)
NMRI	Nuclear magnetic resonance imaging
NPhot	Neutron photolysis
NPV	Normal pulse voltammetry
NR	Neutron reflectivity
NRA	Nuclear reaction analysis
NS	Neutron spectroscopy
NTom	Neutron tomography
PALS	Positron annihilation life time spectroscopy
PAS/TOM	Positron annihilation spectroscopy/tomography
PCM	Phase contrast microscopy
PCS	Photon correlation spectroscopy
PEEM	Photoelectron emission microscopy (under magnetic field)
PERALS	Photon electron rejecting alpha liquid scintillation
PhAS	Photo-acoustic spectroscopy
PHOS	Phosphometry
PhotA	Photoactivation
PhotN	Neutron photoactivation
PIGE	Proton induced gamma emission
PIXE	Proton (particle) induced X-ray emission
PLAP	Pulsed laser atom probe
Rad	Radiography, auto-radiography
RamS	Raman spectroscopy
RBS	Rutherford back-scattering
RIMS	Resonance-induced mass spectroscopy
RIS	Resonance-induced spectroscopy
RMA	Raman microprobe analysis
RNAA	Radiochemical neutron activation analysis
SAED	Selected area electron diffraction
SANS	Small angle neutron scattering
SAXS	Small angle X-ray scattering
SEAM	Scanning electron acoustic microscopy
SEM	Scanning electron microscopy
SEXAFS	Surface extended X-ray absorption fine structure
SEXAS	Secondary electron X-ray absorption spectroscopy
SIMS	Secondary ion mass spectroscopy

SMuM	Scanning muon microscopy
SNMS	Sputtered neutral mass spectroscopy
SPC	Single particle counting
SR	Seismic reflexion
SPICPMS	Single particle ICPMS
SSMS	Spark source mass spectroscopy
SSNMR	Solid-state nuclear magnetic resonance
STEM	Scanning transmission electron microscopy
STM/S	Scanning tunneling microscopy/spectroscopy
SWV	Square wave voltammetry
TAP	Tomographic atom probe
TEM	Transmission electron microscopy
TFMS	Time-of-flight mass spectroscopy
TIMS	Thermal ionization mass spectroscopy
TMuM	Transmission muon microscopy
ToF-MS	Time-of-flight mass spectrometry
Tom	Tomography
TRLIFS	Time resolved laser-induced fluorescence spectroscopy
UPS	Ultraviolet photoelectron spectroscopy
UVS/M	Ultraviolet spectroscopy/microscopy
VIAXAFS	Variable incidence angle XAFS
VASE	Variable angle spectroscopic ellipsometry
Vis/M	Visible spectroscopy/microscopy or optical microscopy (OM)
Vol	Volumetry
Volt	Voltammetry
WAXS	Wide angle X-ray scattering
WDS/WDX	Wavelength dispersive X-ray spectroscopy
XANES	X-ray absorption near edge spectroscopy
XAS	X-ray absorption spectroscopy
XMCD	X-ray magnetic circular dichroism (under magnetic field)
XPS	X-ray photo-electron spectroscopy
XRF	X-ray fluorescence spectroscopy
XRR	X-ray reflectometry
XS	X-ray conversion spectroscopy
XTom	X-ray tomography
αS	Alpha spectroscopy
βS	Beta spectroscopy
γRad	Gamma radiography
γS	Gamma spectroscopy
γTom	Gamma tomography

Materials

CRUD	Chack River undefined deposit
DMA	Dimethyl acetamide
DMF	Dimethyl formamide
FWHM	Full width half maximum
HEU	High enrich uranium
IMF	Inert matrix fuel
LEU	Low enriched uranium
MCA	Multichannel analyzer
MOX	Mixed oxide fuel
ODS	Oxide dispersion-strengthened alloy
PUREX	Plutonium uranium reextraction
RAFM	Reduced activation ferritic/martensitic material
RG	Reactor grade
SIMFUEL	Simulated fuel
SS	Stainless steel
WG	Weapon grade

System and Organization

AGR	Advanced gas-cooled reactor
BWR	Boiling water reactor
GEN	Generation
HTR	High-temperature reactor
IAEA	International Atomic Energy Agency
IUPAC	International Union of Pure and Applied Chemistry
LMR	Liquid metal reactor
LWR	Light water reactor
MSR	Molten salt reactor
NPT	Non-proliferation treaty
SCWR	Supercritical water reactor
VHTR	Very high-temperature reactor
USDOE	United States Department of Energy

Units

Abbreviation	Prefix	Value
f	femto	10^{-15}
p	pico	10^{-12}
n	nano	10^{-9}
μ	micro	10^{-6}
m	milli	10^{-3}
k	kilo	10^{3}
M	Mega	10^{6}
G	Giga	10^{9}
T	Tera	10^{12}
P	Penta	10^{15}
E	Exa	10^{18}

Constants

According to Codata (2002), values in parentheses are estimated standard deviation in the last significant digit units.

Symbol	Constant	Value	Unit
c	Light speed in vacuum	299792458	$m\ s^{-1}$
e	Elementary charge	$1.60217653(14) \times 10^{-19}$	C
F	Faraday constant	96485.3383(83)	$C\ mol^{-1}$
g	Acceleration of free fall	9.80665	$m\ s^{-2}$
G	Gravitational constant	$6.6742(10 \times 10^{-11})$	$m^3\ kg^{-1}\ s^{-2}$
h	Plank's constant	$6.6260693(11) \times 10^{-34}$	J s
k	Boltzman constant	$1.3806505(24) \times 10^{-23}$	$J\ K^{-1}$
N_{Av}	Avogadro constant	$6.022145(10) \times 10^{23}$	$J\ mol^{-1}\ K^{-1}$
R	Gas constant	8.314972(15)	$J\ K^{-1}$

Parameters

Symbol	Parameters	Unit
A	Activity (radiochemical)	Bq
	Activity (chemical)	–
	Surface area	m^2
	Atomic mass	u

(continued)

(continued)

Symbol	Parameters	Unit
B	Magnetic field	–
C	Concentration	M, mol L^{-1}
	Cell constant	–
D	Diffusion coefficient	–
d	Thickness, depth	m, mm, …
	Lattice parameter	m, nm, pm
E	Energy	J
	Electric potential	V
	Young modulus	–
F	Force	N
H	Height	m
I	Intensity	A
K	Partition coefficient	–
k	Spring constant	–
L	Length, distance	m
l	Length, distance	m
m	Mass	kg
M	Mass	kg
N	Object number	–
n	Breaking index	–
p	Pressure	Pa
P	Pressure	Pa
Q	Charge	C
q	Charge	C
R	Reflectance	–
	Resistance	–
r	Radius	m
S	Surface area	m^2
s	Scattering factor	–
	Spin number	–
T	Temperature	(°C), K
t	Time	s
U	Internal energy	J
v	Speed	m s^{-1}
V	Volume	m^3
W	Wave number	cm^{-1}
X	Element symbol	–
x	Dimension	m
y	Dimension	m
Z	Atomic number	–

(continued)

(continued)

Symbol	Parameters	Unit
	Acoustic impedance	kg m^{-1} s^{-1}
z	Dimension	m
	Charge	C

Greek Symbol	Parameters	Unit
α	Angular deviation	rad
β	Prompt neutron factor	–
γ		–
δ	Phase change	rad
Δ	Splitting magnitude (MoBA)	–
χ	Normalised absorption coefficient (XAS)	–
ε	Electric permittivity	Fm^{-1}
ϕ	Electric field	V m^{-1}
η	Viscosity	kg s^{-1} m^{-1}
	isotopic abundancy	–
φ	Scattering angle	rad
κ	Specific resistance	–
λ	Wave length	m, cm, nm,
	Decay constant	a^{-1}, d^{-1}, s^{-1}
μ	Magnetic moment magneton	–
	Absorption coefficient	m^{-1}, cm^{-1}
	Reduced mass	kg, g
ν	Frequency	s^{-1}, Hz
	Sweep rate	V s^{-1}, s^{-1}
θ	Scattering angle	rad
ρ	Density	kg m^{-3}, g cm^{-3}
σ	Cross-section	m^{-2}, cm^{-2}
	Standard deviation	(parameter unit)
τ	Half-life	a, d, s
	Pulse time	s
ϖ	Larmor frequency	s^{-1}
ω	Angular frequency	rad s^{-1}
ξ	Stoechiometric coefficient	–
ψ	Azimuthal angle	rad
ζ	Scattering angle	rad

Chapter 1
Background: Nuclear Materials and Their Analysis Needs

Since the middle of the last century and even before, Mankind has produced nuclear materials and increased the use of these materials for applications as well as for the conversion and deployment of the nuclear energy. The production of these materials is due to the nuclear power industry and the military fabrication of nuclear weapons. These activities have created anxiety about introduction of radioactive contaminations into the environment. Consequently, environmental systems that contain radionuclides or that are exploited for nuclear energy, or are contaminated by radionuclides, must also be investigated and regulations are required for safety and ecological reasons. The analytical techniques introduced in this work are used, after sampling when required, to identify and characterize the nuclear material or to quantify component or material constituent properties.

1.1 Materials

The general concept and description of a material deals with nuclei, atoms, molecules and phases. The multi-scale approach needed to describe the behavior of material requires however more than this restricted consideration. General information may be found in reviews such as Gottstein (2004), Mittemeijer (2010, 2011), Predojević and Mitchell (2015). Ranging from the picometer (pm) to the meter (m) scale or more and from the femtosecond (fs) to the gigasecond (Gs or ka) or more, the study of the system is very demanding and needs extensive analysis. The analyzed items start with the nucleons: protons and neutrons (see Table 1.1), their interactions in the nucleus and their specific properties such as nucleus spin, nucleus stability. The description of the nucleus begins with the proton p^+ that is

C.A. Degueldre, *The Analysis of Nuclear Materials and Their Environments*,
DOI 10.1007/978-3-319-58006-7_1

Table 1.1 Fundamental data on particles, nuclei, ions or atoms and molecules

Particle	**Symbol**	**Mass (u)**	**Mass (MeV c^{-2})**	**Half life (s or a)**
Antineutrino	$\bar{\nu}$	<2 × 10^{-9}	<2 × 10^{-6}	–
Electron	e	(548.57990946 ± 0.00000022) × 10^{-6}	0.510998928 ± 0.000000011	>4.6 × 10^{26} a
Muon	μ	0.1134289267 ± 0.0000038	105.6583715 ± 0.0035	2.197 × 10^{-6} s
Proton	p or ${}^{1}_{1}H^{+}$	1.00727646681 ± 0.00000000009	938.272046 ± 0.000021	>2.1 × 10^{29} a
Neutron	n or ${}^{1}_{0}n$	1.0086649160 ± 0.0000000004	939.5654133 ± 0.0000058	880.3 ± 1.1 s
Nucleus	**Symbol**	**Mass (u)**	**Mass (MeV)**	**Binding energy/nucl**
Deuteron	${}^{2}_{1}H^{+}$	2.01410178	1875.612928 ± 0.000012	2.22452 ± 0.00020 MeV
Triton	${}^{3}_{1}H^{+}$	3.016049	2808.921112 ± 0.000017	2.8272 MeV
Helium-4	${}^{4}_{2}He^{2+}$	4.001506	3727.379	7.095 MeV
Iron-56	${}^{56}_{26}Fe^{26+}$	55.9349	52110.6448	8.791 MeV
Uranium-235	${}^{235}_{92}U^{92+}$	235.0439	218974.0073	7.591 MeV
Atom → ion	**Reaction**	**Electron shell**	**Ionis. E (eV)**	**Ionis. E (kJ mol^{-1})**
Hydrogen	${}^{1}_{1}H \rightarrow {}^{1}_{1}H^{+} + e^{-}$	1 s	13.60	1312.0
Deuterium	${}^{2}_{1}H \rightarrow {}^{2}_{1}H^{+} + e^{-}$	1 s	15.47	1492.4
Iron	${}^{56}_{26}Fe \rightarrow {}^{56}_{26}Fe^{+} + e^{-}$	1 s	7112.00	686098.8
Uranium	${}^{238}_{92}U \rightarrow {}^{238}_{92}U^{+} + e^{-}$	1 s	115606.00	11152578.8
Molecule	**Molecular**	**Bond length**	**Intramolecular bonding E (eV)**	**Intramolecular bonding E (kJ mol^{+1})**
Dihydrogen	${}^{1}_{1}H_2$: H-H	H–H: 74.1 pm	4.52	−458.9
Water	H–O–H	H^{1+}-O: 94.2 pm	2.82/2	−286/2
Iron monoxide	Fe=O	Fe^{2+}–O: 212 pm	2.68	−271.9
Uranium dioxide	O=U=O	U^{4+}–O: 237 pm	10.4/2	−1060/2
Intermolecular		**Distance**	**Interaction energy (eV)**	**Interaction energy** (kJ mol^{-1})
H-bridging	$H_2O^{|}_{|}$ – HOH	~340 pm @RT	0.24	29
H-bridging	H_3N^{I} – HNH_2	–	0.11	13

Note the relative decreases in Energy interaction when the size of the object increase from nucleon to nucleus, atom to molecule to supramolecular structures
Conversion factors: 1 eV at^{-1} = 23.069 kcal mol^{-1} = 96.485 kJ mol^{-1}, 1 eV = 1.6022 × 10^{-19} J; u for atomic mass unit, at for atom, 1 u* = 931.454 MeV (*u for amu Ref IUPAC). Data from Olive (2014) and handbooks

stable over more than the 2.1×10^{29} a (Olive 2014) period, while the neutron, when isolated, decays with a half life of 880.3 ± 1.1 s according to:

$$n \rightarrow p^{+} + e^{-} + \bar{\nu} \qquad \{1.1\}$$

yielding a proton p^+, an electron e^- or β^- (half-life $\tau > 4.6 \times 10^{26}$ a) and an antineutrino $\bar{\nu}$. As a consequence of the energy conservation the beta spectrum is continuous with emission ranging from 0 to a maximum values: E_{Max}. It is consequently more difficult if not impossible to identify the neutron on the basis of the beta activity spectrum alone. An isotope nucleus denominated ${}^{A}_{Z}X$ with X the element symbol, is characterized by Z the atomic number and A the mass number, which equals the sum of $Z + N$, with N the number of neutrons.

The construction including neutrons and protons yields at the pm scale a nucleus that is not only characterized by an atomic number and a mass number but also by a spin number. The nucleus $\left({}^{A}_{Z}X\right)$ is either stable or unstable according to the proton and neutron numbers and their interaction. The stable nuclei or isotopes (e.g. $\tau > 10^{10}$ a) are ranging from ^{1}H to ^{209}Bi (except Tc and Pm for which all isotopes are radioactive). The stable isotope nuclei define a stability zone in the Z versus N or E (average building energy per nucleon) versus A diagram (see Fig. 1.1a). Isotopes above ^{209}Bi are all unstable. They decay toward stable isotopes of Bi or Pb by β and α decay reactions (the latter being treated in Sect. 3.1.4). The stability domain is depicted in Fig. 1.1 by the binding energy per nucleon as a function of the nucleon number or mass number A. It shows a minimum in the surrounding of iron suggesting that both fission and fusion reactions are possible with heavy or light nuclei respectively.

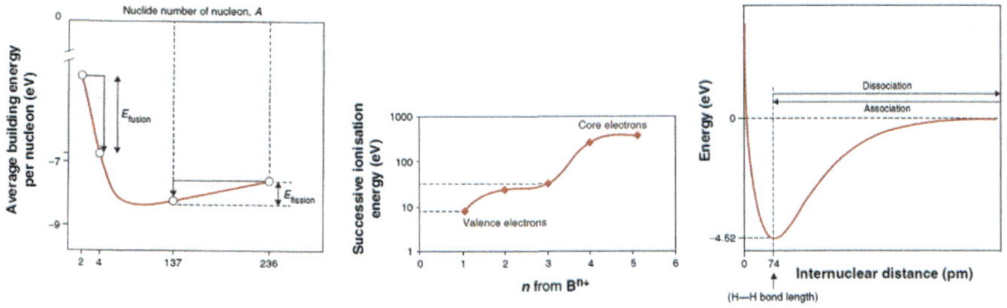

Fig. 1.1 Stability of nuclei (when increasing nucleon number from ^{2}H to ^{238}U), atoms (as ions and electrons when loosing electrons, e.g. for ^{11}B valence vs. core electrons), and, molecules (when splitting, e.g. for $^{1}H_2$). *Note* the energy level ranging from the MeV (for nuclei: fusion −5 MeV per nucleon, fission −1 MeV per nucleon) to the eV (for molecules), $1 \text{ eV} = 1.602 \times 10^{-19}$ J

The decay toward stability is for a nucleus ${}_{Z+1}^{A}X$ driven by a β^{+} (positron: e^{+}) emission such as:

$$ {}_{Z+1}^{A}X \rightarrow {}_{Z}^{A}X + e^{+} + \nu \qquad \{1.2\} $$

It yields the element nucleus ${}_{Z}^{A}X$ and a neutrino ν. For a nucleus ${}_{Z-1}^{A}X$ the decay is driven by a β^{-} (e^{-}) emission following similar reaction with co-emission of an antineutrino $\bar{\nu}$.

The nucleus is either excited or not. When excited, it can decay in energy, emitting one or several gamma photons. Isotopes may be quasi stable when the half life is very much larger that the operational tests e.g. >10^{15} a (^{50}V, ^{100}Mo, ^{113}Cd, ^{144}Nd, ^{148}Sm, ^{174}Hf, ^{180}Ta, ^{186}Os).

To counterbalance the charge of a nucleus, electrons are found in its environment in the 2–200 pm scale forming the ion or the atom of the isotope (or element) either excited or not (with energy difference, see Fig. 1.1). The ions are denominated X^{n+} or X^{n-} according to the charge difference n with the uncharged atom. If these ionic or atomic entities are excited they can lose energy and reach the ground state by emitting one or several X-rays.

The atoms may combine yielding architectural structures that are called molecules (e.g. Fig. 1.1). They are denominated X_x or X_xY_y with $_x$ and $_y$ the stoechiometric coefficients or number of atoms in the molecule. Their formation reaction for binary compound is for example given by:

$$ xX + yY \rightarrow X_xY_y \qquad \{1.3\} $$

Single molecules may be excited or not. When excited they can reach the lower energy state by emitting UV or visible (or near infrared) photon(s).

Molecules may interact together in condensed mater or in solution. They may combine in structures called dimers, trimers or polymers, $(X_xY_y)_p$, with $_p$ the polymerization grade as a mater of intramolecular interaction (polymer generation) or react by intermolecular interactions (colloid aggregate generation) or segregate. In these conditions, intermolecular interactions yield molecule or macromolecule aggregates. Here, again, the energy of interaction is among other functions of the size of the object going from the nm to the µm.

At the level of the material grain the size reaches the µm and the energy deals mainly with vibrations in the system. This is the domain of infrared photons or at the level of the grain of the phonon quanta.

At larger scale the grains assemblies and the multi-phase construction and morphology may reach the mm to the km size (with reaction timescale varying from the s to Ma).

1.2 Nuclear Materials

Nuclear materials are any kind of material that can or did undergo a nuclear process. These nuclear materials are natural and/or man-made (military, civil). Their characterization needs to be performed analyzing their isotopes, elements, species and phases as found in nature or in specific environments. They are also analyzed for properties inherent to their nuclear material nature and utilization. They are found in current natural environment as natural material deposits or repositories as well as in the environment of machines in locations or events that are sketched for actinides (and fission products) in Fig. 1.2. Nuclear materials are inactive: e.g. structural

Fig. 1.2 Natural and man-made (military and civil) actinides with their isotopes (radioisotopes), elements, species and phases, and nuclear materials in natural or anthropogenic environment. Adapted from Degueldre (2006), with permission © Springer. *Note 1* natural major actinides, *2* neptunium and plutonium generation in UO_2 deposits, *3* uranium mining liberates light actinide isotopes, *4* natural actinide contaminations from milling, *5* uranium enrichment and fabrication plants. Anthropogenic actinides as potential contamination; *6* released intentionally below the legal norms from nuclear power plants; or *7* by accident; or *8* reprocessing facilities. Actinides released from military activities, for example during: *9* atmospheric tests; *10* underground detonations; or *11* objects from nuclear naval vessels; or while *12* used as depleted uranium warheads. Waste management activities such as: *13* intermediate; or *14* geologic; and *15* oceanic disposal. Unexpected events with units driven by nuclear power source such as: *16* satellite re-entry in atmosphere; *17* nuclear icebreaker; or *18* nuclear submarine sinking; *19* weapons production; *20* nuclear material research center including a fusion reactor prototype

materials prior irradiation or radioactive such as fuel prior irradiation or after irradiation. Radionuclides numbers N decay with time t according to:

$$N(t) = N(0)e^{-\lambda t} \tag{1.1}$$

with $N(0)$ the initial number of radionuclides and λ the decay constant linked to half life τ by:

$$\lambda = (\ln 2)/\tau \tag{1.2}$$

The natural actinides elements are radioactive. Their isotopes were originally generated in a supernova, condensed from a solar nebula, and incorporated into the condensed material that accreted to form Earth, with Th and U being elements at the 10^{13} ton level in the Earth's crust today as reported by Wasserburg et al. (1964). They are called primordial nuclides formed in the build-up of the terrestrial matter. They are still present today. Naturally occurring uranium contains 99.27% of ^{238}U and only 0.72% of ^{235}U (fissile). They currently produce traces of radioisotopes formed by decay of the major natural actinide isotopes.

1.2.1 Generation 0 Reactors Materials

The specific case of the natural fossil reactors (called "generation 0") in the Oklo region (e.g. Oklo, Oklobonde, Bagombe in Gabon), which underwent spontaneous fission reaction (see reaction {1.4}) and neutron chain reactions some 2×10^9 years ago, was studied extensively and revealed the build-up of large quantities of transuranium elements and fission products during chain reactions (Gauthier-Lafaye et al. 1996). However, most of them have now decayed. Natural plutonium has been produced and has remained in the environment even before man produced it. It remains the heaviest natural element found in the environment at the level of mg per 100 tons of uranium ore residues as reported by Peppard et al. (1951). The only natural fissile isotope is ^{235}U for thermalized neutrons. One of many potential fission reactions writes:

$$^{235}_{92}U + ^{1}_{0}n \rightarrow ^{90}_{37}Rb + ^{143}_{55}Cs + 3\,^{1}_{0}n \qquad \{1.4\}$$

Fission reactions with ^{238}U and ^{232}Th are also possible but they require fast neutrons. Consequently, attention will first be focused on U and Th. The atomic ratio ^{235}U/^{238}U has changed with geological time over the last billion years. Since this ratio used to be around 5–4% 2×10^9 years ago, some U deposits were able to react similarly to today's boiling water reactors, running for several thousand of years. These "generation 0" reactors produced tons of fission products and actinides e.g. Pu, Np and radioactive isotopes that disappeared. They are studied today as analogous to reactors and repository environment. Naturally occurring Np and Pu were present in environmental material at the ultra-trace level generated by nuclear reactions, they may also be residual traces of primordial actinides, or man-made as discussed below.

1.2.2 From Generation I to II Reactor Materials

In the natural environment U and Th deposits are found. Uranium, for example, is exploited from mines where it has been locally concentrated by specific geo-chemical reactions. The ores when exploited are ground and leached producing a concentrated leachate from which uranium is systematically extracted. In a final phase it is transformed in a solid phase (e.g. yellow cake: U_3O_8). The later product is subsequently transformed in UF_6 gas that can be enriched in ^{235}U by diffusion or centrifugation. This enrichment is required for light water reactor (LWR) utilization. After enrichment and transformation in UO_2, the powder is pressed and sintered in the form of pellets that can be used as nuclear fuel and burn in reactor, e.g. LWR. After several years of utilization, irradiated fuel (also called spent fuel containing newly formed fission products and actinides e.g. Pu, Np, Am), may be discharged from the reactor and further reutilized after reprocessing of U and Pu in the form of mixed oxide fuel (MOX), i.e. $(U, Pu)O_2$.

The first experimental fuels were initially tested in the 1940s, 1950s and early 1960s in experimental or pilot reactors (**Generation I**), e.g. Bowker (1960). Following this first generation of reactors, deployment of commercial reactors from **Generation II** started in the 1960s with an apogee in the 1970s and 1980s. Today, after the occurrence of local nuclear reactor accidents, a change towards a safer generation of reactors and fuel is taking place as an evolution rather than a revolution.

Artificial or anthropogenic actinides are those generated by civilian and military activities, see Fig. 1.2. Actinide isotopes that have been produced by Mankind in significant amounts are ^{233}Pa, ^{233}U, ^{236}U, ^{237}Np, ^{238}Pu, ^{239}Pu, ^{240}Pu, ^{241}Pu, ^{242}Pu, ^{241}Am, ^{243}Am, ^{242}Cm, ^{243}Cm and ^{244}Cm as reported by Michel et al. (1999) and Lujaniene et al. (1999) with about 2000 tons Pu produced until now that some groups would like to reuse in a very pragmatic way, burning it using inert matrix fuel such as described by Degueldre and Paratte (1999), while part of it is sued in mixed oxide (MOX) fuel.

Military materials deal mainly with highly enriched uranium (e.g. ^{235}U) and plutonium (e.g. ^{239}Pu) metallic phases used for nuclear bombs. These materials are studied for their composition with additives such as gallium, for their density and for their quality (e.g. affected by traces of other actinides). Recently, attention has also been drawn to depleted uranium and its use in projectiles. Material such as PuO_2 (^{238}Pu) has been used by NASA as fuel for thermo-electric convertors in satellites (see Fig. 1.2).

Examples of nuclear material from electric power utilities during construction, operations and decommissioning are numerous; for example, containments, reactor pressure vessel primary and secondary loop pieces, pipes, pumps, instruments, structural and fuel element materials. At the end of the fuel cycle waste form elements and products are found as low-level (filters, ion exchange resin used for reactor water clean-up), intermediate-level (structural pieces from primary) and high-level waste (spent fuel).

In the ocean, the materials of objects from naval reactor pressure vessels in submarines (Mount et al. 1994) or waste dumping on the seabed (Rastogi and Sjoeblom 1999) are potential sources of contamination. All these cases are schematized in Fig. 1.3 as earlier sketched in Degueldre (2006).

Figure 1.3 locates nuclear materials for traditional light water reactor from Generation II (and III). They include from the periphery of the reactor to the core: the reactor pressure vessel materials (mainly stainless steel), the fuel element materials (mainly zirconium alloys) and in the cladding the fuel material: the uranium dioxide that can be doped with various additives e.g. plutonium dioxide or chromium sesqui-oxide e.g. Degueldre et al. (2011a, b).

During reactor operation, the reactor pressure vessel slightly oxidizes and may undergo slight changes (e.g. hardening) that need to be analyzed; the cladding also oxidizes and may cover by corrosion product deposits, and the fuel undergoes transformation during burn-up due to the build-up of fission products and actinides (trans-uranium elements) e.g. Walker et al. (2005). In addition to the actinides the fuel also produces fission products (see reaction {1.4}) that are analyzed and studied in the frame of fuel performance and safety studies.

The fuel may be classical UO_2 doped with PuO_2 (MOX) with additive such as neutron poison (e.g. Gd from Gd_2O_3). Traditionally, the fuel is analyzed at beginning of life (BOL) or prior to irradiation in the reactor and compared with fuel samples after irradiation or at end of life (EOL) e.g. Degueldre et al. (2011a, b).

Several experimental fuels are also under investigation prior to potential commercial utilization such as Minor Actinide (Np, Am, Cm) Mixed Oxide (MAMOX), e.g. IAEA (2006), Inert Matrix Fuel (IMF) or Accident Tolerant Fuel (ATF). All fuels at Beginning Of Life (BOL) or End Of Life (EOL) can be either real or simulated at the laboratory level (with surrogate elements such as Ce for Pu, and inactive fission product elements) e.g. Lucuta et al. (1991).

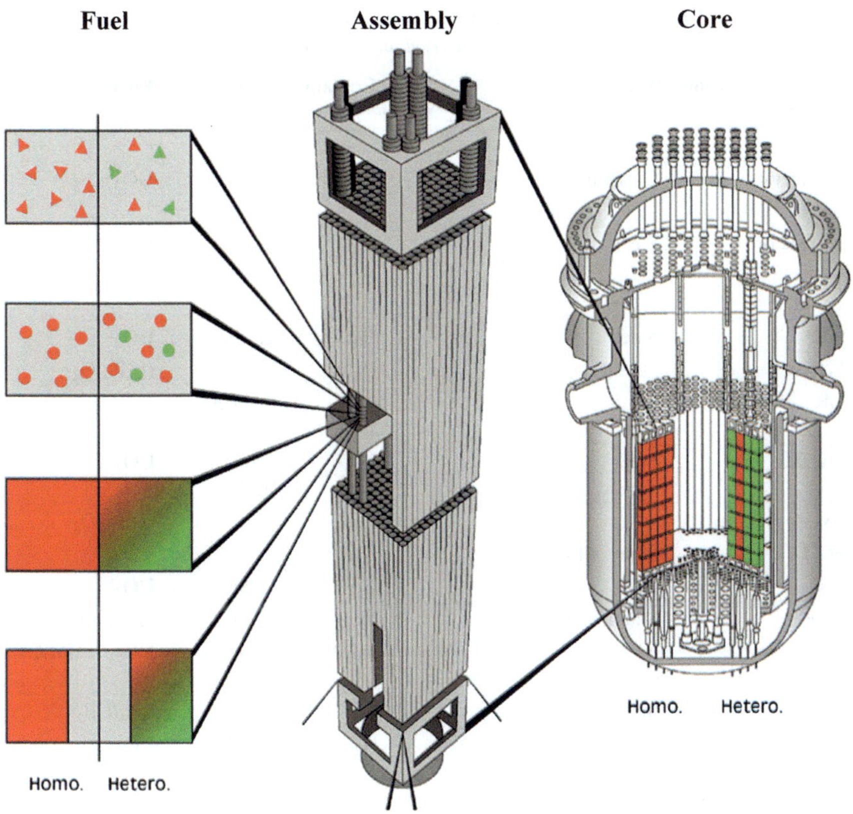

Fig. 1.3 Nuclear materials for Generation II & III reactors UO_2 *green* (PuO_2 and/or ThO_2 *red*), from fuel pellet, to fuel element and to fuel core within the reactor pressure vessel (Degueldre and Paratte 1999, with permission © Elsevier)

1.2.3 Generation III Reactor Materials

The reactors of **Generation III** appears to be the fruit of an evolution rather than a revolution of the Generation II reactors. This evolution is dictated by design changes targeting passive fluid transfers and safety issues. They are more design optimization than new material selection. Risk reduction is the inherent target. Table 1.2 presents a summary of some candidates in term of material for the various generation or type of reactors.

What concerns the nuclear fuel management, the tendency has been to increase the burn-up for economical and ecological reasons, e.g. Degueldre et al. (2011a, b). More recently, fuels with higher safety properties have been identified, and advanced research and development focus on the so called accident tolerant fuel

Table 1.2 Typical nuclear material utilized as vessel, coolant, structural fuel element cladding and fuel classified per generation and type (*abbreviations* see text or abbreviation list)

Reactor generation and type	Vessel	Coolant	Clad	Fuel
Gen. 0				
Geo-reactor	Clay/sand	H_2O	–	UO_{2+x}
Gen. I				
Test reactor	Steel	H_2O	Steel, aluminum	UO_2
Gen. II				
LWR	Steel	H_2O	Zr alloy	UO_2
HWR		D_2O	Zr alloy	$(U,Pu)O_2$
AGR		CO_2	Stainless steel	
Gen. III	Steel	H_2O	Stainless steel	UO_2
EPR, APR				$(U,Pu)O_2$
Gen. IV	Super alloy			
HTR	ODS	He	C, SiC	UO_2
VHTR	Stainless steel	He	C, SiC	$(U,Pu)O_2$
LMR	Ni alloys	Na, Pb	Stainless steel	UO_2, $(U,Pu)O_2$
MSR	Stainless steel	LiF-NaF	–	UF_4/UF_3
HCWR	C/C, SiC/SiC, ODS, ZrO_2 Refractory alloys	H_2O	Stainless steel	UO_2
Gen. V				
Fusion reactor	SiC, ODS, RAFM Steel		Divertor: W, ODS	Plasma ^3H, ^2H

(ATF). These fuels among other display higher thermal conductivity, inertness with the coolant, high melting temperature (e.g. uranium nitride—uranium silicide) and the cladding material display higher resistivity to corrosion such as stain less steel or SiC/SiC.

1.2.4 Generation IV Reactor Materials

The reactors of the next **Generation IV**, that already exist as pilots or prototypes, are

- thermal:
  - the Very High Temperature gas Reactor (VHTR)
  - the Molten Salt Reactor (MSR)
  - the Hypercritical Water Reactor (HCWR)

- or fast:
  - Liquid Metal Reactors (LMR)
  - Sodium Fast Reactor (SFR), lead fast reactor and
  - Fast Gas Reactor (FGR).

The HTR design of the core, assembly (pebble) and element (microsphere) is depicted in Fig. 1.4.

Generation IV reactor vessel structural's for HTR and VHTR may be: the super alloy 9–13% Cr, C/C, SiC/SiC, Oxide Dispersed Strengthened (ODS) steel, ZrO_2 coated refractory alloys, 21/4 Cr -1Mo alloys and graphite. For MSR, nickel is required as being resistant as vessel for (LiF-KF) MSR. The other Generation IV reactors make use of stainless steel.

The fuel elements or assemblies that can be manipulated as group of fuel pieces are made of various materials. In the gas reactor, the cladding material is C and SiC.

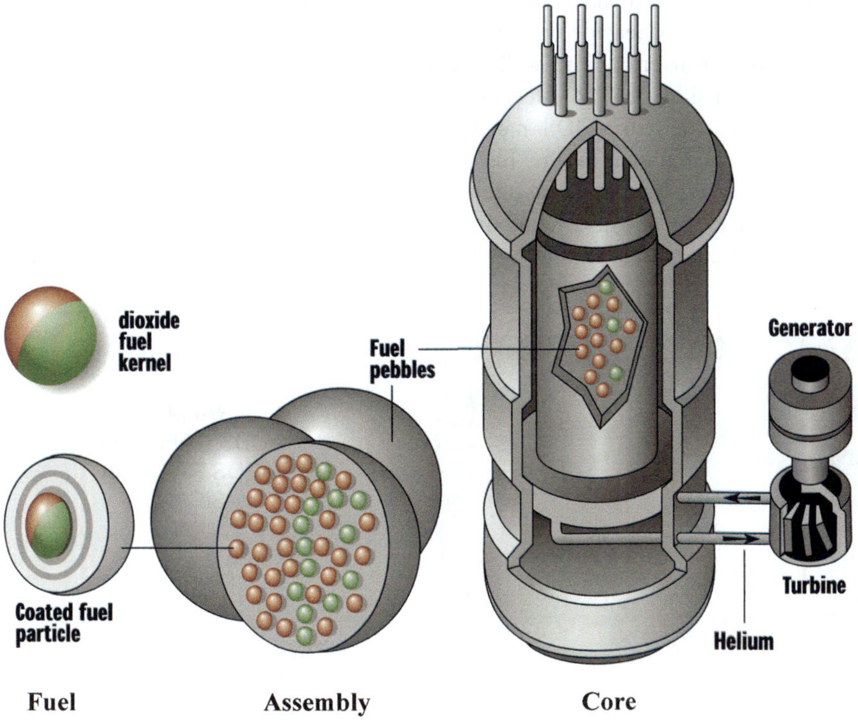

Fig. 1.4 The three levels for fissile material utilization in high temperature reactors (Pebble bed reactor) considering homogeneous vs. heterogeneous systems at the fuel, assembly, and core levels. UO_2 *green* (PuO_2 and/or ThO_2 *red*) (Degueldre 2007, with permission © Elsevier)

Pyrolytic carbon is used in between UO_2 fuel microsphere. The fuel assembly may be spherical (Pebble) or prismatic. After a first layer of carbon around the fuel, a SiC layer improves the retention of fission products. The latter is subsequently covered by a third layer of carbon.

The fuel in the gas reactor is classically uranium dioxide (UO_2) also called UOx. It could also be uranium nitride (UN) or uranium carbide (UC).

1.2.5 Materials from Generation V Reactors and Accelerators

Generation V reactors concern units dedicated to fusion reactions. The Generation V reactor structural materials are those of the plasma vessel wall and the bottom divertor (see Fig. 1.5). To reduce radioactivity induced during operation, low-activation materials have been researched (Zucchetti 1991).

The Generation V reactor structural materials are those of the plasma vessel wall and the bottom divertor. The wall is made of SiC or ODS Reduced Activation Ferritic Martensitic (RAFM) steel in order to resist irradiation damage. For the divertor, a tungsten-yttria ODS–material may be suggested.

The materials found in the fusion reactor environment are low activation materials. The most intensive research concerns the use of C, Si, Ti, Fe, Cr, V and Nb as elements of alloys or compounds ceramics or composites. The alloys are ferritic martensitic steel, Ti alloys or V-Cr–Ti alloys with or without oxide coating

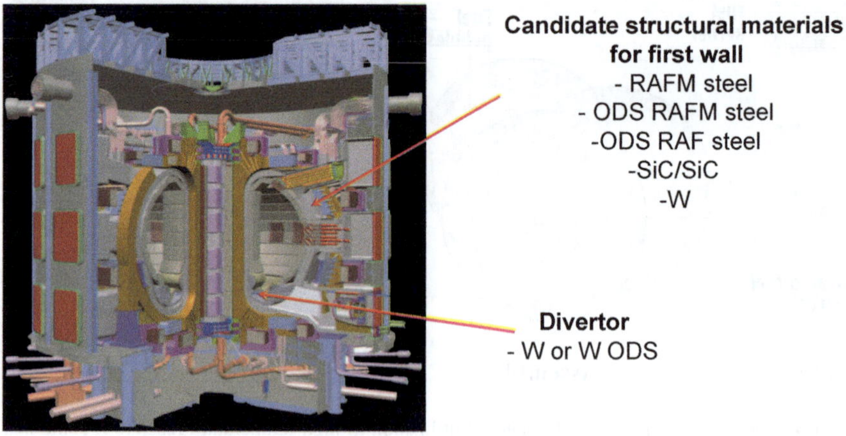

Fig. 1.5 Nuclear structural material for Generation V units, with permission © EUROfusion

e.g. CaO or ZrO_2. Ceramics are for example fiber reinforced SiC - SiC. In these materials the formation of He nano-phase is a key issue e.g. Dai et al. (2012).

Material from the **accelerator** may also be considered as nuclear materials. These are materials from the accelerator ion source, from the vessel wall and from the target. Ion source and vessel materials may be similar to GenV materials while target materials are specific to the experiment or application. The impact of particles on the material during irradiation may be measured as well as evaluated by calculation (e.g. model SRIM). The material is then affected by damage that can be quantified in a number of displacements per atom (dpa).

In operating particle accelerators, charged particles diverging from the beam may interact with accelerator structures and magnetic fields devices activating their components. Magnetic coils are made of metal copper or superconductors, for example Nb-Ti. Beam loss criteria for ion accelerator deal with various phenomena including particle density and scattering, homogeneity of the fields, stability of the geometrical conditions and natural scattering of the particles. Activation of the structural components is due to the nuclear reactions between the incident particles and the elements of the structures. In medical linear accelerators, generated activation products induce β–γ-activities. Material properties such as nuclear reaction cross-sections are required to predict the kind and the quantity of activated products:

- for Al, Fe, Ni, Cu, and Au the activation products cooling may last 1 year;
- for Pb, Al, Cu, W it can be 1000 years.

Target interaction with beam yields the largest activation rate. Divergent particles from the beam may cause activation and remanent ambient activity inside accelerator tunnels and structural elements. Large quantities of activated material usually arise when a whole accelerator part some of them remain active during decommissioning.

Finally, some nuclear designs may contain a radioisotope as energy source for a thermoelectric convertor. These generators make use of ^{238}Pu to provide a long-lasting source of electrical power for the electronics in devices such as satellites.

1.2.6 Military Nuclear Materials

The initial development of nuclear technology was **military**, during the Second World War. Emphasis was placed on research and development of military technology related to U and Pu metal.

After the Second World War, nuclear warheads were produced and tested in various countries.

However, what are military or non-military materials? Materials categorized as military or non-civilian are diverse and can be found in different forms, at different facilities, and for different uses. Some are found at the crossroads between military and non-military.

Weapons-grade uranium is highly enriched, to over 90% ^{235}U. The only current military uses for 80% ^{235}U are found as warheads and require an estimated 15–20 kg of material per warhead. High-Enriched Uranium (HEU) was produced mainly for military use for nuclear weapons. HEU has also been produced for research reactors. Weapon-grade plutonium is also highly enriched with over 93% ^{239}Pu, e.g. Murray and Holbert (2015).

Thermonuclear weapons were also produced in the 1950s. They make use of nuclides interacting by fusion reactions such as:

$$ {}_{1}^{2}H + {}_{1}^{3}H \rightarrow {}_{2}^{4}He + {}_{0}^{1}n \qquad \{1.5\} $$

This arsenal constitutes a potential danger for the world. Accidentally, nuclear missiles have also been dropped and lost by aircraft, contaminating the local environment e.g. Chamizo et al. (2010). Decontamination programs were scheduled when needed or possible.

As a consequence, safeguard programs were designed to prevent diversion of nuclear material. This was successfully performed by increasing control and analysis of the nuclear materials being actually or potentially used for any nuclear military activities.

Nuclear materials such as uranium and plutonium can be used for both peaceful and military purposes. **Nuclear safeguards issues** were established to guarantee peaceful use of nuclear materials.

The Non-Proliferation Treaty (NPT) signed in 1968 was a major step in the regulation of the nuclear weapon stockpile, e.g. Pellechi (2012). Under the NPT, non-nuclear weapon states were prohibited from possessing, manufacturing or acquiring nuclear weapons or other nuclear explosive devices. All signatories, including nuclear weapon states, were committed to the goal of total nuclear disarmament. Nuclear **materials** declared surplus to **military** requirements (e.g. by the USA and Russia) should be utilized in the frame of peaceful programs.

It may be believed that military and other non-civilian materials are under protection, and it is assumed that such materials are better protected than are those in civilian programs. Consequently, the International Atomic Energy Agency (IAEA) and other agencies have developed control and analysis programs to make an, inventory of all fissile and sensitive materials.

The Comprehensive Nuclear-Test-Ban Treaty bans all nuclear explosions in any environments. In this frame, tools such as muon and neutrino tomography can help to track the spread of nuclear material before it is armed into a weapon.

Several types of nuclear missile warheads may be listed. They include:

- Nuclear explosive: an explosive charge is used to disintegrate the target, and damage surrounding areas with a blast wave or a thermal wave. Nuclear explosives are:
  - the nuclear bomb based on nuclear fission reactions of ^{235}U or ^{239}Pu metal, or
  - the thermonuclear weapon based on nuclear fusion reaction with ^{3}H.
- Shaped charge: the effect of an explosive charge is focused onto a specially shaped heavy metal (U) liner to project a hypervelocity jet of metal, to perforate heavy shielding. Depleted uranium, because of its high density, when used to tip bullets, displays such tremendous capacity to open holes in thick iron plates and concrete.

Nuclear materials declared surplus to military requirements by the USA and Russia are being converted into fuel for commercial nuclear reactors. The main material is highly enriched uranium (HEU), containing at least 20% ^{235}U and usually about 90% ^{235}U. HEU can be blended down with uranium containing low levels of ^{235}U to produce low-enriched uranium (LEU), typically less than 5% ^{235}U, fuel for power reactors. It is blended with depleted uranium (mostly ^{238}U) or partially enriched uranium. It could also be used in an inert matrix as a fuel, e.g. Degueldre and Paratte (1999).

World stockpiles of weapon-grade plutonium are reported to be some 260 tons, e.g. Murray and Holbert (2015). For disposal, it could be used blended in depleted uranium oxide in the form of mixed oxide (MOX) fuel to be burnt in LWR. If used in MOX fuel in conventional reactors it would be equivalent to burning a little over one year's world uranium production. After LEU or MOX is burned in power reactors, the spent fuel is not suitable for weapons manufacture. The option of burning Pu excess within an IMF strategy has also been studied (see Fig. 1.6).

It is in military weapons that depleted uranium is used on an extremely large scale, and it is used mainly as penetration body; that is, attached to bombshells for the sake of increasing its penetration capacity, and also as armor for tanks in order to increase its defense capacity.

Waste from **nuclear weapons decommissioning** usually does not contain much β or γ activity other than ^{3}H and ^{241}Am. It is more likely to contain alpha-emitting actinides such as ^{239}Pu which is a fissile material used in bombs.

The β decay of ^{241}Pu forms ^{241}Am. Americium is a gamma emitter (increasing external-exposure to workers) and is an alpha emitter which can cause the generation of heat. The plutonium could be separated from the americium by specific separation; this includes a pyro-chemical process or an aqueous/organic solvent extraction. A truncated PUREX type extraction process could be one possible method of making the separation, e.g. Herbst et al. (2011).

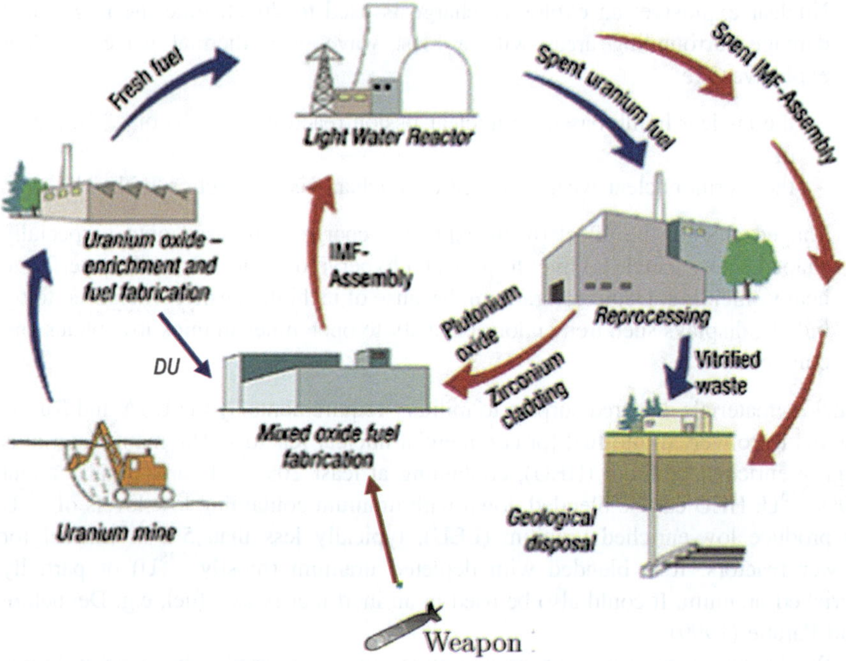

Fig. 1.6 Fuel cycle including once through, and recycling of reactor grade Pu (or use of weapon grade Pu) with inert matrix fuel in LWR: approach at the crossroads between civil and military Pu utilization; DU depleted uranium. Adapted from IAEA-TECDOC-1516 (2006)

Contaminations may be due to operations including nuclear material. As a consequence of an accident involving an aircraft carrying nuclear weapons as reported by Montero and Sánchez (2001) and Mitchell et al. (1995), or during nuclear bomb testing (Kudo 2001) activated or contaminated phases are investigated for their activities and retention properties.

1.2.7 Nuclear Waste Management

Radioactive wastes are usually by-products of nuclear power generation and other applications of nuclear technology, such as research and medicine as well as in the production of nuclear weapons. Radioactive waste is hazardous to most forms of life and the environment, and is consequently regulated by government agencies in order to protect human health and the environment.

Radioactivity naturally decays over time, reducing in principle the radiotoxicity. Radioactive waste has to be isolated in appropriate disposal facilities and confined in specific waste form material for a sufficient period. The time radioactive waste must be stored for depends on the type of waste, activity and radioactive isotopes. It can range from a few days for very short-lived isotopes to millions of years. Current major approaches to managing radioactive waste have been segregation and storage for short-lived waste, near-surface disposal for low and some intermediate-level waste, and deep burial for the high-level waste. The partitioning and transmutation for the high-level waste is also an option with recycling of actinides in the form of advanced fuel and reutilization in commercial or dedicated reactors, e.g. Stanculescu (2006).

Radioactive waste comes from a number of sources. The majority of waste originates from the nuclear fuel cycle and nuclear weapons reprocessing. Other sources include medical and industrial wastes, as well as naturally occurring radioactive materials that can be concentrated as a result of the processing or consumption of coal, oil and gas, and some minerals.

Treatment and conditioning processes are used to convert radioactive waste materials into a form that is suitable for its subsequent management, such as transportation, storage and final disposal, e.g. Ojovan and Lee (2014). The principal aims are to:

- Minimize the volume of waste requiring management via treatment processes.
- Reduce the potential hazard of the waste by conditioning it into a stable solid form that immobilizes it and provides containment to ensure that the waste can be safely handled during transportation, storage and final disposal.

Conditioning processes such as cementation and vitrification are used to convert waste into a stable solid form that is insoluble and will prevent dispersion to the surrounding environment. A systematic approach incorporates:

- Identifying a suitable matrix material (such as cement, bitumen, polymers or borosilicate glass) that will ensure stability of the radioactive materials for the period necessary. The type of waste being conditioned determines the choice of matrix material and packaging.
- Immobilizing the waste through mixing with the matrix material.
- Packaging the immobilized waste in, for example, metallic drums, metallic or concrete boxes or containers, copper canisters.
- Near-field and far-field selection

The concept is depicted in Fig. 1.7. The choice of process(es) used is dependent on the level of activity and the type (classification) of waste. Each country's nuclear waste management policy and its national regulations also influence the selected strategy.

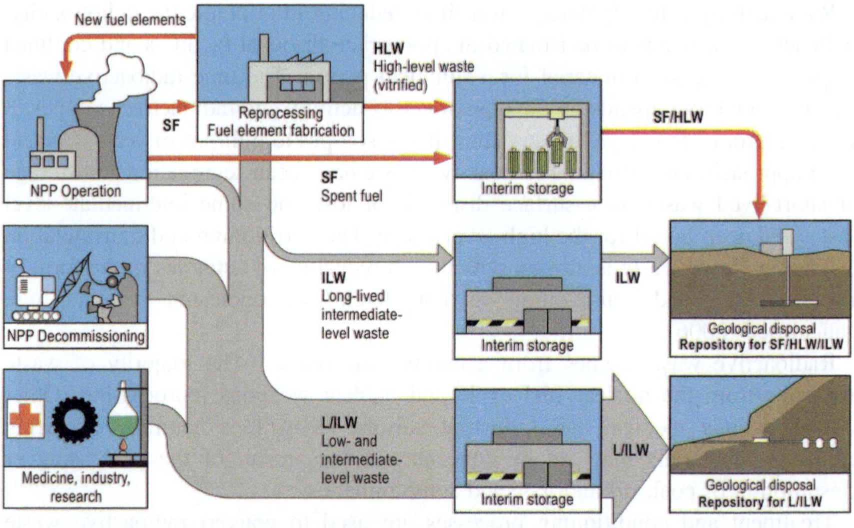

Fig. 1.7 Swiss disposal concept on material disposal, waste form production, interim storage and final repository. Courtesy © Nagra, Switzerland, with permission

1.2.8 Contaminated Materials

Finally, contaminated materials have also to be addressed since their status of anonymous material becomes labeled after contamination. Radioactive contamination is per se the deposition or presence of radioactive substances on surfaces or within solids, liquids or gases, e.g. Chartin et al. (2013). Experimental and modeling investigations of the repository surrounding system are required at the multi-scale level to understand in a comprehensive way contamination behavior through the environment of the nuclear materials, see Fig. 1.8. Contaminations can occur during operation or accidentally at any step of the nuclear fuel cycle, e.g.

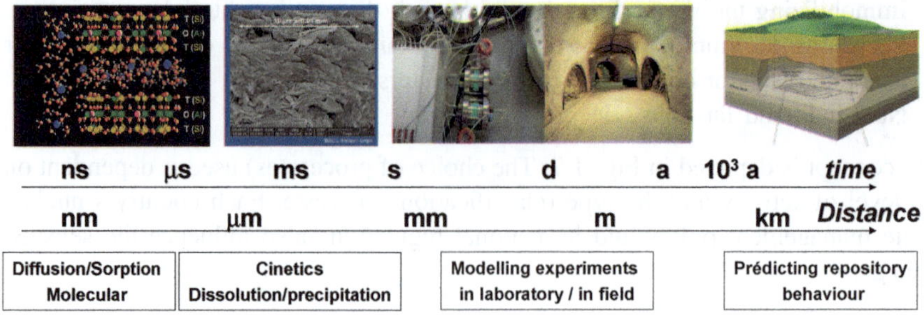

Fig. 1.8 Investigating the repository system at the multi-scale level in a comprehensive way to understand contamination behavior through the environment of nuclear materials. With permission from LES, PSI, Switzerland, adapted from: Churakov et al. (2014)

Déjeant et al. (2016), or during decommissioning of any nuclear installation. Disposals are to be taken by industrial groups and authorities in any cases based on the analysis of the situation.

The analyses of these military and civil nuclear materials target isotopes, elements, species and phases are required in their specific environments. These analyses are needed for many reasons inherent to their production, utilization and management before and after utilization. These analyses are the subject of quality control, licensing, as well as for fissile non-proliferation aspects analysis for safeguard. The analysis of these materials is the core of the topic of this work.

1.3 Analysis Needs and Limitations

On the basis of the previous considerations, it is evident that nuclear materials have to be analyzed. A **need for analysis assessment** is required to plan effectively the analysis of nuclear materials, namely: the analysis manager has to identify priorities, make decisions and solve problems. Its practice affects every aspect of analytical campaigns. This also requires aspects which need to be studied or may be gained by training and the practice of the analysis.

1.3.1 Required Analyses and Analysis Requirement

Definition of the analyzed space, phase and system of the nuclear material must be specified.

Analysis can be performed after responding to some questions:

- What is it about a material that has to be analyzed and what is the analysis required?
- Why is this analysis performed?
- When has the analysis to be carried out (e.g. for fluid sample, or prior/during/after irradiation)?
- Where has the analysis to be done (on or into the sample, in the periphery …)?
- How will the analysis be performed (analytical mode)?

All the questions need be addressed and answered prior to starting the analysis of the nuclear materials. Analysis may concern isotopes, elements and/or chemical species to contribute to the description of the nuclear material, or specific properties, e.g. mechanical, thermophysical, to complete the characterization of the material. The interdependence of the properties has to be respected and cross-checked to contribute to the assessment of the data. However, the analysis performances (detection limit, sensitivity, interference free, lateral resolution) have to be known prior planning the analytical test or campaign.

Table 1.3 System and components (isotopes, elements, and species) basic properties

Property parameter	Symbol	Unit (SI) or (cgs)	Unit
Absolute temperature	T	K	Kelvin
Total pressure	P	Pa	Pascal
Size	$x\ y\ z$	m	Meter
Geometrical volume	V	m^3	Cubic meter
Mass	M	kg	Kilogram
Time	t	s	Second
Component	i	–	–
Partial volume of i	V_i	m^3	Cubic meter
Partial pressure of i	p_i	Pa	Pascal
Concentration of i	C_i	kg m^{-3}, (mol m^{-3})	Mole per cubic meter
Molar fraction of i	X_i	mol mol^{-1}	–
Activity of i	A_i	Bq	Becquerel

Although performing the analysis and recording data are a challenge, a bigger challenge is collecting and analyzing the information effectively and acting on it to upgrade the comprehension of the material properties. Some of the basic properties are given in Table 1.3.

In addition, physical chemical properties are also measured to complete the description of the system. They are thermodynamic: specific mass, heat capacity, enthalpy, entropy, free energy; or physical: elasticity, hardness or thermal conductivity; or, other properties e.g. phase normalized size distribution ($nm^{-1}\ L^{-1}$, $m^{-1}\ m^{-3}$), cumulative size distribution (L^{-1}, m^{-3}), sample activity (Bq) and specific activity (Bq m^{-3}).

1.3.2 Analytical Limitations

Limitations are first dictated by the smallest quantity that can be detected and determined, and by the smallest spot size or sample that can be analyzed (see Table 1.4). Analytical limits may be dictated by:

Table 1.4 Key information provided for the investigated analytical techniques

Technique	Acronym	Detection Limit (unit)	Depth Resolution (unit)	Lateral Resolution (unit)	Provided information

- The beam or sensor size Vs
- The sample or the subsample size.

However, the most important limitation is the detection limit, which, as its name connotes, defines the smallest concentration that can be detected by the used analytical technique.

From the experimental side these concentration and amount limits are given by:

$$C_{\mathrm{DL}}(i) = 3 \cdot \sigma_{\mathrm{BK}} \cdot \kappa(i) \tag{1.3}$$

where $C_{\mathrm{DL}}(i)$ is the concentration at the detection limit (e.g. mol L^{-1}) and σ_{BK}(au) is the standard deviation of the limiting noise signal; the factor 3 is suggested in the definition of detection limit (while 10 corresponds to the quantification limit) by IUPAC, e.g. Curie (1999) see Fig. 1.9, and $\kappa(i)$ the sensitivity of the analysis. The smaller amount of detected component $N_{\mathrm{DL}}(i)$ (mol) is given by:

$$N_{\mathrm{DL}}(i) = V_{\min} \cdot C_{\mathrm{DL}}(i) \tag{1.4}$$

where $V_{\min}$ (cm^3) the minimum volume or 1 mL (arbitrary).

In consequence, the minimum detectable mass, $m_{DL}(i)$ of e.g. an element in a matrix is dependent on the concentration limit $C_{DL}(i)$ (e.g. g g^{-1}) and the minimal analyzable mass of the sample, $m_{\min}$, (g) which just still yields a usable signal.

$$m_{\mathrm{DL}}(i) = m_{\min} \cdot C_{\mathrm{DL}}(i) \tag{1.5}$$

From the theoretical side, a detection limit may be evaluated from the physical-chemical process (background recorded for a given sample) and from the performance of the analytical unit.

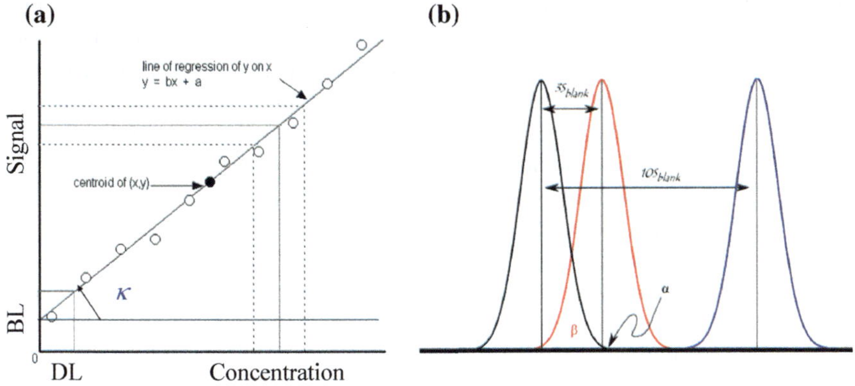

Fig. 1.9 **a** Detection Limit (DL) $3\sigma_{\mathrm{BK}}$ and Quantification Limit (QL) $10\sigma_{\mathrm{BK}}$ in concentration and **b** analysis of both limits in term of Gaussian distribution. *Note BL* blank; depth or lateral resolution depth resolution can be derived from this limitation concept

Two subsequent concepts can be grafted to the detection limits: these are the depth resolution and the lateral resolution. These concepts may be defined the same way as the detection limits, however both are expressed size/dimension unit.

Information required—such as activity (chemical, radioisotopic), amount (mass, number of mole), concentration (fraction, molarity) and structures of the radionuclides in the studied phases—has to be determined at the nuclear (pm), atomic, molecular (nm), microscopic (μm), macroscopic structural (mm), bulk scale (cm), at the component or system scale (m), or at environmental or geographic scale (km) according to the requirements of the study. Identification concerns the radionuclide elements and isotopes, but speciation may not only be understood at the molecular scale but also in a broader sense such as at the environmental scale. Its understanding in the macroscopic scale by plain washing, leaching or extraction tests would be a step for remediation investigations in the case of contaminated material or land.

Although recording accurate **data** artifact-free is a challenge, a bigger challenge is analyzing the data and the information effectively. The question is: What does the data **assessment** mean? The **assessment** of the analytical results **needs to be performed and the data need** to be **analyzed** to learn whether results make sense or not. The criteria on the learning outcomes emerge from a methodological analysis of the analysis results. This is not a problem of precision only but a control of the accuracy of the data and the analytical method. Biases are not only the problem of sample contamination but also analytical signal contamination by artifacts such as extra signals due to physical phenomena. The mode of analysis treated in Sect. 2.1.2 may partially solve the contamination problem.

1.3.3 Analysis Modes and Combinations

Passive and interactive analytical methods will be reviewed (Table 3.1) through Sects. 3.1, 3.2 and 3.3 with examples of their utilization in transmission, injection, diffusion or reflective modes. The sampling area, beam size and reagent quantities are either macroscopic, microscopic or nanoscopic in nature, while spatial-temporal conditions makes excitation vs. detection direction through solid angle, with synchronous detection or with temporal delay. The effect of field on the analysis is treated in Sect. 3.3. In Chap. 4, combinations of techniques are discussed. For example, Seismic Reflection (SR), which cannot be used by itself for locating thorium or uranium deposits, can be used in combination with other techniques as a prospecting tool. Atom force microscopy (AFM) morphological studies also provide useful information; however, they must be complemented with other technique results to provide the required identification result, e.g. (Walther 2003). Similarly, *Eh* electrode (EHE) measurements may contribute to the speciation of redox sensitive actinides such as U, Np or Pu in waters. They are however generally completed by spectroscopic investigations. Chromatography, which is basically a separation technique, must be combined with detectors and is also treated in Sect. 2.2.1.

Separation of elements of interest, which are later analyzed by different analytical techniques, is an important prerequisite of any analytical method as discussed in Chap. 2. The analytical procedure typically includes sampling or sample preparation (e.g. decomposition), separation and/or enrichment prior to analysis in either a passive (Sect. 3.1) or interactive way without (Sect. 3.2) or within (Sect. 3.3) a fixed field.

References

Bowker, H. W. (1960). The Shippingport pressured water reactor. *Journal of Nuclear Energy A, 13,* 38–39.

Chamizo, E., Jiménez-Ramos, M. C., Enamorado, S. M., García-León, M., García-Tenorio, R., Mas, J. L., et al. (2010). Characterisation of the plutonium isotopic composition of a sediment core from Palomares, Spain, by low-energy AMS and alpha-spectrometry. *Nuclear Instruments and Methods in Physics Research Section B: Beam Interactions with Materials and Atoms, 268*, 1273–1276.

Chartin, C., Evrard, O., Onda, Y., Patin, J., Lefèvre, I., Ottlé, C., et al. (2013). Tracking the early dispersion of contaminated sediment along rivers draining the Fukushima radioactive pollution plume. *Anthropocene, 1,* 23–34.

Churakov, S. V., Gimmi, Th, Unruh, T., Van Loon, L. R., & Jurany, F. (2014). Resolving diffusion in clay minerals at different time scales: Combination of experimental and modeling approaches. *Applied Clay Science, 96,* 36–44.

Curie, L. A. (1999). Detection and quantification limits: Origins and historical overview. *Analytica Chimica Acta, 391,* 127–134.

Dai, Y., Odette, G. R., & Yamamoto, T. (2012). The effect of helium in irradiated structural alloys. *Comprehensive Nuclear Materials, 1,* 141–193.

Degueldre, C. (2006). Identification and speciation of actinides in the environment. In *The chemistry of the actinide and transactinide elements, chapter 27*. Berlin: Springer.

Degueldre, C. (2007). Zirconia inert matrix for plutonium utilisation and minor actinide disposition in reactors. *Journal of Alloys and Compounds, 444–445,* 36–41.

Degueldre, C., & Paratte, J. M. (1999). Concepts for an inert matrix fuel, an overview. *Journal of Nuclear Materials, 273,* 1–6.

Degueldre, C., Martin, M., Kuri, G., Grolimund, D., & Borca, C. (2011a). Plutonium—uranium mixed oxide characterization by coupling micro-X-ray diffraction and absorption investigations. *Journal of Nuclear Materials, 416,* 142–150.

Degueldre, C., Bertsch, J., Kuri, G., & Martin, M. (2011b). Nuclear fuel in generation II and III reactors: Research issues related to high burn-up. *Energy & Environmental Science, 4,* 1651–1661.

Déjeant, A., Galoisy, L., Roy, R., Calas, G., Boekhout, F., Phrommavanh, V., et al. (2016). Evolution of uranium distribution and speciation in mill tailings, COMINAK Mine, Niger. *Science of the Total Environment, 545–546,* 340–352.

Gauthier-Lafaye, F., Holliger, P., & Blanc, P.-L. (1996). Natural fission reactors in the Franceville Basin, Gabon: A review of the conditions and results of a "critical event" in a geologic system. *Geochimica et Cosmochimica Acta, 60,* 4831–4852.

Gottstein, G. (2004). *Physical foundations of materials science*. Berlin: Springer.

Herbst, R. S., Baron, P., & Nilsson, M. (2011). Standard and advanced separation: PUREX processes for nuclear fuel reprocessing. In *Standard and advanced separation techniques for nuclear fuel reprocessing and radioactive waste treatment, chapter 6*, 141–175.

IAEA WG chaired by Degueldre, C. (2006). *Viability of inert matrix fuel in reducing plutonium amounts in reactors* (IAEA-TECDOC-1516).

Kudo, A. (2001). *Plutonium in the environment*. Amsterdam: Elsevier.

Lucuta, P. G., Verrall, R. A., Matzke, H., & Palmer, B. J. (1991). Microstructural features of SIMFUEL—simulated high-burnup UO_2-based nuclear fuel. *Journal of Nuclear Materials, 178*, 48–60.

Lujaniene, G., Lujanas, V., Jankunaite, D., Ogorodnikov, B., Mastauskas, A., & Ladygiene, R. (1999). Speciation of radionuclides of the. Chernobyl origin in aerosol and soil samples. *Czechoslovak Journal of Physics, 49*, 107–114.

Michel, H., Gasparro, J., Barci-Funel, G., Dalmasso, J., Ardisson, G., & Sharovarov, G. (1999). Radioanalytical determination of actinides and fission products in Belarussoils. *Talanta, 48*, 821–825.

Mitchell, P. I., Downes, A. B., Condren, O. M., Vintró, L. L., & Sánchez-Cabeza, J. A. (1995). Recent observations on the physico-chemical speciation of plutonium in the Irish sea and the western Mediterranean. *Applied Radiation and Isotopes, 46*, 1175–1190.

Mittemeijer, E. J. (2010). *Electronic structure of the atom; the periodic table, fundamentals of materials science* (pp. 9–36). Berlin: Springer.

Mittemeijer, E. J. (2011). *The microstructure-property relationship using metals as model systems, fundamentals of materials science*. Berlin: Springer.

Montero, R. P., & Sánchez, M. A. (2001). Plutonium contamination from accidental release or simply fallout: study of soils at Palomares (Spain). *Journal of Environmental Radioactivity, 55*, 157–165.

Mount, M. E., Sheaffer, M. K. & Abbott, D. T. (1994). Kara Sea radionuclide inventory from naval reactor disposal. *Journal of Environmental, 25*, 1–19.

Murray, R. L., & Holbert, K. E. (2015). Nuclear weapons, nuclear energy chapter 27 (pp. 495–513).

Ojovan, M. I., Lee, & W. E. (2014). *Principle of nuclear waste management, an introduction to nuclear waste immobilisation* (2nd ed., pp. 75–97) (Chapter 7).

Olive, K. A. (2014). Particle data group, the review of particle physics. *Chinese Physics, 38*, 090001 (update).

Pellechi, M. S. (2012). Non-proliferation safeguards in nuclear power programmes.In *Infrastructure and methodologies for the justification of nuclear power programmes* (pp. 421–464) (Chapter 13).

Peppard, D. F., Studier, M. H., Gergel, M. V., Mason, G. W., Sullivan, J. C., & Mech, J. F. (1951). Isolation of microgram quantities of naturally-occurring plutonium and Examination of its Isotopic Composition. *Journal of the American Chemical Society, 73*, 2529–2531.

Predojević, A., & Mitchell, M. W. (Eds.). (2015). *Engineering the atom-photon interaction controlling fundamental processes with photons, atoms and solids*. Berlin: Springer.

Rastogi, R. C., & Sjoeblom, K. L. (1999). *Inventory of radioactive waste disposal at sea, IAEA-TECDOC-1105*.

Stanculescu, A. (2006). IAEA activities in the area of partitioning and transmutation. *Nuclear Instruments and Methods in Physics Research Section A: Accelerators, Spectrometers, Detectors and Associated Equipment, 562*, 614–617.

Walker, C. T., Rondinella, V. V., Papaioannou, D., Van Winckel, S., Goll, W., & Manzel, R. (2005). On the oxidation state of UO_2 nuclear fuel at a burn-up of around 100 MWd/kgHM. *Journal of Nuclear Materials, 345*, 192–205.

Walther, C. (2003). Comparison of colloid investigations by single particle analytical techniques—a case study on thorium-oxyhydroxides. *Colloids and Surfaces A: Physicochemical and Engineering Aspects, 217*, 81–92.

Wasserburg, G. T., McDonald, G. J. F., Hoyle, F., & Fowler, W. A. (1964). Relative contributions of uranium, thorium and potassium to heat production in the earth. *Science, 143*, 465–467.

Zucchetti, R. (1991). Criteria for defining low activation fusion reactor applications. *Fusion Engineering and Design, 15*, 235–244.

Chapter 2
Sampling, Handling, Treatment and Separation

Prior to analysis, the sample must be defined, described and subsequently extracted from the considered material, if necessary, and prepared for the analysis. A rigorous approach has to be followed before initiating any expensive analysis.

2.1 Sample Collection and Analysis Mode

In a first step, sample collection and analytical protocol need to be addressed.

2.1.1 Sample Selection and Collection

Prior to sample selection and data collection, a minimum of information is required on the sample to perform the sampling. Is the sample:

- Solid, liquid or gaseous?
- Heterogeneous or homogeneous?
- 1 m^3, 1 mm^3… or 1 nm^3 in size?
- A sub-sample?
- A sample series in space and/or time?
- A major, a minor or a trace/radio-trace component in terms of its analyte?
- Analyzed for basic properties, e.g. mass, volume, pressure (total and partial), temperature, gradients and location (coordinates) fixed in the system?
- Representativeness defined?
- Homogeneous or with spatial variation?
- Size larger than the phase's size?

For practical and safety reasons it will be interesting to reduce sample size in order to reduce sample activity. For example, sampling will pass from 1 fuel element

C.A. Degueldre, *The Analysis of Nuclear Materials and Their Environments*,
DOI 10.1007/978-3-319-58006-7_2

(assembly) to 1 fuel rod, to 1 fuel pellet to 1 fuel fragment; the sample needs however to be representative, which requires careful sample selection, otherwise series of samples need to be selected.

In statistics, quality assurance and survey methodology, sampling is concerned with the selection of a sub-set of individual objects within a statistical set of samples to estimate characteristics of the material. Each observed properties (such as location, weight, color) are noted for independent or individual objects. In survey sampling, weights can be corrected to adjust the data for the sample design, particularly for stratified sampling. Probability and statistical theories are applied on the results to guide the analyst.

The sampling process comprises several stages:

- Defining a population of concern in the investigated material
- Specifying a set of samples possible to measure
- Specifying a sampling method for selecting samples from the material
- Determining the sample size
- Implementing a sampling plan
- Sampling and data collecting/recording
- Selecting data which can be handled.

In the most straightforward case, such as the acceptance of a batch of material from the production line, one could identify and measure every single specimen in the population and to include any one of them in the sample. However, this is generally not possible. The imprecise populations are not amenable to sampling in any of the ways below to which statistical theory could be applied.

Sampling method

Probability sampling is carried out when a sample from a population has a given probability of being selected from the set of samples, and when this probability can be accurately determined.

Non-probability sampling is any sampling method where some elements of the population have *no* chance of selection, or where the probability of selection cannot be accurately determined. Hence, because the selection of elements is non-random, non-probability sampling does not allow the estimation of sampling errors.

Advantages over other sampling methods:

- Focuses on important sub-populations and ignores irrelevant ones
- Allows use of different sampling techniques for different sub-populations
- Improves the accuracy/efficiency of estimation
- Permits greater balancing of statistical power of tests of differences between strata by sampling equal numbers from strata varying widely in size.

Disadvantages over other sampling methods:

- Requires selection of relevant stratification variables, which can be difficult
- Is not useful when there are no homogeneous sub-groups
- Can be expensive to implement.

Oversampling may be performed to minimize error. Choice-based sampling is one of the stratified sampling strategies. In choice-based sampling, the data are stratified on the target and a sample is taken from each stratum so that the rare target class will be more represented in the sample. The model is then built on this biased sample. The effects of the input variables on the target are often estimated with more precision with the choice-based sample even when a smaller overall sample size is taken compared to a random sample. The results usually must be adjusted to correct for the oversampling.

In imbalanced datasets, where the sampling ratio does not follow the population statistics, one can resample the dataset in a conservative manner called **minimax** sampling. Gaussian distribution may be selected to classify the worst case error over all the possible sample population.

Accidental sampling is a type of improbable sampling which involves the sample being drawn from a specific part of the population. The researcher using such a sample cannot scientifically make generalizations about the total population from this sample because it would not be representative enough but specific to a given area where sampling is carried out.

Finally, in a breakthrough or tailing, sampling may be performed on a logarithm scale frequency to reduce the population of samples.

2.1.2 Sampling from Analysis Mode

The material targeted for analysis may flow in a pipe or a channel. Several modes of analysis are possible to perform the assay on a fluid, i.e. gas, liquid, powder fluid bed (see Fig. 2.1), see Callis et al. (1987):

- Non-invasive
- In-line
- On-line
- At-line
- Off-line.

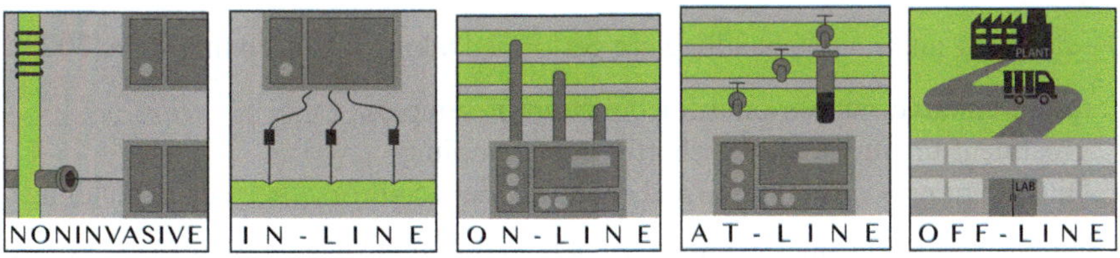

Fig. 2.1 Representation of five analysis modes based on sample location. Adapted from concept described in Callis et al. (1987), with permission © ACS

- The **non-invasive** analysis may be performed within a field through the pipe and be applied continuously or intermittently. Another possibility is to adapt a window and perform the investigation in transmission or in reflection. In both cases the information is provided without intrusion of a sensor in the pipe and without change in the thermodynamic conditions (*P*, *T*). The sample analyzed is defined by the beam, or the size of the sample passing in the field and back analyzed.
- The **in-line** analysis requires intrusion of a sensor, e.g. electrode or optrode in the pipe. There may be modification of the sampled material flow but no significant change in the thermodynamic conditions (*P*, *T*). The analysis may be performed in a continuous way; however, flow change may locally modify traces e.g. colloid's distributions, dissolved gases' fractions.
- The **on-line** analysis requires a deviation of the sample flux toward local analytical unit. Again, analysis may be performed in a continuous way; however, flow changes may locally occur, modifying the sample. The thermodynamic conditions (*P*, *T*) may consequently be slightly affected, potentially altering the sample.
- **At-line** analysis implies the sampling becomes intermittent with amount, mass (*m*) or a volume (*V*) of material collection. Here the thermodynamic conditions —including pressure and temperature of the sample—may significantly change, consequently affecting the sample.
- **Off-line** analysis also implies intermittent sampling, with potential significant change of thermodynamic conditions (*V*, *P*, *T*). Transport of fuel pins to hot laboratory with transfer in hot cell line is needed for non-destructive testing and sample preparation of irradiated fuel pins or irradiated structural materials. This would be the situation from one fuel rod in the reactor fuel pool (at-line) transferred to a hot cell of a hot laboratory to sample one fuel fragment from one fuel pellet from one fuel segment, from one fuel pin for example.

The concept of destructive versus non-destructive analysis may be introduced at this point, the latter allowing multiple analysis on the same sample.

2.2 Sub-sampling

The sampling and sample handling are performed taking into account (Salbu 2000):

- Representative samples and fractionation of samples
- Treatment in situ, at-site or shortly after sampling
- Dilution or pre-concentration and chemical yield control (efficiency of handling).

Sampling, pre-treatment, shipment to laboratory and analysis are areas where contaminations, losses or speciation changes can occur (Harvey et al. 1987). Corrections for these artifacts must be applied by use of isotopic tracers or specific handling conditions.

2.2.1 *Fluid Samples*

Samples of fluids may be gaseous and or liquids. Fluid samples may need physical or chemical treatments.

Gaseous samples are generally homogeneous, except in stratified large-volume samples with gases of high density differences. Some of them may include aerosols. They may be handled as delivered or after pretreatment for separation. Sampling is generally carried in loops (on-line) or in syringes (at-line). Gases are typically separated using chromatographic techniques. Analysis is then performed after separation if needed using analytical techniques described in Chap. 3.

In the cases of liquid samples, the sample may include gases, liquids and/or solids dissolved in the solvent. When gas, liquid and/or solid are dissolved forming a solution, segregation can occur falsifying the analysis of the solution (considered as a single phase). Re-homogenization is then mandatory prior to analysis. For apparent homogeneous fluid with dissolved species, such as reactor water, a separation technique such as ultra-filtration (colloids) or chromatography (ions) is needed prior to analysis of the separated species.

For heterogeneous fluid, such as suspension of organic phase in water (reprocessing), phase separation is also required prior to analysis in each single phase. Leached samples require full digestion (analytical leaching) using optimal leaching process as described in Sect. 2.2.2. In all cases an analytical sampling plan is required prior to determining the analyte properties.

- Treatments may be required at-site, shortly after sampling e.g. spiking acid for trace analysis (e.g. nitric acid prior to ICPMS analysis, see Sect. 3.2.18.4).
- Dilution of a suspension may be required prior analysis e.g. single particle analysis.
- Pre-concentration with chemical yield control (efficiency of handling) may be required prior to analysis, e.g. radiochemical analysis of a suspension.

- **Physical** separation may be performed:
  - field-free: by (ultra)-filtration, by gel-permeation or by volatilization, or
  - within a field: flow field fractionation, centrifugation, capillary or gel electrophoresis.

- **Chemical** separation: (co)-precipitation, ion exchange, extraction, electroplating, or chromatography; e.g. High Performance Liquid Chromatography (HPLC).
- Conductivity (Sect. 3.3) or spectrometer (Sect. 3.2) detectors are commonly used with ion chromatography.

2.2.2 *Solid Samples*

Sub-sampling can easily be performed at the macroscopic level on solid samples. The reduction of the sample size implies also the reduction of its activity, which is a great advantage for its handling. The reduction in size may be performed by cutting, milling polishing collection of fragments (replicates), chemical etching, or physical milling by Focused Ion Beam (FIB). The produced macro- (for mechanical test), micro- (for spectroscopy test), nano- (for microscopy test) sub-sample can be prepared prior to its specific analysis.

Sub-sample preparations for transmission electron microscopy (TEM) analysis were recently reported and compared for nuclear materials by Aitkaliyeva et al. (2015). The preparation may include the following:

- Mechanical cutting and milling/polishing are physical subtractive processes. **Cutting** goes from crystal cleavage (using hard edge and a hammer shock) to cutting using a saw, e.g. a diamond powder coated disk or wire-saw. These two last techniques may require an inert cooling phase flow that reduces temperature excursion during cutting. The inert phase used for cooling must definitively be inert in a physical and chemical sense to avoid element dissolution or reactivity within the fluid flow. A further abrasive action including a polishing phase may be required prior to sample analysis. On the other side, **milling** may be used to produce powder that can be the sample to analyze. The milling material must be selected to avoid/reduce contamination of the powder.
- **Chemical milling** or **etching** is a subtractive reaction process. Using temperature-regulated baths allows chemical etching to remove material and to create an object with the desired shape. It is mostly used on metals, although other materials can also be treated. The process essentially involves bathing the cutting areas in a corrosive chemical known as an **etchant**, which reacts with the material in the area to be cut. Inert substances known as **maskants** are used to isolate the coated surface from the specific areas etched on the material.

For steels the etchant may be:

- Hydrochloric and nitric acids
- Ferric chloride for stainless steels
- Nital (a mixture of nitric acid and ethanol or methanol) for mild steels.

For ceramic the etchant may be:

- Hydrochloric and nitric acids
- Basic leaching with molten carbonate or hydroxide.

Protocol: the solid powder is treated by acid or basic contact. Temperature treatment by microwave (in situ) or heating (ex situ) affects the grain size distribution. Optimal size depends on type of analysis foreseen, e.g. macro- sub-sample for mechanical test, micro- sub-sample for spectroscopy test, nano- sub-sample for microscopy test.

Ion milling and etching

The use of **Focused Ion Beam (FIB)** is required for production of micron-sized samples with regular dimensions. Currently, this can be performed on low active

Fig. 2.2 FIBing a lamella sub-sample from a fuel "pie" piece. **a** set-up in the FIB unit, **b** fuel material excavation around the sub-sample, **c** approach of the pin for sub-sample transfer, **d** freed sub-sample and its transfer to sample carrier, **e** sub-sample fixed on sample carrier by carbon welding, **f** lamella obtained after fine abrasion. Details from Degueldre et al. (2013a), with permission © Elsevier

nuclear material (e.g. Degueldre et al. 2013a), or at very few other hot laboratories on irradiated fuel and active nuclear material samples, e.g. Teague and Gorman (2014). The sub-sample preparation is depicted through various production steps in Fig. 2.2. Such prepared samples can then be used for direct analysis in the FIB by scanning electron microscopy or synchrotron investigations (X-ray absorption or diffraction techniques), or for TEM, Degueldre et al. (2013b). A coating with thin film conductive material may be needed.

Other treatments

- Replicate sub-samples may be obtained by surface polishing, contact with adhesive tape and transfer of particles. To reach representativeness, several sub-samples from various places on the sample may be necessary. For example, the sampling requires cutting of the fuel rod, section handling and sub-sample preparation such as specimen cutting, cross-sectioning, possibly embedding in resin, and polishing with abrasive sand paper. Replicate (grain samples detach/attach) production may be performed using rotational grinding (detach) a fuel cross-section and transfer (attach) of fuel particles on adhesive tape pieces, e.g. Degueldre et al. (2011). The production steps are depicted in Fig. 2.3.
- Abrasion without field, may be upgraded by abrasion within a field, e.g. gravity (centrifugal), magnetic.

The size of the sample or sub-sample to analyze must be compared to the size of the spot analyzed.

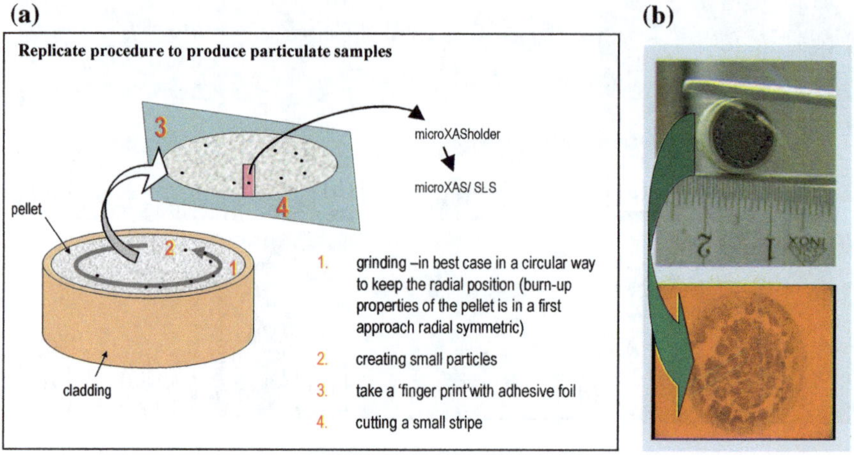

Fig. 2.3 Replicate procedure to produce particle samples on fuel cross-section. **a** Schematic procedure for sample preparation. **b.** *Top* fuel specimen prior to grinding, bottom: sub-sample obtained by contacting the adhesive Kapton foil on the grinded fuel section. A piece of sub-sample may be cut for activity reduction (below dose limit); Degueldre et al. (2016), with permission © Elsevier

2.3 Reference Materials and Calibration

2.3.1 Reference Materials

Certified Reference Materials are standards used to control the quality of the analysis, to validate analytical methods or for the calibration of instruments. Reference materials are important for analytical chemistry and nuclear material analysis. Since most analytical instrumentation is comparative, it requires a sample of known composition (stable, fixed fraction of components) for accurate calibration. These reference materials are produced under stringent preparation and manufacturing procedures. Certification and traceability of the data must be documented in detail. Quality management for laboratory accreditation requires metrological traceability and use reference materials for calibration (Calli 1979).

The certified reference material is "a material characterized by a metrologically valid procedure for one or more specified properties, accompanied by a certificate that provides the value of the specified property, its associated uncertainty, and a statement of metrological traceability".

These reference materials may be:

- Pure substances; stoechiometric chemicals, characterized for major constituents.
- Chemical substances; characterized for major chemical and trace impurities.
- Standard solutions and gas mixtures, often prepared from pure substances.

- Matrix reference materials characterized for the composition of specified major, minor or trace chemical constituents. Such materials may be prepared from matrices containing the components of interest, or by preparing synthetic mixtures.
- Reference objects or artifacts, including microscopy specimens characterized for properties ranging from object type to sized specimens.

2.3.2 Calibration with Internal or External Standards

Internal standard methods are used to improve the precision and accuracy of results where volume errors are difficult to predict and control. A systematic approach has been used to compare internal and external standard methods in high-performance analytical techniques. The precision may be determined, with both analyte and internal standard combined. Precision using three methods of adding the internal standard to the analyte before final dilution may be examined. The internal standard method is used more frequently and more performing than the external standard methods in all instances.

A systematic approach was used to compare **Internal Standard** (**ISTD**) and **External Standard** (**ESTD**) methods used in high-performance analysis.

An internal standard in analytical chemistry is a chemical substance that is added in a constant amount to samples, the blank and calibration standards in a chemical analysis. This substance can be used for calibration by plotting the ratio of the analyte signal to the internal standard signal as a function of the analyte standard concentration. This is done to correct analyte losses during sample preparation. The internal standard is a compound that must show similar behavior to the analyte.

The external standard approach is similar but the assay on the standard is carried out independently, the analyte in one batch, the standard in a separate batch. This practice could however be slightly affected for its precision compared to the internal standard approach.

References

Aitkaliyeva, A., Madden, J. W., Miller, B. D., Cole, J. I., & Gan, J. (2015). Comparison of preparation techniques for nuclear materials for transmission electron microscopy (TEM). *Journal Nuclear Materials, 459,* 241–246.

Calli, J. P. (1979). The role of reference materials in the analytical laboratory. *Fresenius' Journal of Analytical Chemistry, 297,* 1–3.

Callis, J. B., Illman, D. L., & Kowalski, B. R. (1987). Process analytical chemistry. *Analytical Chemistry, 59,* 624A–637A.

Degueldre, C., Martin, M., Kuri, G., Grolimund, D., & Borca, C. (2011). Plutonium—uranium mixed oxide characterization by coupling micro-X-ray diffraction and absorption investigations. *Journal of Nuclear Materials, 416,* 142–150.

Degueldre, C., Borca, C., & Cozzo, C. (2013a). Curium analysis in plutonium uranium mixed oxide by x-ray fluorescence and absorption fine structure spectroscopy. *Talanta, 115,* 986–991.

Degueldre, C., Schaeublin, R., Krbanjevic, J., & Minikus, E. (2013b). Electron energy loss spectroscopy investigation through a nano ablated uranium dioxide sample. *Talanta, 106,* 408–413.

Degueldre, C., Bertsch, J., & Martin, M. (2016). Post irradiation examination of nuclear fuel: Toward a complete analysis. *Progress in Nuclear Energy, 92,* 242–253.

Harvey, B. R., Lovett, M. B., & Boggis, S. J. (1987). Some experiences in controlling contamination of environmental materials during sampling and processing for low-level actinide analysis. *Journal of Radioanalytical and Nuclear Chemistry, 115,* 357–368.

Salbu, B. (2000). Speciation of radionuclides in the environment. In R. A. Meyers (Ed.), *Encyclopedia of analytical chemistry* (Vol. 1, pp. 2993–3016). Chichester: John Wiley

Teague, M., & Gorman, B. (2014). Utilization of dual-column focused ion beam and scanning electron microscope for three dimensional characterization of high burn-up mixed oxide fuel. *Progress in Nuclear Energy, 72,* 67–71.

Chapter 3
Characterization Using Passive or Interactive Techniques

Principles of passive and interactive techniques

Analytical techniques can be distinguished between passive ones, taking profit of the inherent activity of the sample, and those which are interactive in character. Based on the methodological classification proposed in Table 3.1, a list of techniques is discussed in Sects. 3.1 and 3.2. Passive techniques are performed without excitation while interactive techniques require excitation. Detection is carried out in both cases. These analytical techniques are applied for characterization of isotopes, elements, or species. Excitation (when needed) and detection are performed with phonons, photons, leptons (electrons, positrons, muons and neutrinos), neutrons, ions (or atoms or molecules) considering the particles (plain, solvated, or as clusters) or their associated waves. The approach takes into account the absorption (A), the emission (E) or fluorescence (F) or phosphorescence (Ph) as depicted on an energy diagram in Fig. 3.1.

Each technique is presented according to the following scheme:

- Principle and history
- Excitation (if any) e.g. by absorption (A)
- Detection of emitted (E), fluorescent (F) or phosphorescent (Ph) signal
- Resolution (lateral and/or in depth)
- Detection limit (DL)
- Examples for nuclear materials and their environment.

After description of the techniques, their advantages and drawbacks as well as their combinations are applied and discussed.

Application mode

- Bulk analysis (0D)
- Linear scan (1D)
- Static radiography (2D)
- Static tomography (3D)
- Dynamic tomography (4D).

C.A. Degueldre, *The Analysis of Nuclear Materials and Their Environments*,
DOI 10.1007/978-3-319-58006-7_3

Table 3.1 Detection—excitation: passive versus interactive analytical techniques

Detection exitation	**Phonon**	**Photon**	**Lepton: neutrino electron positron muon**	**Neutron**	**Ion atom**
–	AE	XS, γS, γRAD, γTOM MOES	νS, EH, RAD, LSC, βS PET	NS, NRAD, NTOM	ISE αS, αRAD, LSC
Phonon	SR AAFM	–	STM	–	–
Photon	LIPAS, (LIPDS), LIBD	GPR IRS, DRS, AAS NIR-VIS-UVS, COL, RAMS, TRLITS PHOS, SPC, PCS, XAS, XRD XRF, STXM, MOAS, XγTOM,	UPS, XPS, SEXAS	NPHOT	TIMS, RIMS, LAMMA, LAICPMS
Lepton: neutrino electron positron muon	SEAM	EDS, EMPA, WDS, PAS, SPM,	EAFM, SEM, TEM, ED AES, EELS, μRAD, μTOM, SMUM, TMUM	–	ESMS, SSMS
Neutron	–	NAA	νS NAA	DNAA NRAD, NTOM	NAA RAD NERDA
Ion	–	ICPOES, PIXE, PIGE	–	NRA	VOL, AFM, NI, ICPMS, SIMS, AMS, RBS, ERDA, NRA

Analytical techniques including excitation (if any) and detection for isotope, element, or species characterization (**see list of abbreviations in the Glossary**), *Note* Excitations or detections are performed with phonons, photons, electrons, neutrons, ions (or atoms or molecules) considering the particles (plain, solvated or cluster) or their associated waves. *Note* FTIR with IRS, XAS for XAFS, XANES and EXAFS. Specific techniques used within a field are presented Table 3.17

The bulk analysis (0D) is in principle carried out with a detector (and a beam) of the same size as the sample. The whole sample volume is analyzed and the analysis yields an average measurement of an analyte property in the sample.

With the scan (1D), analysis is generally carried out with a detector (and a beam) whose size is smaller than or equal to the sample width and consequently is smaller than its length. The analytical set-up includes a fixed sample and mobile detector

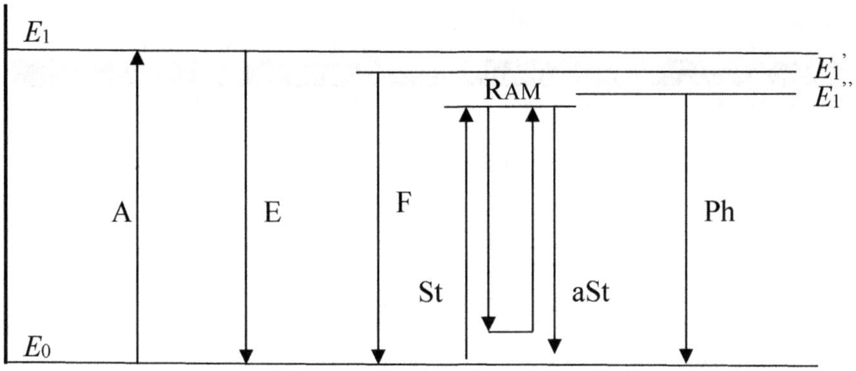

Fig. 3.1 Energy transitions including: Absorption (A)—Emission (E)—Fluorescence (F)—Phosphorescence (Ph)—Raman (RAM)—Stoke (ST)—anti-Stoke (aST)

(and beam) orthogonal to the sample with a scan (stepwise or continuous) along the sample length. Or the detector (and beam) is fixed and the sample is stepwise or continuously shifted along a given direction. Alternatively the detector (and beam) is larger than the sample and the signal is recorded with the detector as a picture. The 1D information is then obtained without being a real scan.

The 2D scan and radiographies (2D) may be performed in a scan mode with a detector (and/or beam) smaller than the sample. Record is carried out row (1D scan) per row, with the set-up described above in scan mode. However, the classical way to perform radiography is to record the 2D picture of the sample as "seen" by the detector in passive or interactive mode.

For the static tomograms (3D), the sample may be analyzed by rotation of the sample on itself (or the detector) and the image data over the angles going at least from 0° to 180° need to be collected in passive or interactive mode. Data require processing for the 3D image reconstruction.

For the dynamic tomography (4D) the 3D reconstruction needs to be repeated as fast as required or as possible. The 3D film may be displayed in real time (fast data processing) or deferred.

The application as bulk (0D) analysis, scanning (1D), radiography (2D) and tomography (3D and 4D) may be applied in-line, on-line at line or off-line. In general, the aim of computerized tomography is to obtain information regarding the nature and position of materials inside a sample. Transmission tomography, using the absorption of a physical entity (gamma-ray, X-ray, light, ultrasound) crossing a body, is able to assess the internal morphology of an object. In the specific case of application to nuclear fuels, gamma-ray transmission tomography can be used to investigate the fuel rod density as well as to determine the within-pin spatial distribution maps of linear attenuation coefficients (to be used in emission tomography application).

Frequency	EHz	PHz	THz	GHz	MHz				
Wavelength	pm	nm	μm	mm	dm	m			
	γ ray	X-ray	farUV	UV	VIS	NIR	IR	farIR	Radio
Particle energy	MeV	keV	eV	meV	μeV				

Fig. 3.2 Indicative spectral range of radiation wave explored

In all techniques, geometrical parameters such as size of the system analyzed, size of the detector and object–detector distance are key parameters, which, together with acquisition time, energy level etc. rule the detection limit of the analyzed isotopes, elements and species.

In a vacuum, electromagnetic waves travel at the speed of light, c, that is linked to its wavelength, λ, and frequency, v, by:

$$c = \lambda \cdot v \tag{3.1}$$

Techniques are normally classified according to their particle or wave energy going from low energy to high energy (see Fig. 3.2).

3.1 Material Characterization by Field-Free Passive Techniques

The material may emit particles that are detected for their analysis during a fixed acquisition time. The detectors are any kind of system sensitive to the emitted particles or associated waves, e.g. piezoelectric sensor, optical sensor, scintillator, semi-conductor, electrode etc. High-efficiency detectors record data or spectra in less time than low-efficiency detectors. However, the detector efficiency may reduce its resolution, which is a key issue in the analysis of nuclides for example. The detection limit as defined in Sect. 1.3.2 is function of the background but also of the efficiency of the detector and of the occurrence of potential interferences.

3.1.1 Techniques with Phonon Detection

Phononic or acoustic emission measurement

Acoustic Emission (AE) refers to the stochastic generation of transient elastic waves produced by a sudden redistribution of stress in a material. This is a non-destructive testing technique. The detection and analysis of AE signals can supply valuable information regarding the origin of the investigated material.

Acoustic emission is the phenomenon of radiation of acoustic (elastic) waves in solids that occurs when a material undergoes irreversible changes in its internal structure, for example as a result of crack formation or plastic deformation due to aging, past or present temperature gradients or external mechanical forces. Information on Acoustic Emission Testing may be found in the work edited by Grosse and Ohtsu (2008).

Acoustic emission can be detected using piezo detector in frequency ranges under 1 kHz. AE has been reported at frequencies up to 100 MHz, but most of the released energy is within the 1 kHz–1 MHz range. Stress-releases generate a spectrum of acoustic waves starting around 1 Hz, and falling off above some MHz.

The three major applications of AE techniques are to:

- Determine the locations where a stress source occurred during material production;
- Evaluate material mechanical performance; and
- Monitor operational safety of a structure, i.e. pressure containers, pipes, etc.

AE can be related to an irreversible release of energy. It can also have been generated from sources not involving material failure including intrinsic stress, cavitations and impact of insignificant P or T variations.

The technique may be used, for example, to observe the formation of cracks after welding. In a material under stress, transducers can record signals, and locate the area of their origin by measuring the time required for the sound to reach the transducers. The technique is valuable for detecting cracks initiation and formation in pressure vessels and pipelines transporting liquids under high pressures. In addition to non-destructive analysis, AE has applications in process monitoring and is also used for investigation of corrosion in structure. Information on practical acoustic emission testing may be found in the work edited by the Japanese Society for Non-Destructive Inspection (2016).

3.1.2 Techniques with Photon Detection

3.1.2.1 X-Ray and γ-Ray Counting and Spectroscopy

In **X-ray** and **γ-ray counting and Spectroscopy** (XS, γS) the relaxation from excited stage of an atom or a nucleus yields emission of one or several photons of well-defined energy. The X-rays are due to the relaxation of atoms or ions after excitation to a higher electronic energy state. The γ-rays are due to the energy decay of nuclei to lower nuclear energy level. The discrimination between γ rays and X-rays may be difficult. X-rays refer to photon emission from atoms or ions, up to about 120 keV i.e. for actinides (K line see Sect. 3.2.3.8), whereas the lowest photon energy nuclei emissions (γ-rays) may be less than 10 keV. Radionuclides emitting γ-rays radiate photons in the energy range from a few keV to ~10 MeV anticipating an overlapping between the two types of emission. Such radionuclides produce γ-ray line spectra with photons emitted at well-defined discrete energies.

More detailed information on γS and XS can be found in Gehrels et al. (1988), and Friedlander et al. (1981).

The gamma emission of the isotope ${}_{Z}^{A}X$ in its exited state ${}_{Z}^{A}X^{*}$ undergoes the decay reaction:

$$ {}_{Z}^{A}X^{*} \rightarrow {}_{Z}^{A}X + \gamma \qquad \{3.1\} $$

yielding the nucleus ${}_{Z}^{A}X$ in its ground state or in a lower energy state and a γ-ray photon. This reaction is usually associated to another nuclear reaction or decay.

In γ-spectroscopy detectors γ photons interact in the detector material volume. The interaction mechanisms are the photoelectric effect, the Compton effect and the pair production. The photoelectric effect is preferred since all the energy of the incident γ-ray is absorbed within the detector volume. When a γ-ray undergoes a Compton interaction or pair production a fraction of the energy escapes from the detector material volume without being absorbed. The background rate in the spectrum is locally increased by one count. Because of this energy manco, this count will be recorded in a channel below the full energy of the γ-ray. Larger detector volumes reduce this effect.

Gamma spectroscopy systems are selected to take advantage of several performance characteristics. Two of the most important issues include detector resolution and detector efficiency. Reduction of the background is a key issue in low activity analysis. Important developments include the production of high efficiency coaxial and well-type detectors operating with anti-cosmic ray or anti-Compton shielding. Detection is currently carried out using a semiconductor crystal or by scintillation.

The gamma-rays recorded in a spectroscopic system display signals in the spectrum in form of peaks or lines. The peak width, a very important parameter, defines the detector resolution. High resolution allows separation of two close γ peaks. The peak may usually be fitted as a Gaussian count distribution. In most spectra the maximum of the line is fixed by the γ-ray's energy that is a characteristic of the emitting radionuclide. The gamma line is currently used to identify unknown radionuclides in an investigated sample. The peak area is determined by the detector efficiency and the fluency of γ-photons being itself a function of the acquisition time and of the amount of nuclei. To quantify detector resolution evaluation of the full width at half maximum (FWHM) of peaks is required. This is the width of the γ-ray peak at half of the highest intensity. Resolution figures are given for reference energies. Resolution can be expressed in absolute, e.g. eV, or relative values. For a sodium iodide mono-crystal detector the FWHM may be about 10% or 10 keV at about 100 keV, and about 80 keV at 800 keV. A germanium detector gives a resolution of 0.5 keV only at 100 keV (0.50%) However, its efficiency is weaker for crystal of similar sizes. More detailed information may be found in Friedlander et al. (1981) and Ehmann and Vance (1991). The detector efficiency is determined by the photon–detector material interaction. The probability that a γ photon interacts with the detector and produces a count is the detector **efficiency**. In general, larger detectors have higher efficiency than smaller detectors, although the shielding effect of the detector material is important. It is estimated by the fraction of the number of event registered with respect to the number of events emitted from a source of defined activity $A(t)$:

$$A(t) = N(t)\,\lambda_{\mathrm{i}} \tag{3.2}$$

with $N(t)$ the present number of radioactive nuclides i of the source and λ_{i} their decay constant (see Eqs. 1.1 and 1.2). The two main kinds of counters are the scintillator and the semiconductor detectors.

Scintillation detectors use crystals converting γ photons in visible light flashes generated by their interaction in the crystal. The flash intensity is proportional to the γ photon energy deposited in the crystal. The counting is carried out by a photomultiplier that transforms each flash as an electric pulse. The signal amplification and analysis are subsequently performed together using a Multi Channel Analyzer (MCA). Common crystals include thallium-doped sodium iodide, called sodium iodide (NaI) detectors, and bismuth germanate (BGO). Since photomultipliers are sensitive to visible light, the detector is designed as a light-tight system. Detailed information may be found in Friedlander et al. (1981) and Ehmann and Vance (1991).

Semiconductor detectors, also called solid-state detectors, are based on the detection of charge carriers including electrons, generated by γ photon interactions (see Friedlander et al. 1981; Ehmann and Vance 1991). In these detectors, an electric field is applied trough the crystal volume. Consequently, electrons in the conduction band move in the electric field toward the cathode. The collection of electrons and electron-deficient sites called holes migrating respectively to the positive and negative electrodes generates an electrical signal that is amplified and treated by a MCA prior to spectrum display. Semiconductor detectors used currently are made of silicon, germanium, cadmium telluride, and cadmium zinc telluride. The principle of the semiconductor detector is similar to that of a gas-filled detector such as ionization chambers.

Si(Li) detectors consist essentially of a 3–5 mm thick silicon diode with a bias of 1000 V across it (see Szöghy et al. 1981). The lithium-drifted center part forms a non-conducting depleted layer at the p-n junction. When an X- or γ-ray photon passes through, it causes a cloud of electron-hole pairs to form, and this produces an electric pulse. To obtain sufficiently low conductivity, the detector must be maintained at low temperature, and liquid-nitrogen cooling must be used for the best resolution. With some loss of resolution, the much more convenient Peltier cooling can be employed. More recently, high-purity silicon wafers with low conductivity have become routinely available. Cooled by the Peltier effect, this provides a cheap and convenient detector; however the liquid nitrogen cooled Si(Li) detector still has the best resolution.

To keep the geometry of the sample-detector assembly constant, the sample is prepared as a flat disc of diameter 2–5 cm placed at a fixed small distance from the counter window. The production of sample discs may be metal shape machined, finely ground minerals, pressed tablet, or cast glasses that emit γ-rays with a wide variety of energies (for energy and efficiency calibration). A source of ^{241}Am is frequently used for low energy calibration. A flat and representative sample surface is needed because secondary X-rays are generally only emitted from the top micrometer layer of the sample. To reduce the surface effect, the sample is usually spun at 5–20 rpm. It is necessary to ensure that the sample is sufficiently thick to absorb the entire primary beam. For higher-*Z* materials, a few millimeters thickness is adequate, but for a light-element material such as coal, a thickness of 3–4 cm is needed.

Since the position on the channel axis is proportional to the γ ray energy, the MCA channel axis in γ- and X-ray spectroscopy may be converted to energy. The energy spectrum must be calibrated using the peaks of known emitters, such as ^{137}Cs or ^{60}Co. For a crystal detector (volume) an intensity calibration has also to be performed, so that the peak intensities of an unknown radionuclide doped sample can be determined together with its energies. The identification of the detected nuclides consequently can then be performed together with the determination of its amount.

The detection limits (DL) may then be estimated for a given nuclide and the associated energy for its characterized γ line, for the type of detector, geometry of the system and background in the spectrum. The DL is currently of the order of one mBq. The detection limits can be calculated in term of moles on the basis of the half-life of the isotope. Typical analytes include natural actinides such as ^{234}Th or anthropogenic actinides with detection limits of the order of 1 mBq. Based on this activity limit, the detection limits for actinide isotope amounts are calculated for relevant isotopes in Table 3.2. A comparison of detection limits is reported in Sect. 4.1.

More specific fissile material detection for safeguard issues may be gained by coincidence counting due to annihilation. Quantification of ^{235}U and ^{238}U activities for undeclared nuclear materials by a digital gamma–gamma coincidence spectroscopy has been reported by Zhang et al. (2011).

Table 3.2 Passive analytical techniques used for actinide isotope, element or species identification. Detection limit (DL) in **mol** recalculated from A_{DL} in Bq

Detection	Goal	Sample	^{A}An(*Y*)	N_{DL} (mol)	Remarks
Photon: XS γS	Determination of isotope activity and identification of isotope	Solid or liquid bulk or film	^{227}Ac ^{230}Th ^{232}Th ^{231}Pa ^{234}U ^{238}U ^{237}Np ^{239}Pu ^{240}Pu ^{241}Am ^{242}Cm ^{244}Cm	2×10^{-14} 1×10^{-11} 5×10^{-7} 2×10^{-14} 2×10^{-11} 2×10^{-8} 2×10^{-12} 7×10^{-12} 1×10^{-12} 1×10^{-16} 5×10^{-15} 1×10^{-16}	γS, A_{DL}: 1 mBq Bojanoswki et al. (1987) Guillot (2001)

3.1.2.2 From Scanning 1D to Mapping 2D

Uranium and thorium ore body prospection

Gamma ray spectroscopy (γS) has been systematically used to detect ^{232}Th or ^{238}U from environmental samples. The detection of these isotopes may be performed using γ photons from daughter nuclei. In situ determination of uranium in surface soil was carried by γ spectroscopy measuring ^{234}Th and ^{234m}Pa using a high resolution γ-ray spectrometer and assuming secular equilibrium (Miller et al. 1994). On the other hand, uranium and thorium were also detected in soil samples by measuring respectively ^{208}Tl and ^{214}Bi as reported by LaBreque (1994) which were also assumed to be in secular equilibrium with their respective parents.

A gamma-logging (γS) probe has been used to monitor thorium and uranium as a function of depth in a borehole (Mwenifumbo and Kjarsgaard 1999; Nagra 1991) as presented in Fig. 3.3a. The technique is used for uranium exploration; it discriminates between valuable uranium ore and other radioactive material of less value.

The determination of the specific activity of these major natural actinides may be carried out by airborne gamma spectroscopy using the above key nuclides, or other

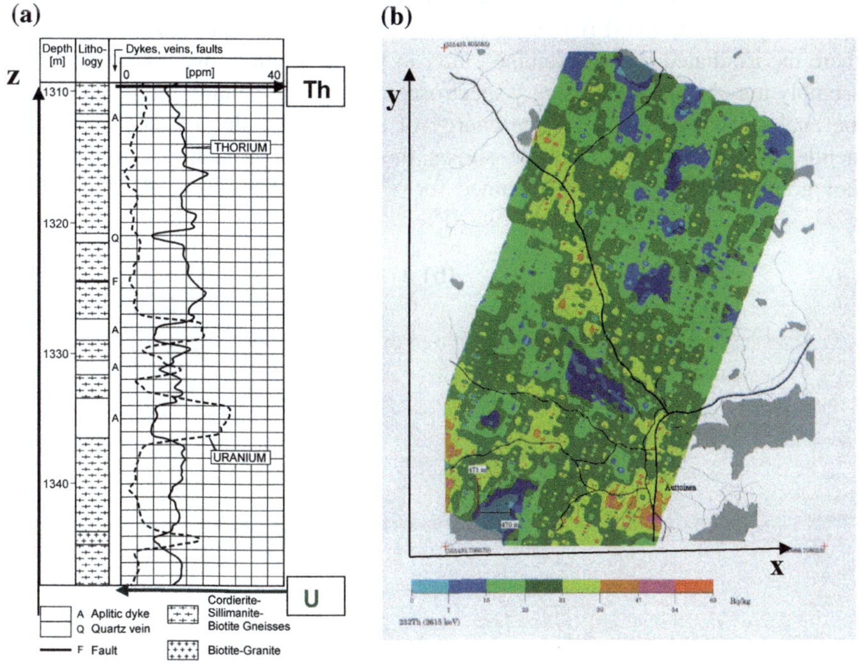

Fig. 3.3 Gamma survey, from 1D scanning to 2D mapping. **a** Th and U profiles from gamma spectroscopic (γS) instrumental analysis in Leuggern borehole (north Switzerland). Note the uranium (opposite scale) depletion through defined faults in formation. Nagra (1991) with permission © Nagra. **b** ^{232}Th maps (from exercise in Finland), area processed by the filtering and window methods. Guillot (2001) with permission © Springer

nuclides, e.g., U by Ra as reported by Kerbelov and Rangelov (1997). This method enables analysis during fixed-wing aircraft or helicopter flight (Guillot 2001). The sensitivity of the spectral analysis of windows at 2615 and 1764 keV for ^{232}Th (by ^{208}Tl) and ^{238}U (by ^{214}Bi) respectively was optimized by subtraction of the Compton continuum in the detection window. The detection of ^{232}Th and ^{238}U is possible in their natural background of 33 Bq kg^{-1} in a large-volume NaI detector (16 L) and a short sampling time (1–5 s) at 40 m ground clearance. The calculation of the concentrations is then simple and reliable. A quantitative estimate of radioactive anomalies can then be obtained easily. The spectral profile analysis is of great interest and has been applied within the framework of environmental monitoring studies. Figure 3.3b shows a map (2D γ radiography) obtained for ^{232}Th during a mapping exercise. Similarly aerial measurements above uranium mining and milling areas have also been reported (Winkelmann et al. 2001; Guillot 2001).

3.1.2.3 From Scanning 1D to Tomography 3D

Irradiated nuclear fuel analysis

Nuclear fuel analysis may be performed in the frame of the so called Post Irradiation Examination (PIE) of the irradiated material. Currently, in a reactor pool where the irradiated fuel is examined after in pile irradiation, fuel rod from a fuel assembly may be analyzed using γ spectroscopy in a 1D scanning mode along rod. The analysis is performed for the energy of a selected γ line of an investigated nuclide. Figure 3.4a for example shows the analyzed irradiated CANDU rod photograph and the 1 D scan obtained for ^{134}Cs, (see Ionescu et al. 2009). The

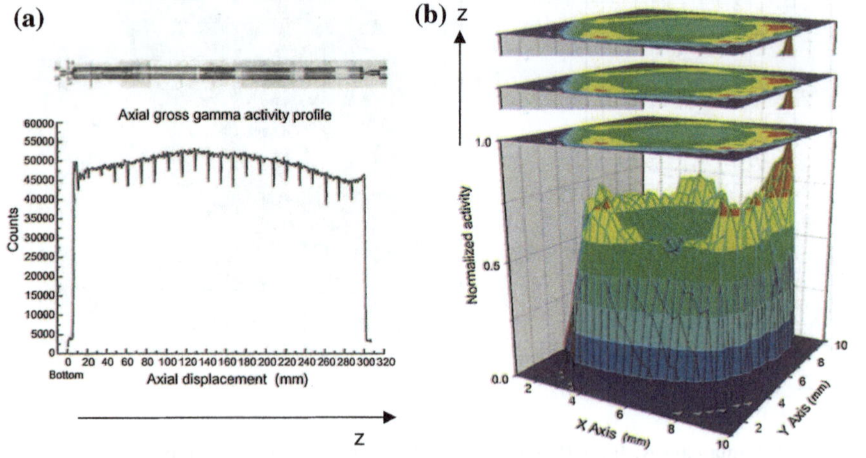

Fig. 3.4 Axial gamma scanning (1D) (**a**) and tomography (3D) (**b**) of ^{137}Cs on a CANDU fuel rod irradiated in the INR–TRIGA reactor in a power ramping test, conditions: normalized activity $A(x, y)/A_{max}$; adapted from Ionescu et al. (2009), with permission © Elsevier

abscissa here is the 30 cm long CANDU rod, the activity slightly increases up to the middle of the pin then decrease corresponding to higher burn up at mid assembly, sharp minima can also be observed that on the record corresponding to the 22 single fuel pellets filling the rod.

The 3D tomography reconstruction for the same isotope of Cs is reported on Fig. 3.4b. For the given position rotating the pin, the activity is recorded and the series of angular activity data treated by an algorithm for conversion in a 3D plot of activity. The activity is normalized to the maximum activity recorded per space unit (pixel). The tomogram shows that the activity of Cs has concentrated in the periphery (colder) while the center (hotter) is depleted in Cs and that the maximum observed is to be found where the burn-up was higher (e.g. higher moderation). These analytical results deliver precious information on the fuel behavior during irradiation.

3.1.2.4 From Static Source to Dynamic Source Counting

Mössbauer emission spectroscopy

A radiometric technique alone is not effective for radioelement speciation. **Mössbauer Emission Spectroscopy** (MoES), as field-free emission of γ photons, can however provide molecular information.

Conservation of momentum induces radionuclide recoil during γ-emission. If a nucleus in a mater emits a γ-photon, its energy is slightly less than the transition energy, the difference being due to recoil. Consequently, nuclear resonance, corresponding to the reabsorption of emitted γ photons of the same energy by identical nuclei, is not observed with free nuclei e.g. in gas or liquid, because the shift in energy is too large.

Nuclei in a solid crystal, however, are fixed in the crystal lattice and not free to recoil. When a nucleus in a solid emits a γ photon, some energy can still be lost as recoil energy, but in this case it always occurs in discrete clusters of phonons (crystal lattice vibrations). Any whole number of phonons can be emitted, including zero, corresponding to a recoil-free event (see May 1971).

Mössbauer found that a fraction of emitted photons are recoil-free, e.g. Mössbauer et al. (1968). This makes Mössbauer absorption spectroscopy also possible, because it means γ photons emitted by one nucleus can resonantly be absorbed by a sample doped by the same isotope nuclei.

The detectors utilized for this specific γ spectroscopy are described in Sect. 3.1.1. The detectors used are the semi-conductor detectors. In the emission methodology the detection system includes also the absorber that acts as a "filter" (see Fig. 3.5). Emission studies are less numerous than absorption investigations that are treated in Sect. 3.2.3.15.

The difference in chemical environments, however, causes the nuclear energy levels to shift in a few different ways, as described below. Although these energy shifts are tiny (often <1 MeV), the extremely narrow spectral line widths of γ-rays for some radionuclides make the small energy shifts correspond to large spectral

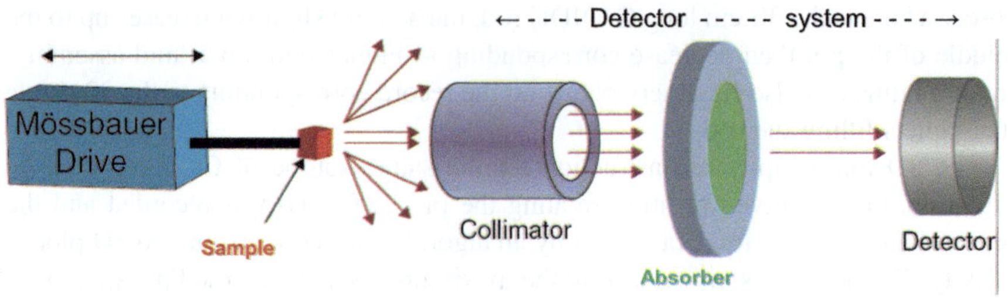

Fig. 3.5 Schematic illustration of the experimental set-up required for Mössbauer emission spectroscopy (MoES)

changes. During Mössbauer emission spectroscopy, the source or sample is accelerated through a range of velocities using a motor to produce a Doppler effect (see Fig. 3.5). The γ photon energy is consequently scanned through a given velocity range. The Doppler effect may be described here by the equation:

$$E_v = E_\gamma (1 - v/c) \tag{3.3}$$

where v is the recoil velocity of the emitter nucleus and c the light velocity. Consequently the energy of the transition E_γ is shifted by the recoil to E_v with $E_{rec} = E_\gamma - E_v$. Conservation of momentum induces radionuclide recoil during γ–emission and the energy shift is given by:

$$E_{rec} = \frac{E_\gamma^2}{2mc^2} \tag{3.4}$$

Velocities are usually of some 10 mm s^{-1}. Equations 3.3 and 3.4 allow calculation of the transition energy of the daughter radionuclide for a velocity. One calculates 1 mm s^{-1} = 48 neV for ^{57}Co (β^-, γ-emitter) with Eγ 14.41 keV, and, 1 mm s^{-1} = 198 neV for ^{241}Am (α-, γ-emitter) with E_γ 59 keV.

Suitable γ source samples consist of a radioactive parent that decays to the absorber isotope. For example, the source for the absorber ^{57}Fe consists of ^{57}Co, which decays by electron capture to an excited state of ^{57}Fe, which in turn decays to a ground state emitting a γ-ray of defined energy. The radioactive cobalt sample is prepared on a foil. Ideally the parent isotope will have a convenient half-life. Also, the γ-ray energy should be relatively low, otherwise the system will have a low recoil-free fraction resulting in a poor signal-to-noise ratio and requiring long collection times.

Table 3.3 indicates elements from nuclear materials having an isotope suitable for Mössbauer emission spectroscopy. Among them, ^{57}Co is the element studied using the technique, although fission product isotopes e.g. ^{129m}Te (Ohtani 1986) are also occasionally studied. Mössbauer emission spectroscopy may also be used to

Table 3.3 Elements from nuclear materials with suitable optimize for Mössbauer spectroscopy; **Emitter** → absorber, conditions: E_γ emitter gamma energy, τ emitter half life

Decay	τ (d∩a)	E_γ (keV)	Species	References
^{57}**Co** → ^{57}Fe	271.8 d	121, 136	Co(II,III)	Fitch and Rees (1981)
^{99}**Rh** → ^{99}Ru	16 d	354	Rh(III,IV)	Ohkubo et al. (1995)
^{119m}**Te** → ^{119}Sn	4.7 d	24	Te(-II,III, IV)	Ambe et al. (1974)
^{121m}**Sn** → ^{121}Sb	18.25×10^3 d	6	Sn(II,IV)	Ambe and Ambe (1975)
^{129m}**Te** → ^{129}I	33.6 d	33.6	Te(-II,IV)	Ohtani (1986)
^{242}**Pu** → ^{238}U	3.75×10^5 a	44.91	Pu(III, IV, V, VI)	Tutsi et al. (1998)
^{241}**Am** → ^{237}Np	4.32×10^2 a	59.57	Am(III,IV)	Smith et al. (2015)
^{247}**Bk** → ^{243}Am	1.38×10^3 a	84	Bk(III,IV)	Smith et al. (2015)

analyze actinide isotopes e.g. ^{242}Pu and ^{241}Am. While actinide isotopes may be excellent Mössbauer nuclei, little speciation has been done for nuclear material samples, perhaps because Mössbauer spectroscopy requires macro-concentrations.

The nuclear material example deals with the speciation of a short life safety relevant nuclide. Cobalt-60 is one of the relevant nuclides that may contribute to human doses in contaminated environment. Zeolite is used as exchanger to decontaminate waters loaded with radionuclides. $^{57}Co^{2+}$ ions have been exchanged into zeolite A. Even at ppm concentrations Co^{2+} ions were found to be located only in the sodalite window site with trigonal coordination to lattice oxygens (see Fitch and Rees 1981). The temperature dependence of the Mössbauer parameters was measured in the range 77−573 K from which the source recoil free fractions for Co^{2+} ions located in sodalite site was derived.

3.1.3 Techniques with Lepton Detection

In this section, lepton detection includes neutrino, electron and positron detection and spectroscopy. Electron measurements are also considered in a broad sense. It goes from the potential in an oxido-reduction reaction to the energy of electrons emitted by radioactive decay: β^- particles. Positrons are finally considered for their application in tomography.

3.1.3.1 Neutrino Counting and Spectroscopy

The neutrino or more precisely the antineutrino is the antiparticle co-produced during the beta emission of the isotope ${}^{A}_{Z-1}X$ as follows:

$$ {}^{A}_{Z-1}X \rightarrow {}^{A}_{Z}X + e^{-} + \bar{\nu} \qquad \{3.2\} $$

yielding a next element nucleus ${}^{A}_{Z}X$ and an antineutrino $\bar{\nu}$. This antiparticle is coproduced as a consequence of the parity conservation during the nuclear reaction. The antineutrino is the antiparticle corresponding to the neutrino, with a spin of ½ and representative of the lepton family, such as the electronic antineutrino $\bar{\nu}_e$. All antineutrinos are particles of zero charge.

Antineutrinos are omnipresent in nature. In one second, 10^{10} $\bar{\nu}_e$ pass through one square centimeter of any part of individual body without interaction. Consequently, they are extremely difficult to detect because their interaction cross-sections are very small. Electron antineutrinos emitted in β decay of a radionuclide naturally occurring in the Earth are also called **geoneutrinos**. Matter is virtually transparent to neutrinos and consequently they travel at near light speed through the Earth from their point of emission.

Information on *Neutrino geophysics* may be read in the monograph of Dye (2007) and on neutrinos in Close's work (2012). Some recent specific details may be found in the work of Kikawa (2016) mainly on the neutrino interaction in mater and on the neutrinos three flavors (i.e. oscillation between the 3 neutrino types: ν_e, ν_μ and ν_τ).

Geoneutrinos carry integrated information about the abundances of their radioactive sources inside the Earth. A major objective of the emerging field of **neutrino geophysics** involves extracting geologically useful information (e.g. such as the occurrence of formation or deposit geoneutrino-producing elements and their distribution in Earth's interior) from geoneutrino measurements. Geoneutrinos originate mainly from β^{-} decay of ^{40}K, and from β^{-} decay-branches of ^{232}Th and ^{238}U. A smaller fraction is also due to β^{-} decay-branches of ^{235}U. Only geoneutrinos from ^{232}Th, ^{235}U and ^{238}U decay chains are detectable by the inverse beta-decay mechanism because these have energies above the corresponding threshold (1.8 MeV).

Detection of geoneutrinos

The geoneutrinos detector applies the reverse beta decay reaction: an electron antineutrino reacts with a proton to produce a positron and a neutron:

$$ {}^{1}_{1}\mathrm{H} + \bar{\nu} \rightarrow {}^{1}_{0}n + e^{+} \qquad \{3.3\} $$

To react the antineutrinos must have energies above the threshold of 1.804 MeV (energy difference between the masses of neutron plus positron and proton). The positron annihilates with an electron:

$$ e^{+} + e^{-} \rightarrow \gamma + \gamma \qquad \{3.4\} $$

with a delay of a few tens to a few hundred microseconds the neutron combines with a proton ^{1}H to form a deuteron ^{2}H nucleus:

$$ {}_0^1n + {}_1^1\mathrm{H} \rightarrow {}_1^2\mathrm{H} + \gamma \qquad \{3.5\} $$

The two light flashes yield by reaction {3.4} and {3.5}, coincident in time and space, provides a powerful method to discriminate single-flash background events in the liquid scintillator. Antineutrinos produced in man-made nuclear reactors (higher energy spectrum see Sect. 3.2.10) may partially overlap in energy with geologically produced antineutrinos and are also counted by detectors but can however be discriminated for their energies.

Existing detectors

The volume of an **antineutrino detector** must be very large in order to detect a significant number of antineutrinos because they only weakly interact with other particles of matter. Antineutrino detectors need to be built underground, to shield the detector from cosmic rays and other background radiations.

KamLAND (Kamioka Liquid scintillator Anti-Neutrino Detector) is a 1.0 kiloton detector located at the Kamioka Observatory in Japan. Results based on an acquisition-time of 749 days and presented in 2005 report the first detection of geoneutrinos e.g. Araki et al. (2005). The total number of counted antineutrinos was 152 while 4.5 to 54.2 were geoneutrinos. The analysis estimated a 60 TW upper limit on the Earth's power due to ^{232}Th and ^{238}U decays. In 2011, an updated result including 2135 days acquisition time, a new scintillator and a reduced reactor background shut down several Japanese reactors. Of 841 detected antineutrinos, 106 were identified as geoantineutrinos. It was found that ^{232}Th and ^{282}U together generate 20.0 TW of radiogenic power.

Borexino is a 0.3 kiloton detector at Laboratori Nazionali del Gran Sasso, Italy. Results published by Collaboration (2010) for an acquisition time of 537 days corresponded to 15 antineutrinos and 9.9 identified as geoneutrinos. The zero geoneutrino hypothesis was rejected (99.997% with (4.2σ) confidence level).

3.1.3.2 From Neutrino Counting to Tomography

Recently, a team of physicists and geologists has generated the first global map of antineutrino emissions. The map (Usman et al. 2015a) provides a series of images of reconstruction of Earth's antineutrino sources (see Fig. 3.6) and existing human-made sources of radiation.

The interior of Earth is difficult to investigate, even with modern technology. Locating the activity of antineutrinos sources allows creation of their distribution images. The map should be particularly useful for future studies of processes within the lower crust and mantle. Antineutrino Global Map 2015 (AGM2015) is an experimentally informed model of Earth's surface electron antineutrino flux over the 0–11 MeV energy range. The open source AGM2015 provides fundamental data for experiments, assists in strategic detector placement to determine neutrino mass hierarchy, and aids in identifying undeclared nuclear reactors (e.g.

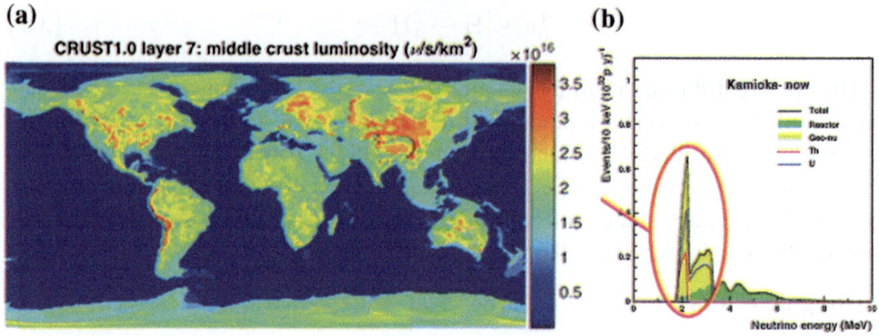

Fig. 3.6 **a** Worldwide antineutrino flux map combining geoantineutrinos from natural ^{238}U and ^{232}Th decay in the Earth's middle crust. Adapted from Usman et al. (2015a), neutrino flux $cm^{-2}\ s^{-1} \Longrightarrow 10^{-2}\ cm^{-2}\ s^{-1}$. Usman (2015b), with permission © Scientific Report. **b** Geoneutrinos detected by KamLAND following the Japan reactor shut down. Shimizu (2013), with permission © APS

underground facilities). The current composition of the lithosphere is confirmed by the estimated flux from the mantle, as measured by the 2 underground detectors, to calculate **the planetary antineutrino** ($\overline{\nu}_e$) luminosity at 3.4 (+2.3/−2.2) × 10^{25} s^{-1} (see Usman et al. 2015a). These data were used by Degueldre and Fiorina (2016) to revisit the hypothesis of a central Earth georeactor which was found unlikely.

3.1.3.3 Measurement of Redox Potential

The redox potential (E or pe) is a fundamental parameter linked in solution to the redox reagents reaction:

$$\mathrm{Ox} + n\,\mathrm{e}^- \Leftrightarrow \mathrm{Red} \qquad \{3.6\}$$

by Nernst (1888) law:

$$E = E_0 + \mathrm{R}T/n\mathrm{F}\ \ln C_{\mathrm{Ox}}/C_{\mathrm{Red}} \qquad (3.5)$$

with E and E_0 the reagents and the standard potentials respectively, C the concentration of the redox species, n the number of electron exchanged during the redox reaction {3.6} F the Faraday constant, R the gas constant and T the absolute temperature of the reagents. A normalized potential is given by:

$$\mathrm{pe} = n\mathrm{F}(E - E_0)/\mathrm{R}T \qquad (3.6)$$

The potential measurement is carried out with an inert working electrode (without fixing the potential) coupled to a reference electrode and corrected by the potential difference with the standard hydrogen electrode. Optimized measurements require clean electrode. This status can be obtained by electrochemical, chemical or mechanical cleaning, e.g. Degueldre et al. (1999).

Redox potential measurements in reactor water have been performed extensively on the water from the primary; however, little has been done under in situ conditions (e.g. 300 °C, 100 Atm). The redox potential also called electrochemical potential (ECP) is measured vs. a reference electrode that is usually at room temperature an Ag/AgCl electrode, and the potential is finally reported to the standard hydrogen electrode (SHE) for comparison.

In PWR system, redox potential measurements on platinum and Type 304 stainless steel in equilibrium (H_2 or O_2) and non-equilibrium ($H_2 + O_2$) boric acid and lithium hydroxide solutions at 275 °C are reported by Macdonald et al. (1981). For equilibrium systems containing hydrogen, good agreement was obtained between the measured potentials on platinum and calculated equilibrium potentials for *p*H (275 °C) = 5.49, 5.60 and 8.02. For highly basic systems (0.1 M LiOH), however, the observed potentials tended to be more positive than the calculated values; the difference was probably due to trace amounts of oxygen remaining in the system. In oxygenated equilibrium systems, good agreement was found between the observed and calculated potentials for the 0.1 M LiOH solution for $C_{O2} \leq 7$ ppm, but the observed potentials tended to be less positive than the calculated values for more acidic solutions and for lower oxygen concentrations. Steady-state measurement on non-equilibrium ($H_2 + O_2$) boric acid solutions at 275 °C showed that the redox potential on platinum becomes more negative when the $[H_2]/[O_2]$ ratio increases. Transient studies on non-equilibrium ($H_2 + O_2$) boric acid solution revealed that the redox potential can vary over a wide range on cycling the relative concentrations of oxygen and hydrogen. The experimental findings were interpreted in terms of mixed potential model. In PWR water a pH of 7.3 ± 0.2 and a redox potential as low as −0.770 ± 0.015 V versus SHE may be measured.

In BWR, the ECP of the primary reactor water at the sampling point is in the range +0.100 to +0.300 V Vs SHE under normal conditions i.e. without any specific chemical addition in the pure water (see Sarott 2005). In order to minimize inter-granular stress cracking risk of stainless steel welds in BWR, hydrogen may be injected into the reactor feed water with a minimum concentration of 2 ppm. This addition reduces drastically the ECP as seen in Fig. 3.7 which also reduces corrosion.

The positive effect of hydrogen water chemistry is unfortunately accompanied by unwanted side effect increase of the main steam dose rate attributed to ^{16}N (and its β, γ decay's) volatile species generated under reducing conditions. The main disadvantage of hydrogen injection may be mitigated by applying 'noble metal addition'. This method is based on the fact that platinum group metals are catalysts which help to recombine oxygen and hydrogen peroxide with hydrogen on the metal surface. With noble metal addition the ECP remains below −0.230 V versus SHE fixed by the specifications.

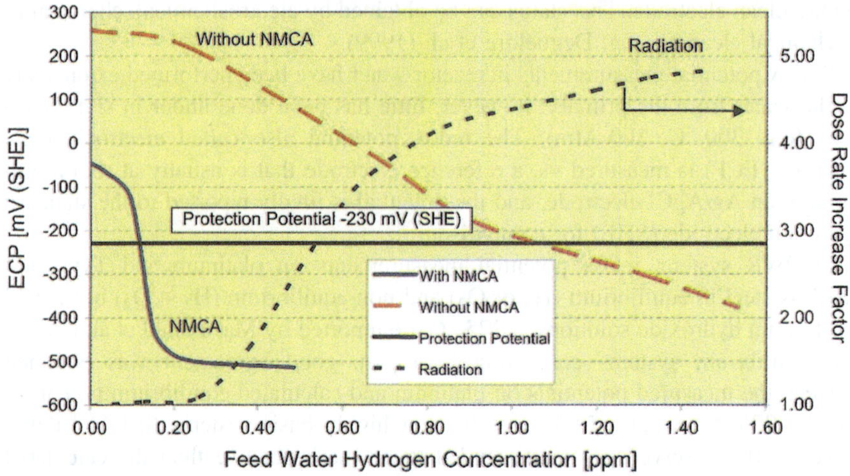

Fig. 3.7 BWR water redox analysis: effect of hydrogen addition on the electrochemical potential (ECP) of stainless steel in the feed water, with and without noble metal chemical addition (NMCA) and main steam line dose rate which increase is due to ^{16}N activity in the steam. Sarott (2005), with permission © Chimia

3.1.3.4 Electron Counting and Spectroscopy

The **beta** (e^-) **emission spectroscopy** (βS) of the isotope ${}^{A}_{Z}X$ undergoes the decay reaction:

$$ {}^{A}_{Z}X \rightarrow {}^{A}_{Z+1}X + e^- + \bar{\nu} \qquad \{3.7\} $$

yielding the next element nucleus ${}^{A}_{Z+1}X$, an electron and an antineutrino $\bar{\nu}$. As a consequence of the energy conservation the beta spectrum is continuous with emission ranging from 0 to E_{max} a maximum values. Figure 3.8 shows the E_{max} of the Cs β^- emitters as a function of their half-life. It is consequently more difficult if not impossible to identify the nucleus on the basis of the beta activity spectrum. However beta activity analysis may be useful in specific cases. The beta spectrometer is a 'C'-shaped vacuum chamber with magnet coils on top and bottom to produce a uniform magnetic field perpendicular to the radius of the chamber and perpendicular to the plane of the 'C' shaped chamber. A source of ^{137}Cs is permanently mounted inside the chamber on a rod top. The detector is mounted at the other end through a slit and a mylar aluminum coated foil over a plastic scintillator detector with light pipe connected to a PhotoMultiplier Tube (PMT).

Liquid Scintillation Counting (LSC) is the measurement of activity of a liquid solution or a gel of colloid sample of radioactive material. It uses mixing the active material with a liquid scintillator, and counting the resultant photon emissions. The purpose is to carry out a more efficient counting using the intimate contact of the

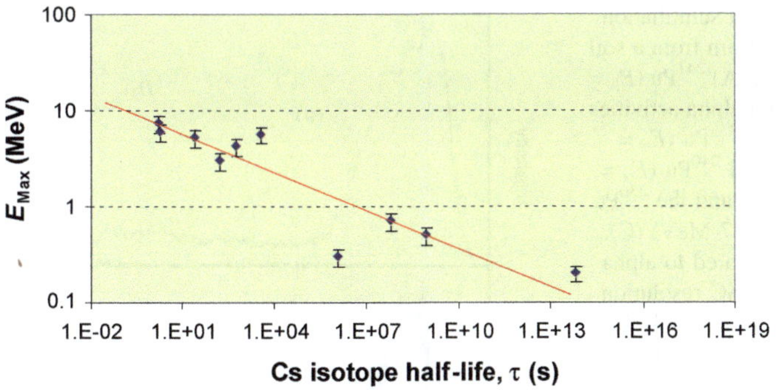

Fig. 3.8 Plot of the maximum energy $E_{\max}$ of the Cs β^- emitters as a function of their half-life. Note the β (and $\bar{\nu}$) energy increase when τ decreases. ^{134}Cs to ^{143}Cs, data from Pfennig et al. (1995)

scintillator with the activity. LSC has been used for beta (and alpha) particle detection, e.g. Rapkin (1964).

The radioactive samples and cocktail are placed in small transparent glass or translucent plastic vials that are loaded into an instrument known as a liquid scintillation counter. Counters usually have two photomultiplier tubes connected. The coincidence counting discriminates and records genuine light pulses, which reach both photomultiplier tubes, while single pulses due to background only detected on one of the tubes are ignored.

Counting efficiencies range from about 30% for ^{3}H ($E_{\max}$ 0.018 MeV) to nearly 100% for ^{137}Cs ($E_{\max}$ 0.5 MeV). Chemical compounds such as chlorine or highly light-absorbing reagents interfere. This interference (quenching) can be corrected or reduced through specific sample preparation.

High-energy beta emitters, such as ^{137}Cs, can also be counted in a scintillation counter without the cocktail, using an aqueous solution. This technique, called Cherenkov counting, is carried out by direct detection of the radiation using the photomultiplier tubes. This simple counting approach is normally used for quick, rough measurements.

The energy resolution is weak and of the order of several 100 keV making the identification of the nucleus difficult. The efficiency is however large (about 100% for the high energy emitters), making this spectroscopy useful for sample doped with a single β emitter.

In the field of ***β*-spectroscopy** (βS), utilization of a very low background **liquid scintillation counting** (LSC) spectrometer enables the analysis of soft beta emitters such as ^{241}Pu (see Fig. 3.9) with detection limits of the order of 10 mBq as reported by Yu-fu et al. (1990).

This makes it possible to estimate the detection limit for β-spectroscopy in Table 3.4. However, since the beta spectrum is continuous, application of the beta

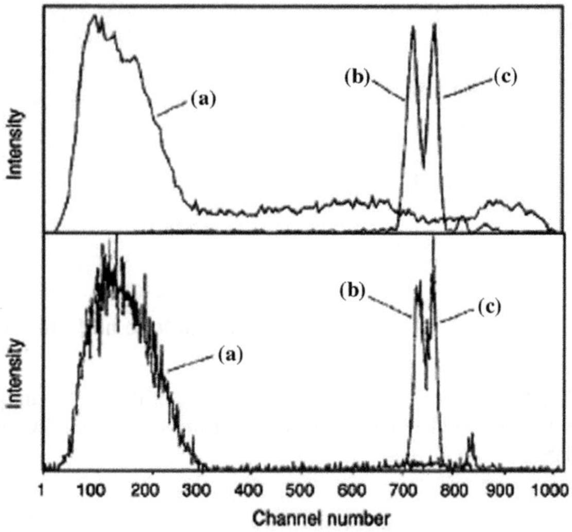

Fig. 3.9 Liquid scintillation counting spectrum from a soil layer showing (A) ^{241}Pu (E_β = 0.02 keV) and alpha activities including both ^{239}Pu (E_α = 5.14 MeV) and ^{240}Pu (E_α = 5.17 MeV) (B) and the ^{236}Pu tracer (E_α = 5.77 MeV) (C). Note that compared to alpha spectrometry LSC resolution is lower. Yu-fu et al. (1990) with permission © Springer

Table 3.4 Passive analytical techniques used for actinide isotope, element or species identification

Detection	Goal	Sample	^{A}An(Y)	N_{DL} (mol)	Remarks
Electron: βS LSC	Concentration or activity determination	Solid film electroplated or liquid bulk scintillation	^{228}Ac ^{234}Th ^{233}Pa U Np ^{241}Pu Am Cm	5×10^{-22} 5×10^{-20} 5×10^{-20} – – 1×10^{-17} – –	βLSC A_{DL}: 10 mBq in 1 mL (Yu-fu et al. 1990)

Detection limit (DL) in **mol** recalculated from DL in Bq

spectroscopy cannot be directly used for the identification of actinides in environmental samples without the use of specific separation techniques. The counting yield of β scintillation counting is always smaller than 100%. The DL is currently of the order of the mBq. The detection limit in term of mole may be estimated on the basis of the half-life of the isotopes. Information on LSC may be found in the "Advances in Tracer Methodology" monograph by Rothchild (1963).

Comparing electron counting and spectroscopy detection limits

Radiometric techniques dominate the analyses of short-lived and medium-lived beta emitter actinide nuclides. The passive techniques currently used for actinide detection are summarized in Table 3.4. A comparison of detection limits is reported in Sect. 4.1.

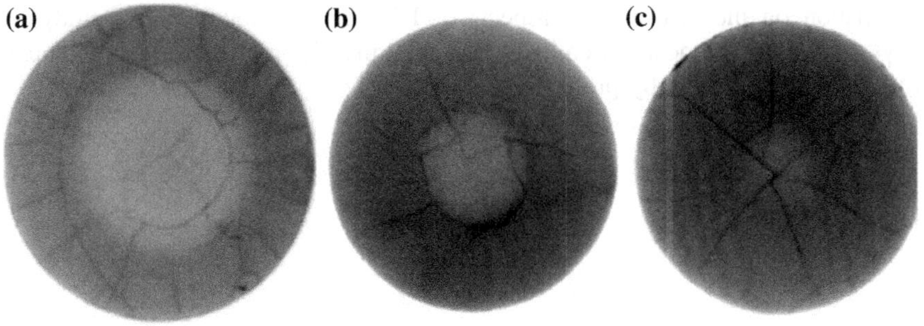

Fig. 3.10 β–γ **autoradiography of fuel pins a** outer, **b** intermediate and **c** central pin of a high burn-up bundler. Sah et al. (2008). with permission © Elsevier

3.1.3.5 β–γ 2D Autoradiography

Autoradiography delivers images from a distribution of the β and/or γ emitters on a nuclear emulsion or an X-ray film. The autoradiograph is also available as a digital image, scintillation gas detectors or rare earth phosphor-imaging systems (see Rogers 1979). The film is apposed to the sample section to obtain the autoradiograph. The *auto*-prefix refers to the passive aspect of the technique, as distinguished from the case of microradiography, in which the sample is X-rayed. Some autoradiographs can be investigated using optical microscope for localization of silver grains in which the process is termed micro-autoradiography.

The **β–γ autoradiography** of fuel pins of a high burn-up fuel element are compared in Fig. 3.10 for the outer, intermediate and central pin positioning the assembly Sah et al. (2008).

The central white region in the autoradiograph indicates that the radioactive fission products had migrated from the central region towards the periphery of the fuel section, resulting in lower activity in the central region and a higher activity in the outer region of the fuel cross-section.

3.1.3.6 Positron Emission Tomography

Positron Emission Tomography (**PET**) is an imaging technique that may be utilized to observe processes occurring in an object. The PET unit detects pairs of γ-rays indirectly generated by a β^+ emitter nuclide, which was introduced into the object to investigate. These γ-rays are emitted by β^+ annihilation with a local electron according to the reaction:

$$\beta^+ + e^- \rightarrow 2\gamma \qquad \{3.8\}$$

The 2γ are recorded by detectors located around the sample. The 3D images of the β^+ tracer concentration in the object are reconstructed by using an algorithm.

Information on the instrumental aspects of PET may be found in the study of Clayton (1990). The positron emitters that are currently used in PET are isotopes produced at a cyclotron by proton p addition:

$$ {}^{A}_{Z}X + p \rightarrow n + {}^{A}_{Z+1}X \qquad \{3.9\}$$

or

$$ {}^{A}_{Z}X + p \rightarrow 2n + {}^{A-1}_{Z+1}X \qquad \{3.10\}$$

The isotopes used traditionally in the industrial applications are presented in Table 3.5 together with their characteristics.

Outside clinical uses, PET has been developed for industrial applications and is expected to become an accepted technology in the chemical engineering industry (see Clayton 1990). The most important current developments are in the environmental sciences, and especially in minerals exploration, process control and contaminant transport. The reason lies in elemental specificity and the penetration of the radiation from nuclear processes. Porosity and pathway determination in crystalline rock by positron emission tomography have been demonstrated experimentally thanks to calibration tests (Degueldre et al. 1996d). Lateral resolution was found to be of the order of 1 cm for a source of 1 MBq ^{68}Ge, a 17 min acquisition time and for a cylindrical granitic sample of 21.5 cm diameter and 20 cm high, with activity in cylindrical holes (of 2.5, 2.0. 1.4 and 1.2 cm diameter).

PET has been used as non-invasive tools to obtain dynamic 3D images of radioactive tracer concentrations through porous columns during infiltration. Dynamic PET obtained using ^{18}F was used for following flow (through a 5 cm diameter × 20 cm high) with and without an impermeable obstacle (see Boutchko et al. 2012). A fluid dynamics model has been used to calculate the observed flow dynamics. The tracer distributions observed in the simulations, in the smaller column uniformly packed with sand and in the column with an obstacle, are similar to the reconstructed images in the PET experiments.

Table 3.5 Characteristics of selected PET nuclides that can be used for analysis of nuclear materials or their environment

^{A}X	**Half life**	β^{+} **decay (%)**	β^{+} E_{max} (MeV)	**Reaction**	**Abundance (%)**
^{76}Br	16.2 h	58.2	3.98	^{76}Se + p → n + ^{76}Br	9.1
^{124}I	4.18 d	22.0	2.15	^{124}Te + p → n + ^{76}Br	4.8
^{66}Ga	9.49 h	56.5	4.15	^{66}Zn + p → n + ^{66}Ga	27.8
^{64}Cu	12.7 h	17.8	0.66	^{64}Ni + p → n + ^{64}Cu	1.16
^{89}Zr	78.5 h	22.8	0.40	^{89}Y + p → n + ^{89}Zr	100
^{18}F	1.83 h	97.0	0.60	^{18}O + p → n + ^{76}Br	99.8
^{68}Ga	1.13 h	88.0	1.90	^{68}Ge + e^{-*} → ^{68}Ga	–

e^{-*} for electron capture
Source Quaim et al. (2007)

PET was also applied by Kulenkampff et al. (2008) for direct visualization of solute transport in order to overcome the limitations of conventional methods for measuring advection and diffusion properties. At intervals from minutes to days the 3D distribution of the tracer was determined. This spatio-temporal profile of the tracer activity can be used as experimental data for clarification of the relevant transport processes, derivation of transport parameters, and model verification. Here, ^{18}F and ^{124}I in 0.01 M carrier solution of KF and KI have been selected from the limited number of PET-tracers, based on their decay time and the required experimental time. The sample was a granite core from the Äspö Hard Rock Laboratory which carries an axial fracture. The experiments showed that the transport path through the fracture is modulated by the flow rate.

3.1.4 Techniques with Neutron Detection

The neutron, *n*, is a non-charged particle (see data Table 1.1). This particle, that was discovered by Chadwick (1932), can be detected using specific spectrometers.

Neutronic decay occurs for radionuclides containing excess neutrons. Neutron emission is one of the ways a nuclide reaches its stability. The instability of the nucleus is the result of nuclei having extra neutrons such as ^{13}Be (mean half-life 2.7×10^{-21} s) and ^{5}He (7×10^{-22} s). Spontaneous fission happens when an atom's nucleus splits into two smaller nuclei and generally one or more neutrons.

Neutron Spectroscopy (NS) is used to record the fluxes and energies of neutrons. This is carried out to measure neutron emitters in material.

In **neutron time-of-flight scattering**, a form of inelastic neutron scattering, the initial position and velocity of a pulse of neutrons is fixed, and their final position and the time after the pulse the neutrons are detected and measured. By the principle of conservation of momentum, these pairs of coordinates may be transformed into momenta and energies for the neutrons, and the experimentalist may use this information to calculate the momentum and energy transferred to the sample. Inverse geometry spectrometers are also possible. Information on neutron applications in materials for energy may be found in the work edited by Kearley and Peterson (2015).

Neutron spectroscopy can be applied in a plain counting mode to detect spontaneously fissile actinides in the environment and in the framework of trafficking. Plutonium-239 is hard to detect by means of its α, X-rays or γ-rays, but neutrons are more penetrating and can be more easily and specifically detected. Recently, sensitive neutron detectors including ^{3}He proportional counter tubes moderator and integrated electronic have been developed. They are based on the reaction:

$$^{3}\mathrm{He} + {}^{1}n \rightarrow {}^{4}\mathrm{He} + \gamma \qquad \{3.11\}$$

It is used to detect ^{239}Pu down to the g (5×10^{-3} mol) level at 20 cm in 5 s (see Klett 1999).

3.1.5 Techniques with Atom or Ion Detection

The ion energy measurement is approached here in a broad sense. It goes from the potential in an oxidation-reduction reaction in solution to the energy of an α particle.

3.1.5.1 Ion Sensitive Electrodes

An **Ion-Selective Electrode (ISE)** is a sensor that converts the activity of a specific dissolved ion in a solution, which can be measured in a passive way using a voltmeter. For detailed information, the monograph "Working with Ion-Selective Electrodes Chemical Laboratory Practice" by Cammann (1979) and "Ion-Selective Electrodes in Analytical Chemistry" by Freiser (1980) may be consulted.

The voltage is theoretically dependent on the logarithm of the ionic activity, according to the Nernst (1888) equation. The equation of the potential $n\mathrm{F}$ (E) is dictated by the ion diffusion from a reference state where the ion activity is A_1 to that being measured A_2:

$$\mathrm{R}T\ln(A_1/A_2) = n\mathrm{F}(E) \quad \text{or} \quad E = (E^*) + \mathrm{R}T/n\mathrm{F}\ln(A_1/A_2) \qquad (3.7)$$

with E positive for a cationic species and negative for an anionic species, n the charge of the ionic species, T the absolute temperature, R the ideal gas constant, F the Faraday constant and E^* a standard potential for $A_1 = A_2$.

Measurements are usually made using both an ion-specific membrane electrode and reference electrode. Ion-selective electrodes can be used in-line in systems where measurements of the ionic concentration in solution are required, usually on a real-time basis. Because of the logarithm form of the equation, the analysis of the activity of the species can be measured over several orders of magnitude. The precision is of the order of the %, and the response is relatively rapid (s to min).

The most serious problem limiting use of ion-selective electrodes is interference from other, undesired, ions. No ion-selective electrodes are completely ion-specific; all are sensitive to other ions having similar chemical properties, to an extent which depends on the degree of similarity. Most of these interferences are weak enough to be ignored, but in some cases the electrode may actually be much more sensitive to the interfering ion than to the desired ion, requiring that the interfering ion be present only in relatively very low concentrations, or entirely absent.

There are three main types of ion-selective membrane used as ISEs: glass, solid state and liquid-based electrode.

Glass membranes are made from an ion-exchange type of glass (silicate or chalcogenide). These ISE's have good selectivity, but only for several single-charged cations; mainly H^+, Na^+, and Ag^+. Chalcogenide glass also has selectivity for double-charged metal ions, such as Pb^{2+}, and Cd^{2+}. The glass membrane has excellent chemical durability and can work in aggressive media at

Fig. 3.11 Schematic presentation of the cell including the ion selective liquid membrane for M^{2+} (Ni^{2+} or Cu^{2+}) analysis

room temperature. A very common example of this type of electrode is the pH glass electrode.

Crystalline membranes are made from mono- or polycrystalline single compound. They have good selectivity, because only ions which can introduce themselves into the crystal structure can interfere with the electrode response. Selectivity of crystalline membranes can be for both cation and anion of the membrane-forming substance. An example is the fluoride selective electrode based on LaF_3 crystals.

Liquid-based electrodes

At room temperature liquid membrane ion-selective electrodes (see Fig. 3.11) have been successfully tested for potentiometric dosage of divalent copper and nickel in solution (see Pleniceany et al. 2005). N-[2-thienylmethylidene]-2-aminoethanol (TNAHE) in nitrobenzene was used as selective liquid. The analysis detection limit was around 10^{-6} M. Signal response was of the order of 2 min for pH going from 2.5 to 9.5.

pH electrodes are glass electrodes utilized on a pH meter: a potentiometer. The electrode potential depends on the activity of H^+ ion and consequently on the pH of the solution. The pH is measured after calibration of the pH meter with standard solution (pH buffer).

The reactor water pH is currently measured on line at the sampling line (Room temperature, P = 1 Atm) in PWR unit. In BWR pH is more difficult to measure because of the high resistivity of the reactor water.

High-purity water quality monitoring based on ion-selective electrode technology has been applied to power plant instrumentation. Measurement of Cl^-, Ca^{2+}, S^{2-} in high-purity water has been carried out successfully on-line by Diggens et al. (1981).

More recently, in order to improve the lithium monitoring in the primary coolant of PWR plants, a new concept using lithium ion selective electrode (Li-ISE) was investigated. Costarramone et al. (2000), measured lithium in PWR water using lithium ion selective electrodes.

A pH-meter is also used for acid titration in extraction phase processes for example (see Sect. 3.2.16.1).

3.1.5.2 α Counting and Spectroscopy

The alpha (α) Spectroscopy (αS) makes use of the α emission of the isotope ${}^{A}_{Z}X$ according to the decay reaction:

$$^{A}_{Z}X \rightarrow ^{A-4}_{Z-2}X + ^{4}_{2}He \qquad \{3.12\}$$

yielding a nuclide $^{A-4}_{Z-2}X$ and a helium $^{4}_{2}He$. As a the alpha spectrum is discontinuous with defined emission of alpha energy lines and energy values for the maximum intensity: E_{max}.

The detection of the alpha particles can be made in vacuum using a silicon diode coupled to a MCA, in a solution/suspension using a fluorescent reagent and photomultipliers also coupled to a MCA, or for a flat sample by autoradiography in contact with a flat detector (see Sect. 3.1.2).

Usually for **α spectroscopy** analysis a drop of the test solution is placed on a metal disk and dried out to give a uniform coating on the disk. The loaded disk is then used as the test sample; a spectrum of the α activity in the coating may be recorded. If the thickness of the coating is too thick then the lines of the spectrum are broadened to lower energies. This is because some of the energy of the alpha particles is lost during their trajectory through the active material coating.

When the solution is too salt concentrated the evaporation technique is replaced by electroplating the α emitters on the metal disc. Several solution additives are suggested to allow better coating (e.g. actinide oxo-hydroxides). Details on this technique may be found in the monograph by Aggarwal (2016a, b).

Standardization for energy scale together with activity calibration are performed for example using standard ^{241}Am sources. Peak resolution is of the order of 60 keV (FHWM). The high-resolution silicon detectors have proved to be sensitive down to 10 μBq levels for analysis of both natural Th and U isotopes and daughter nuclides, as well as for the anthropogenic actinides. In natural samples ^{234}U and ^{238}U may be close to the equilibrium and similar detection limits are observed in term of activity for both. The DL for ^{235}U is better than for ^{234}U and ^{238}U (see Jia et al. 2002). The isotope ^{227}Ac was quantitatively determined in environmental samples after sample treatment and after electrodeposition a first α count at 4.85–4.95 MeV for the 1.38% α decay of ^{227}Ac, and a second for 5.5–6.1 MeV after ^{227}Th build-up to equilibrium (see Bojanowski et al. 1987; Martin et al. 1995).

An alternative method is to use internal **liquid scintillation counting**, where the sample is mixed with a scintillation cocktail as described in Sect. 3.1.2.3. The light emissions are then counted. Some units record the amount of light energy per radioactive decay event. Due to absorption in the scintillation cocktail with the sample, quenching can reduce the number of photons generated per radioactive decay, a broadening of the alpha spectra may be obtained through liquid scintillation. It is likely that these liquid scintillation spectra will be subject to a Gaussian broadening, rather than the distortion exhibited when the layer of an active material on a disk is too thick.

A calibration is required because MCA does not work on energy, it works on voltage. To relate the energy to voltage one must calibrate the detection system. Here different alpha emitting sources of known energy need to be placed under the detector and the full energy peak has to be recorded.

Alpha liquid scintillation counting (LSC) is attractive because it offers a nearly 4π geometry and because the counting yield for an actinide α-emitter is about 100%, but with a lower energy resolution than for alpha spectroscopy (αS). Improvements for α energy resolution and background reduction are key needs. An improvement of α energy resolution for determining low-level plutonium has been achieved using combined solvent extraction-low level liquid scintillation counter (Yu-fu et al. 1990) and can also be applied for $^{239+40}$Pu and ^{241}Pu activity measurements in sea water (Irish Sea and North Sea) and soils (Cumbria and Belorussia) (Yu-fu et al. 1990). Resolution of the order of 275 keV for liquid scintillation spectra can be achieved, which allows low level determination of plutonium,

On a LSC spectrum of α emitters the peaks due to ^{209}Po, ^{210}Po, ^{239}Pu and ^{241}Am are observed from left to right. The fact that isotopes such as ^{239}Pu and ^{241}Am have more than one alpha line indicates that the nuclide has the ability to be in different discrete energy levels.

Comparing ion counting and spectroscopy detection limits

Radiometric techniques dominate the analyses of short-lived and medium-lived alpha emitter actinide nuclides. The passive techniques currently used for actinide detection are summarized in Table 3.6. A comparison of detection limits is reported in Sect. 4.1.

An **autoradiograph** is an image on a nuclear emulsion produced by the pattern of decay emissions from a distribution of a radioactive substance. The film or emulsion is apposed to the labeled tissue section to obtain the autoradiograph. Methodology is as described in Sect. 3.1.2.

As an example, analysis of a thorium MOX fuel pellet cross-section reveals precious information on the actinide distribution into the nuclear ceramic material.

Table 3.6 Passive analytical techniques used for actinide isotope, element or species identification. Detection limit (N_{DL}) in **mol** recalculated from A_{DL} in **Bq**

Detection	Goal	Sample	^{A}An (Y)	N_{DL} (mol)	Remarks
Ion: αS LSC	Determination of mass or of isotope activity and identification of isotope	Solid film electroplated or liquid bulk scintillation	^{227}Ac	4×10^{-18}	αS, A_{DL} : 10 μBq (Bojanoswki et al. 1987; Yu-fu et al. 1990; Degueldre et al. 1994)
			^{228}Th	1×10^{-21}	
			^{230}Th	6×10^{-17}	
			^{232}Th	1×10^{-11}	
			^{231}Pa	2×10^{-17}	
			^{234}U	2×10^{-21}	
			^{235}U	4×10^{-17}	
			^{238}U	3×10^{-11}	
			^{237}Np	2×10^{-15}	
			^{238}Pu	7×10^{-20}	
			^{239}Pu	2×10^{-17}	
			^{240}Pu	5×10^{-18}	
			^{241}Am	3×10^{-19}	
			^{242}Cm	3×10^{-22}	
			^{248}Cm	1×10^{-20}	

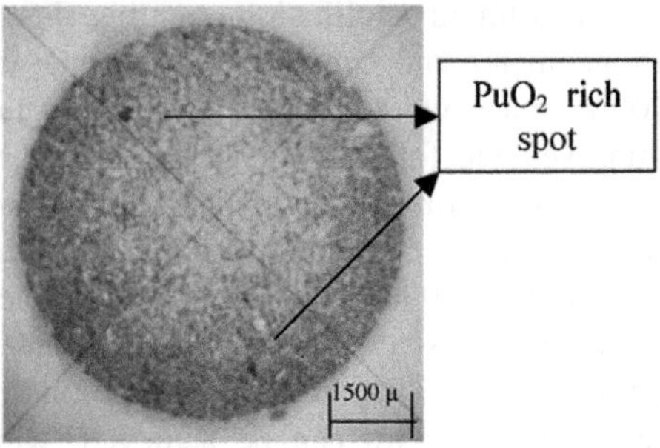

Fig. 3.12 Alpha autoradiograph of a heterogeneous MOX fuel pellet (ThO_2—4 wt% PuO_2) showing PuO_2-rich spots. Rao et al. (2003), with permission © Elsevier

The fuel is made of ThO_2 doped with 4 wt% PuO_2 and was obtained by powder mixing, pelletizing and sintering. Alpha autoradiography (Fig. 3.12) of the MOX fuel pellet cross-section reveals PuO_2-rich spots that remained undisclosed during sintering (see Rao et al. 2003). Microscopic PuO_2 particles may be visualized using this passive technique.

3.2 Material Characterization by Field-Free Interactive Techniques

The characterization of an isotope, element or species can be based on its wave or particle absorption properties. The approach is dictated by the duality particle-wave absorption laws. The incident intensity I_0 of wave is attenuated by absorption through a sample of thickness x according to:

$$\log_{10}(I/I_0) = -\mu x \tag{3.8}$$

where μ is the absorption coefficient of the wave (see Beer 1852), or:

$$\log_{10}(I/I_0) = -\varepsilon C x \tag{3.9}$$

where ε is the specific absorption coefficient and C the concentration.

Particle absorption is described by:

$$\log_{10}(I/I_0) = -N \sigma x \tag{3.10}$$

where σ is the absorption cross section and N the number of absorber or density (see Einstein 1917). In reality, the beam through a sample may interact according to modes going from transmission to reflection as sketched in Fig. 3.13.

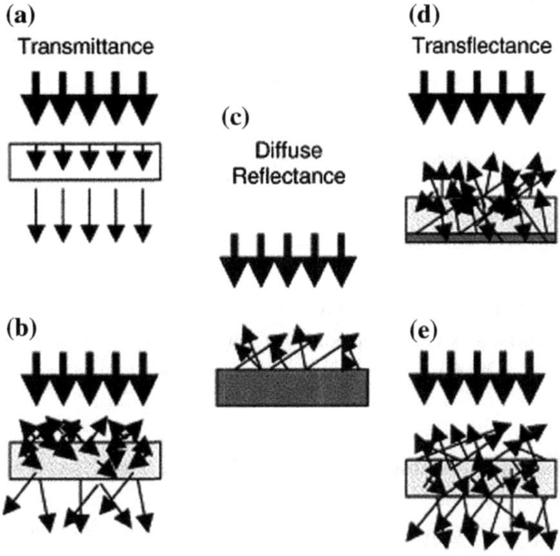

Fig. 3.13 Beam behavior through a sample: from transmission to reflection. **a** Pure transmittance. **b** Transmittance with scattering. **c** Reflectance with scattering. **d** Transflectance. **e** Transflectance with partial transmittance

3.2.1 Techniques with Phonon Interaction and Phonon Detection

Analysis of sound and acoustics is important in engineering for tasks such as product design, production processing, quality performance and production control e.g. Hamstad (1986). The work might involve design fine-tuning to meet research requirements. Designers are also using acoustic instrumentation to reduce the noise generated in power plant. Acoustical measurements and instrumentation range from a handheld sound level meter to a 1000-microphone phased array. Most of the acoustical measurement and instrumentation systems can be broken down into three components:

- source
- sensors
- data acquisition and analysis system.

The most common acoustic sensor used for measurement is the microphone. It must be reactive along the largest range of frequencies starting below 20 Hz–200 MHz (see Fig. 3.14). Measurement microphones are different from typical microphones because they can provide a detailed calibration for their response and sensitivity. Other sensors include hydrophones for measuring sound in water or accelerometers for measuring vibrations causing sound. The three main groups of microphones are pressure, free-field, and random-incidence, each with their own correction factors for different applications.

Audio and acoustic analysis includes: fractional-octave analysis, sound-level measurements, power spectra, frequency response measurements, and transient

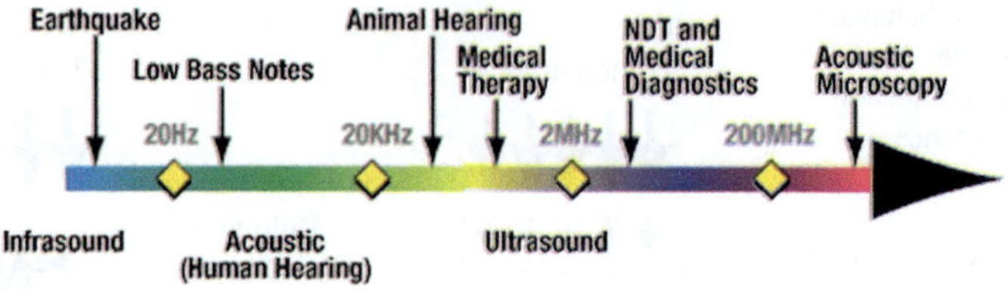

Fig. 3.14 Illustration of the sound spectrum

analysis. Results are viewed on waterfall displays, color map displays, and octave graphs. Acoustic applications for material characterization were reviewed by Zheng et al. (2000).

3.2.1.1 Phononic Acoustic Measure and Spectroscopy

The way sound interacts within materials is very complex. Acoustic wave behaviors include absorption, diffraction, reverberation, and refraction. Absorption refers both to the sound that transmits through and the energy that is dissipated in a material. Sound waves bend around surfaces when reflecting. This bending is called diffraction. Reverberation is caused by boundary reflections. It is the persistence of sound after stopping the acoustic signal. Refraction another sound wave bending is caused by changes in the medium through which the wave is passing. Temperature gradients, for example, cause bending in sound waves. Phonons or acoustic waves are used to investigate various materials. Details on acoustic properties for the analysis of material behavior are available in the monograph by Wevers (1997).

Applying a mechanical load to a composite material can result in many types of damage: matrix cracking, debonding between matrix material and various phases in a composite material, failure and delamination in the case of layered composites. A deeper understanding of the initiation, the growth and interaction of the damages during mechanical loading is important and requires damage monitoring. Acoustic emission is the only non-destructive test technique capable of detecting all of the above mentioned damage types in composites (see Hamstad 1986; Grosse and Ohtsu 2008).

The nuclear material examples deal with concrete and integrity, reactor pressure vessel, internals as well as reactor pipe. Reinforced concrete integrity is an important issue for reactor confinement structures. Many issues are not visible on the surface of consolidated concrete material. Therefore, an investigation of reinforced concrete components should include techniques which allow peering into its internal state (see Scott 2013). Ultrasonic shear wave tomography is used to determine presence and extent of voids, cracks perpendicular to the surface, and delamination. This, or similar, technique has been utilized in the nuclear industry to determine structural conditions. Development and application of mitigation techniques for reactor pressure vessels, reactor internals, steam generators, pressurizer nozzles, control rod drive mechanisms, and secondary piping are also described, (see Jeong and Hwang 2013). Mitigation and repair methods for degradation of PWR components are also suggested.

3.2.1.2 Acoustic Microscopy

Acoustic Microscopy (AM) is a microscopy that makes use of very high or ultra high frequency ultrasound. Acoustic microscopes operate non-destructively and penetrate most solid materials to image internal features, including defects such as cracks, delaminations and voids. The most recent basic types of acoustic microscope developed are the scanning acoustic microscope (SAM), Confocal scanning acoustic microscope (CSAM), C-mode scanning acoustic microscope (C-SAM) and the acoustic-atom force microscope (A-AFM), see Fig. 3.15. This discussion will be limited to these instruments, e.g. Korpel and Kessler (1971).

Ultrasounds are generally defined as having frequencies above 20 kHz. This is the highest frequency detectable by humans. The acoustic microscopes emit ultrasound ranging from 5 to 400 MHz and resolutions of micrometer size are achievable. The ultrasound that penetrates a sample may be scattered, absorbed or reflected by the internal features or the material itself. Ultrasounds that have reflected from an internal feature, or that have traveled through the entire thickness of the sample, are used to make acoustic images. Information may be found in "Advances in Acoustic Microscopy" edited by Briggs (1995).

Samples need no special treatment before acoustic imaging, but they should be able to withstand at least brief exposure to water or to another fluid, since air is a

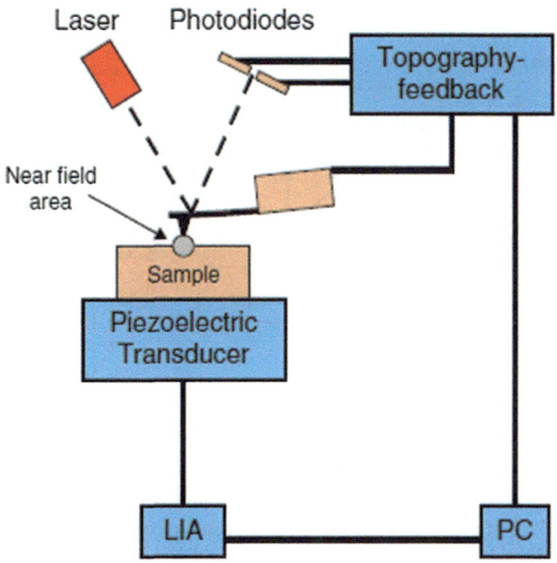

Fig. 3.15 Acoustic-atom force microscope (A-AFM) set-up

very poor transmitter of high frequency acoustic energy from the transducer. The sample may be completely immersed in the water. Alternately, other fluids can be used to avoid contamination of the sample. Samples typically have one planar surface that is scanned. An ultrasound pulse from the transducer travels ns to μs to reach an internal interface to be back reflected to the transducer. When several internal interfaces reflect the acoustic signal at different depths, the echoes reach the transducer after different times. Planar acoustic images do not often use all return echoes from all depths to make the visible acoustic image. Instead, a time window is created that accepts only the return echoes from the fixed depth. Detail information may be found in the work "Acoustic Scanning Probe Microscopy" edited by Marinello et al. (2013).

The nuclear material example deals with hydride detection in Zircaloy cladding. This is a safety-relevant issue because hydride formation can lead to cladding rupture. Figure 3.16 presents the focusing of a low-frequency acoustic microscope and the result about H concentration of a hydrided Zircaloy sample. The determination of hydrogen concentration in Zircaloy claddings using AM is possible in the 0 to 1000 ppm range (see Yang and Huang 2004).

There are also numerous applications in nuclear industries such as products that involve tubing, ceramic materials, composite materials or some types of welds that may be imaged acoustically.

Determination of high burn-up nuclear fuel elastic properties with acoustic microscopy has been reported by Laux et al. (2012). The fuel was a non-irradiated UO_2, simulated spent fuel with additives (inactive elements as fission products mimicking burn-up) and irradiated fuel studied by focused acoustic microscopy. A parametric study was conducted by performing measurements on depleted

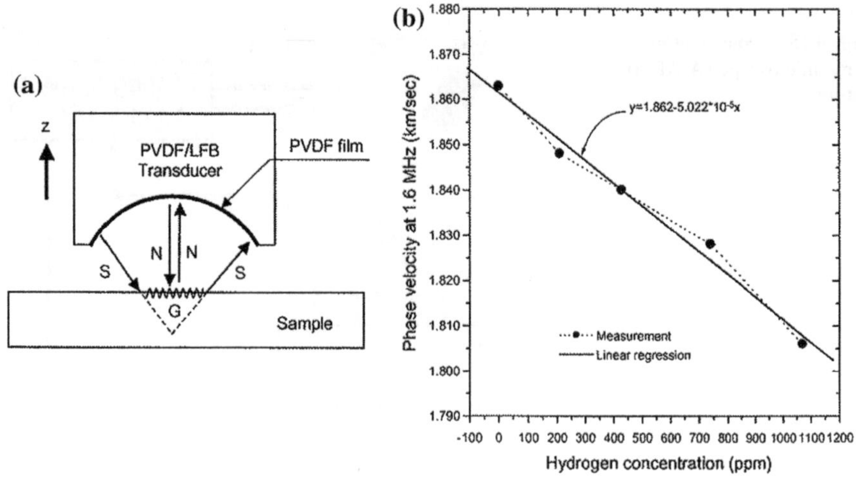

Fig. 3.16 Acoustic microscopy **a** Schema of the principle used for material characterization. **b** Plot of phase velocity versus hydrogen concentration, PVDF and LFB stand for polyvinylidene difluoride and line-focus-beam. Yang and Huang (2004), with permission © Elsevier

uranium oxide for various volume fractions of porosity, oxygen-to-metal ratios, grain sizes and SIMFUEL. It was shown that the main parameters influencing the elastic moduli are the amount of fission products in solid solution, and the pore shape and density. The other parameters irradiation defects, oxygen-to-metal ratio and grain sizes mainly increase the attenuation of the ultrasonic wave but do not change the wave velocity. Measurements on irradiated fuels were subsequently performed. A global decrease of 25% of the elastic modulus was observed between 0 and 100 GWd t^{-1}.

3.2.1.3 Seismic Reflection

Seismic Reflection (SR) (or **reflection seismology**) is a geophysical method of investigation that uses the seismic or shear (S) waves to estimate the properties of the Earth's subsurface from transmitted and reflected waves. Reflection seismology is similar to sonar and echo-location.

Seismic waves are mechanical perturbations that travel in the Earth at a speed governed by the acoustic impedance of the medium in which they are travelling, (see Chapman 2004). The acoustic impedance, Z, is defined by the equation:

$$Z = v\rho \tag{3.11}$$

where v is the wave velocity and ρ is the density of the rock. The wave equation is the relationship between its velocity v, the wavelength λ and frequency ν:

$$v = \lambda \nu \tag{3.12}$$

For a frequency of 50 Hz the seismic wavelength is of the order of 60 m and the resolution limit would be ¼ of the wave or 15 m. In sedimentary rock the wave velocity is of 3000 m s^{-1} while in crystalline rock it can reach 6000 m s^{-1}. Table 3.7 gives the main properties of the S wave for the traversed rocks.

The method requires "controlled" seismic energy sources, such as explosive charges, a specialized air gun or vibrators. Vibrators (see Fig. 3.17) are typically programmed to produce waves with for example increasing frequency from 10 to

Table 3.7 Main properties of the investigated rock and used S wave (data from Mavko et al. 1998)

Rock	Density (t m^{-3})	Velocity (m s^{-1})	Frequency (Hz)	Wavelength (m)
Sand	1.5–1.7 (2.6)	100–500	50	2–10
Clay	2.0–2.4	200–800	50	4–16
Limestone	2.4–2.7 (2.7)	2000–3000	50	40–60
Granite	2.5–2.7	2500–3300	50	50–66
Basalt	2.7–3.1	2800–3400	50	56–67

Corrections for pressure, porosity and humidity are required

Fig. 3.17 Seismic wave source and detector, with permission © Nagra, Switzerland. **a** On sufficiently stable paths and roads, seismic waves are generated. **b** A geophone works like a highly sensitive microphone, converting ground vibrations into electrical signal. The long metal spike ensures good coupling with the ground

120 Hz in about 20 s. The seismic wave travels through the Earth and when it encounters an interface between two materials with different acoustic impedances, some of the wave energy reflects off the interfaces (see Fig. 3.18) and some refracts through the interface. At its most basic, the seismic reflection technique consists of generating seismic waves and measuring the time taken for the waves to travel from the source, reflect off an interface and be detected by an array of receivers (piezosensors or geophones) at the surface (see set-up Fig. 3.17). They are typically disposed every 30 m by groups of 12 for example. Knowing the travel times from the source to various receivers, and the velocity of the seismic waves, a geophysicist attempts to reconstruct the pathways of the waves in order to build up an image of the subsurface, e.g. Fig. 3.18.

The monograph "Seismic reflection processing with special reference to anisotropy" by Upadhyay (2004) may be used as information source.

Interpretation of reflections

The time it takes for a reflection from a particular boundary to arrive at the geophone is called the travel time. If the seismic wave velocity in the rock is known, then the travel time may be used to estimate the depth to the reflector. For a simple vertically traveling wave, the travel time t from the surface to the reflector and back is called the Two-Way Time (TWT) and is given by the formula:

$$t = 2\frac{d}{v} \tag{3.13}$$

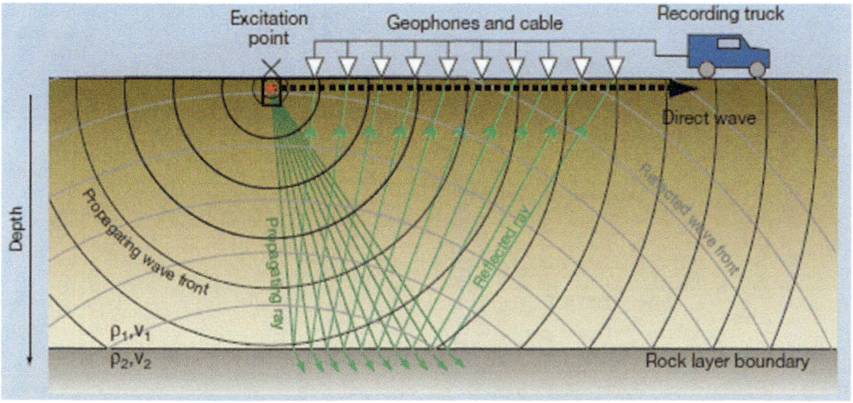

Fig. 3.18 Principe of the reflection seismic method with seismic wave propagation, (see Nagra 2000, with permission © Nagra, Switzerland. Conditions: $\rho_{1,2}$ density of the rock, $v_{1,2}$ wave velocity)

where d is the depth of the reflector and v is the wave velocity in the rock. A series of apparently related reflections on several seismograms is often referred to as a reflection event. By correlating reflection events, a seismologist can create an estimated cross-section of the geologic structure that generated the reflections. Interpretation of large surveys is usually performed with programs using high-end 3D computer graphics, (see Sheriff and Geldart 1995). The work "Aspects of Seismic Reflection Data Processing" edited by: Marschall (1990) may also be used as a source of information.

Three examples related to the nuclear fuel cycle are presented together with an illustration of a local campaign. The first one is related to the prospection of a uranium ore body, the second and third deal with the monitoring of water-bearing features or host rock formation reconstruction and fine analysis in planned radioactive waste repository sites.

The first example deals with the 3D seismic reflection campaigns that have been conducted to assist mining plan at the Millennium uranium deposit, Canada (Juhojuntti et al. 2012). The deposit is located within the crystalline basement, separated from the overlying sediments by an unconformity with significant fluid flow. The first objective of the survey was to image the unconformity and possible deformation structures in and around the deposit. The seismic interpretation shows large variations in the unconformity depth (from ~430 to ~650 m), indicating a pronounced basement depression that coincides with a gravity low. Although the **ore body is not directly visible** in the seismic image, there is a lack of reflectivity coincident with the alteration surrounding the mineralization. The seismic data further indicate faults in the vicinity of the ore body. Based on the initial seismic interpretation, the depth of the crown pillar was adjusted and the mine infrastructure moved away from areas interpreted to be affected by the intense hydrothermal

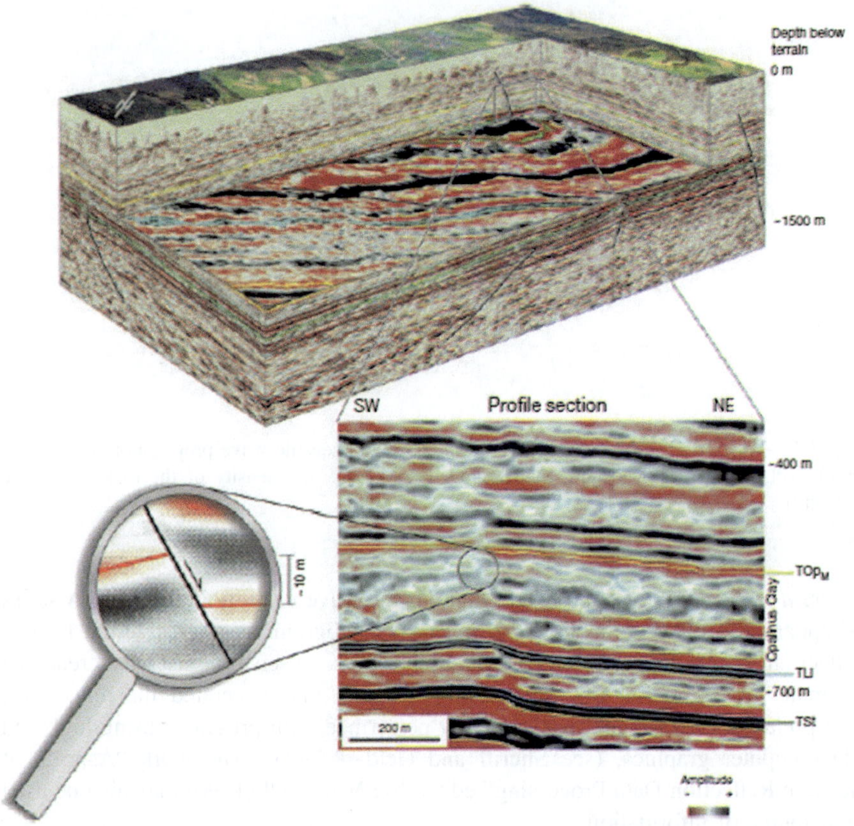

Fig. 3.19 3D seismic reconstructed image of a sedimentary formation system. The location of the rock layers can be visualized. Displacement of ~ 10 m in the largely undisturbed Opalinus Clay can be recognized directly. The enlargement shows one of the few faults in the studied area; it is part of a local flexure. Nagra (2000) with permission © Nagra, Switzerland

alteration surrounding the deposit. The capability to image the system also highlights the potential use of seismic surveys in uranium exploration.

The second example concerns the planning of a final repository in a granitic formation. A Scandinavian company has been carrying out extensive studies at the location of the planned repository for spent nuclear fuel. Identification of sub-horizontal to gently dipping, seismic reflections are especially important since these features may represent transport routes for radionuclides. Studies have shown that such reflections can be generated by water-filled fracture zones that have lower velocity than the surrounding bedrock, (see Zhang and Juhlin 2014). At the site, it is difficult to distinguish fracture zones from mafic sills in the standard reflection seismic processed sections. However, the two possibilities could be differentiated when reconstructing the underground velocity field. Seismic full waveform inversion has the potential to perform this reconstruction, discriminating between fractures zones and mafic sills. In this study, a 2-D waveform inversion code was applied on crooked line data sets acquired at the site.

The third example deals with investigations in Opalinus Clay: a potential host rock for a geological repository of radioactive waste. As illustration of a local campaign, the 3D seismic profile is displayed in Fig. 3.19. The reconstructed image of a sedimentary formation system 5000 $\times$ 2500 m through depth down to 1750 m reveals from the seismic signal the location of the sedimentary rock layers above crystalline rocks. At depths of 550 to 650 m a formation of Opalinus Clay is found. Displacement of $\sim$10 m in the largely undisturbed Opalinus Clay can be recognized directly. The enlargement shows one of the few faults in the studied area; it is part of a local flexure. These flexures need to be identified in the system to be eventually used in models.

3.2.2 Techniques with Photon Interaction and Phonon Detection

These techniques concern acoustic detection of material investigated by laser pulses. Phenomena induced by light absorption are summarized by the reaction scheme:

$$\begin{aligned} &\text{Laser irradiation on sample} \rightarrow \text{excitation} + (\text{chemical reaction \& radiation emission}) \\ &\qquad \rightarrow \text{Heat} + (\text{Phase transition, heat diffusion}) \rightarrow \text{sound wave.} \end{aligned}$$

3.2.2.1 Laser Induced Photo-Acoustic Spectroscopy

In Induced Photo-Acoustic Spectroscopy (LIPAS), a photon–phonon or photo-acoustic method, the absorption of photons induces heat that results by sample dilatation in a small detectable choc wave. With LIPAS the need for sample treatment is minimal. The liquid or solid (solid piece or powder) sample is placed in a cup which is inserted into the photo-acoustic cell sealed for measurement. For example, a piece of ceramics can be introduced in the cup prior to measuring the spectrum. The spectra obtained from different sample preparation protocols will be slightly different due to differences of the samples. Information on photo-acoustic and photo-thermal phenomena may be found in the proceedings of the 5th International topical meeting edited by Hess and Pelzl (1988).

LIPAS has been applied for speciation studies of uranium and transuranium species in aqueous solution as well as in precipitates. LIPAS has been adapted for the UV-VIS and NIR photon energy domain as well as in the infrared. The technique applies the absorption of light that creates locally a thermal shock wave that is detected through a piezo-electric crystal using a dual beam system (see Pollard et al.

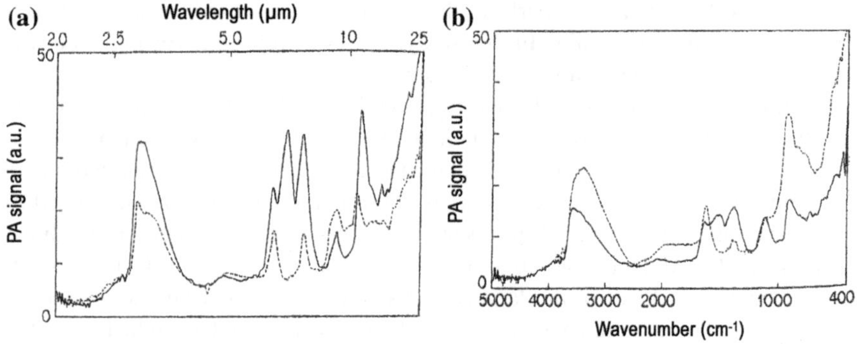

Fig. 3.20 LIPAS IR-spectrum of uranium precipitate **a** in 1 M $NaHCO_3$ (*bold line*) and 1 M $NaClO_4$ (*broken line*) solutions, **b** in 0.05 M (*bold line*) and 0.02 M (broken line) $NaHCO_3$/$NaClO_4$ solutions. Kimura et al. (1992), with permission © Springer

1988; Klenze et al. 1991). Modern photo-acoustic detectors (see e.g. Kosterev 2014) still rely on the same principles as Bell's instrument. The following modifications have however been made to improve the sensitivity:

- Use of intense lasers to illuminate the sample since the intensity of the generated sound is proportional to the light intensity.
- The ear has been replaced by sensitive microphones. The microphone signals are further amplified and detected using lock-in amplifiers.
- By enclosing the gas sample in a cylindrical chamber, the sound signal is amplified by tuning the modulation frequency to an acoustic resonance of the sample cell.

LIPAS permits identification of U, and its speciation may be performed down to 10^{-6} M. It may be used to achieve speciation of uranium ions in mill tailing water (Geipel et al. 1998). This technique may be used for the speciation of uranium in solution or in precipitates (Kimura et al. 1992), see Fig 3.20. LIPAS has also been used to characterize plutonium species as Pu(VI) in slightly acidic solutions (Okajima et al. 1991) and in neutral solutions (Neu et al. 1994) as well as the generation of americium colloids (Buckau et al. 1986). This was carried out for various actinide speciation tests in solution that simulate environmental conditions: e.g. see Fig. 3.20. Using an optical heterodyne interferometer, the LIPAS unit can be modified in a **photo-thermal displacement** instrument, which can be used for the speciation of lanthanides and actinides (Kimura et al. 1992).

3.2.2.2 Laser Induced Breakdown Spectroscopy (LIBS)

Laser Induced Breakdown Spectroscopy (**LIBS**) **in acoustic mode** is a type of atomic excitation spectroscopy which uses a highly energetic laser pulse as the excitation source. The laser is focused to form a plasma, which atomizes and excites samples. In principle, LIBS can analyze any matter regardless of its physical state:

solid, liquid or gas. Because all phase differences emit sound of characteristic frequencies when excited to high temperatures, LIBS can (in principle) detect phases, limited only by the power of the laser as well as the sensitivity and wavelength range of the acoustic detector. If the constituents of a material to be analyzed are known, LIBS may be used to evaluate the relative abundance of each constituent element, or to monitor the presence of impurities. Theory and details on LIBD may be found in the work edited by Miziolek et al. (2006).

LIBS operates by focusing the laser onto a small area at the surface of the specimen; when the laser is discharged it ablates a very small amount of material, in the range of ng to pg, which generates a plasma plume with temperatures in excess of 100,000 K. During data collection, typically after local thermodynamic equilibrium is established, plasma temperatures range from 5000 to 20,000 K. At the high temperatures during the early plasma, the ablated material dissociates (breaks down) into excited ionic and atomic species. During this time, the plasma emits a dilatation wave. Laser Induced Breakdown Detection (LIBD) application to colloids or glass materials using piezo detector has a speciation potential more sensitive than VIS, e.g. Hrdlička et al. (2009). In practice, detection limits are a function of: (a) the plasma excitation temperature; (b) the phonon detection system; and (c) the line strength of the viewed transition.

When the laser intensity is increased, phase breakdown occurs. Instead of a simple thermal shock, the material is atomized (breaks down). LIBS is used to detect colloidal particles (Gutmacher et al. 1987) and allows the detection of actinide colloids or actinide-associated particles. LIBS was applied to determine the solubility product of Th(IV) colloids (Bundschuh et al. 2000; Bitea et al. 2003).

3.2.2.3 Performance of the Interactive Photon–Phonon Techniques

The photon–phonon interactive analytical techniques include LIPAS and LIBD. They have similar performances. Comparison of LIPAS detection limits applied for actinide elements or species for their characterization is given in Table 3.8. A comparison of detection limits is reported in Sect. 4.1.

Table 3.8 Comparison of detection limits for photon–phonon interactive analytical techniques applied for actinide isotope element or species characterization

Detection	Goal	Sample	^{A}An(Y)	C_{DL} (M)	Remarks
LIPAS, (LIPDS) LIBD	Speciation of complexes or redox species	Liquid bulk	Ac(III) Th(IV) Pa(V) U(VI) Np(V) Pu(VI) Am(III) Cm(III)	– – – $5\ 10^{-6}$ $1\ 10^{-7}$ $2\ 10^{-5}$ $1\ 10^{-8}$ –	LIPAS Geipel et al. (1998) Pollard et al. (1988) Okajima et al. (1991) Klenze and Kim (1988)

3.2.3 Techniques with Photon Interaction and Photon Detection

The analysis using photon illumination and photon detection starts with the classical observation in the visible range using white, pink or monochromatic light. In this section the same scheme is suggested, starting with low-energy illumination and investigating the techniques when increasing the incident photon energy.

3.2.3.1 Ground Penetrating Radar

Ground Penetrating Radar (**GPR**) is an interactive photon-photon technique using micro-waves to locate objects in a solid phase which can be soil, rock, concrete, bituminous material or non-metallic phases. The protocol is:

- Emit a radio-wave pulse in the investigated system utilizing antenna
- Record the echoes and their time delays
- Build 1D, 2D or 3D images from the echo data.

GPR is the term applied to techniques that use radio waves, in the frequency range 1 to 1000 MHz, to map shallow structures in the ground or in man-made systems (see Jol 1995). GPR was primarily utilized for mapping underground structures. However, GPR has been used in non-destructive testing of non-metallic structures. Information on civil engineering applications of ground penetrating radar may be found in a work edited by Benedetto and Pajewski (2015).

The concept of probing underground formations by applying radio waves is not new. The most successful early work in this area was the use of radio echo sounders to map the thickness of ice sheets in the Arctic and Antarctic. Investigations with GPR in geological environments started in the 1970 s. The methodology is depicted in Fig. 3.21.

The basic GPR equations are:

$$z = vt/2 \qquad \{3.14\}$$

with z the depth where the waves reflected, v the velocity of the wave in the medium, and t the time required for the pathway (including return). The velocity in the medium v is linked to the speed of light in vacuum c by:

$$v = c/\varepsilon_r^{1/2} \qquad \{3.15\}$$

where ε_r is the relative dielectric constant of the material. Values of these key parameters are reported in Table 3.9 for selected materials.

GPR applications are limited only by the imagination and availability of suitable instrumentation. These days, GPR is being used in various areas including location search of buried utilities, mine site evaluation, archaeological digs, searching for

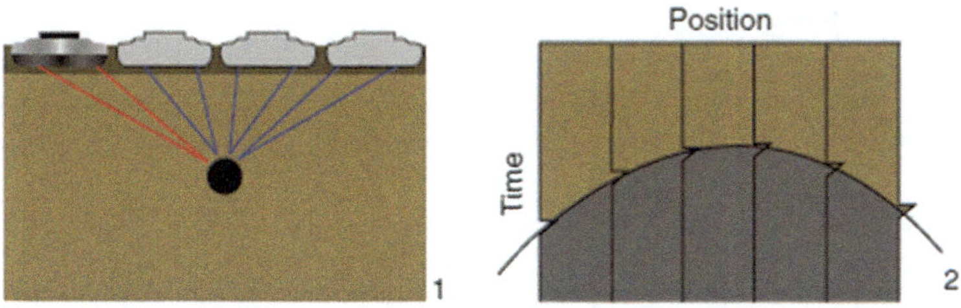

Fig. 3.21 GPR analytical protocol: *1* Emit weak radio frequency signals, and detect the echoes sent back and use them to build an image, *2* Displays signal time delay and strength, and rebuild the 2D or 3D picture on the basis of recorded data

Table 3.9 GPR properties of some related materials

Material	ε_r (–)	*v* (mm ns^{-1})
Air	1	300
Water	81	33
Sand (dry)	3–5	150
Sand (wet)	20–30	60
Limestone	4–8	120
Granite	4–6	130
Shale	5–15	90
Slit	5–30	70
Clay	4–40	60
Salt (dry)	5–6	130

Data from Davis and Annan (1989). *Note* 1 mm ns^{-1} = 10^6 m s^{-1}

buried landmines and unexploded ordnance. To improve accuracy and lateral resolution, strong sources of THz radiation are suggested, e.g. >1 mW (see Dworak et al. 2011). In these conditions the resolution may reach the mm.

The nuclear material examples concern granitic underground investigations. The in situ strength and failure mechanisms of migmatitic gneiss and pegmatitic granite at a nuclear waste disposal site in Finland have been reported by Siren et al. (2015). The outcome of the campaign was that rock failure occurred from structural origin rather than dictated by the expected location of the maximum stress. The experiment also showed that the onset of fracture initiation in the rock occurs at 40 MPa, and the rock mass strength is ~90 MPa compared to the mean laboratory value of 104 MPa.

The second example is due to an experiment performed in granodiorite formation at the Underground facility in Grimsel. Figure 3.22 shows the plot gained from the radargram as well as the results obtained by calculation for ε_r = 5.65 which

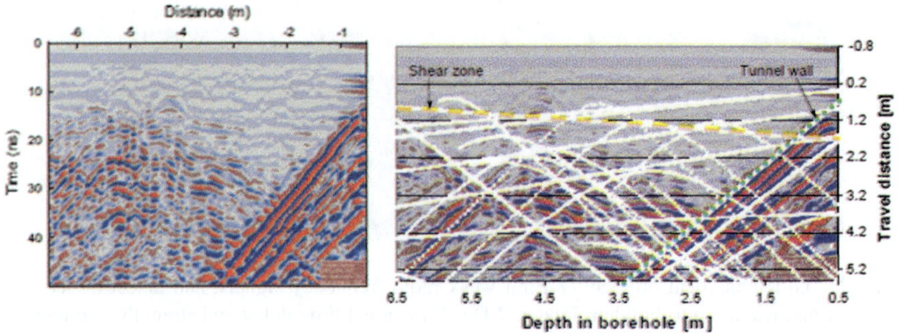

Fig. 3.22 Baseline radar measurements along borehole. *Right* stacked time section. *Left* comparison measurements and calculations for antenna in borehole; the tunnel is indicated in *green dots*, the shear zone by the *yellow dashes*. Lateral resolution: 1 pixel = 3 cm. Data from Marschall and Lunati (2011), with permission from © Nagra, Switzerland

yield a wave velocity of 1.2×10^8 m s^{-1}. The plot exhibits the pattern of reflections and diffractions. It is however difficult to assign all reflector and diffractor to defined structures (see Marschall and Lunati 2011).

3.2.3.2 Infrared Spectroscopy

InfraRed Spectroscopy (IRS) operates by absorption in the infrared spectrum. It is used to identify and study chemicals, e.g. McDonald (1986). The sample may be solid, liquid, or gaseous. The technique uses an **infrared spectrometer** to produce an **IR spectrum**: a record of the infrared light absorbance (see Eq. (3.9)) or transmittance versus frequency or wavelength on the horizontal axis. Units of IR wavelength (λ) are given in μm, a reciprocal of the wave numbers (W) given in cm^{-1} and used as frequency.

$$W = 1/\lambda \tag{3.16}$$

The technique also utilizes Fourier Transform (FT) treatment in the IR domain with the so-called Fourier Transform InfraRed (FTIR) spectrometer. Two-dimensional IR is also possible in a radiography mode, for example.

The infrared portion of the electromagnetic spectrum is usually divided into three regions; the near-, mid- and far-infrared, referencing their proximity to the visible spectrum.

- The near-IR, ~14,000 to 4000 cm^{-1} (0.8–2.5 μm) to study harmonic or overtone vibrations.
- The mid-IR, ~4000 to 400 cm^{-1} (2.5–25 μm) to study atom vibrations and rotations.
- The far-IR, ~400 to 10 cm^{-1} (25–1000 μm) to study low-energy rotations.

The IR spectrum of a sample is recorded by absorption of an IR light beam through the sample. When the energy of an IR photon corresponds to the vibrational energy of a bond, absorption of the photons occurs. Record of transmitted light reveals how much energy was absorbed. This can be achieved by scanning the wavelength using a monochromator over a fixed spectral range. In addition, the whole wavelength range can also be measured using a Fourier transform procedure; transmittance and absorbance spectra are recorded. The maximal energy, shape and intensity of peaks in the spectrum provide information about the sample molecular structure. The work of Havenith (2002) may be used as an information source for more details.

The A–B bond is generally considered as a spring. The frequency of the absorbed photon, v can be calculated:

$$v = \frac{1}{2\pi c}\sqrt{\frac{k}{\mu}} \tag{3.17}$$

where k is the spring constant for the bond, c is the speed of light and μ is the reduced mass of the A–B system given by:

$$\mu = \frac{m_A m_B}{m_A + m_B} \tag{3.18}$$

with (m_i the mass of atom i).

Solid samples are prepared through various means.

- The usual method is to mill the sample with an oily agent (e.g. Nujol) in an agate mortar. This method uses a thin film of the paste that may be coated onto a planar sample carrier in salt and then measured.
- The second method is to grind a sample with a purified salt (usually KBr) finely (to avoid scattering from large crystals). This powder mixture is then pressed to form a pellet through which the IR beam passes.
- The third technique is the “cast film” technique, used mainly for organic materials. The sample is first dissolved in a suitable, non-hygroscopic solvent. A drop of this solution is deposited on the surface of a near-infrared or infrared transparent KBr or NaCl cell. The solution is then evaporated to dryness and the film formed on the cell is analyzed. Care is important to ensure that the film is not too thick otherwise IR light cannot pass through.
- The final method is to use microtome to cut a thin (20–100 μm) film from a solid sample. Experimental spectra are compared to a reference.

Fourier Transform InfraRed (FTIR) spectroscopy is a measurement technique that allows recording of infrared spectra. Infrared light is guided through an interferometer and then through the sample (or vice versa). A moving mirror inside the apparatus alters the distribution of the infrared light that passes through the interferometer. The signal directly recorded, called an “interferogram”, represents

Table 3.10 Vibrational peaks data for waters and actinyl ions

Bond	λ (nm)	*W* ($\times 10^3$ *cm*$^{-1}$)	*ν (THz)*	Refs.
H–O–H	760 and 960	13.16 and 10.42	395 and 312	e.g. Gaunt (1956)
D–O–D	1000 and 1300	10.00 and 7.15	300 and 231	
U^{6+}=O	870	11.49	345	
Np^{6+}=O	860	11.63	349	Basile et al. (1978)
Np^{5+}=O	767	13.04	391	
Pu^{6+}=O	835	11.98	359	

light output as a function of mirror position. For detailed information the work "Fourier transform infrared spectroscopy industrial chemical and biochemical applications" edited by Theophanides (1984) can be referenced.

The nuclear material examples deal first with the isotopic effect. The different isotopes in a particular species may exhibit different fine details in their IR spectra. For example, the spectra of H_2O and D_2O are different. H_2O absorption displays a band centered at 760 nm and stronger band at 960 nm. D_2O absorption spectrum has similar structure to H_2O but shifted in the IR spectra, with a peak at 1000 nm and a stronger one at 1300 nm (see Table 3.10). This difference could be used to control the fraction of heavy water in light water. This is a way to analyze the fraction of light water in the heavy water reactor moderator. This technique was implanted for example on line (e.g. Gaunt 1956) for CANDU heavy water analysis.

Infrared spectroscopy has occasionally been used to the study of actinides under environmental conditions. It has been used in transmission as well as in diffuse-reflectance mode (see Fig. 3.13), both applying Fourier transform or not. IRS may provide useful information on the speciation in submicron size phases of an actinide when it is present in relatively large concentration. For example, the process of complexation and reduction of uranium by lignite was determined with site-specific material (Nakashima 1992). This phenomenon occurs during the build-up of a uranium repository and specific vibrational frequencies assigned to AnO_2^{i+} yield peaks near 870 nm for U(VI), 860 nm for Np(VI), 767 nm for Np(V) and 835 for Pu(VI) (see Table 3.10) and for actinide concentrations above 10^{-3} M, (see Basile et al. 1978; Maya and Begun 1981). Spectroscopy may be applied at the macro-scale (mm^3) and could be reduced down to the micrometric scale by using a focused beam.

3.2.3.3 Diffuse Reflection Spectroscopy

Diffuse Reflection Spectroscopy (DRS) may be used to characterize the sample surface and subsurface. Since light cannot penetrate opaque (solid) samples, it is reflected on the surface of the samples. As shown in Fig. 3.13c, incident light not reflected symmetrically with respect to the normal line but scattered in different directions is called diffuse reflection. In addition, part of the incident light penetrates the sample Fig. 3.13d and a shallow layer scatters the light. The plain reflection must be completed by an absorption scattering as described in the Kubalka and Munk (1931) theory.

$$F(R) = \frac{(1-R)^2}{2R} = \frac{\mu}{s} = \frac{\varepsilon c}{s} \tag{3.19}$$

where, R is the reflectance (for infinite thickness); μ the absorption coefficient; s the scattering coefficient; c the concentration of the absorbing species and ε the absorbance. It gives a correlation between the reflectance and the concentrations of absorbing species in weak absorbing samples, something like the Lambert-Beer law for absorbance.

Detection may be carried out with integrating spheres. Measurement is performed by placing the sample in front of the incident light window, and concentrating the light reflected from the sample on the detector using a sphere coated inside with a barium sulfate. The obtained value becomes the reflectance (relative reflectance) with respect to the reflectance of the reference standard white board, which is taken to be 100%. When light is directed on the sample at an angle of 0°, specular reflected light exits the integrating sphere and is not detected. As a result, only diffuse reflected light is measured.

DRS may be used to characterize the metal surface and subsurface during corrosion process under LWR conditions. This method has been adapted for the study of the corrosion of stainless steel. The system has been designed for non-invasive record of diffuse reflection spectra directly on stainless steel coupon under boiling water reactor conditions (T = 298 °C, P = 100 Atm); see Fig. 3.23a, b; Degueldre et al. (1996c). The technical set-up with optical fiber probe and loop part includes an inert window that isolates the optical fibers from the hot water. Optical scheme for in situ investigations shows that the investigation may be initially performed in a specular reflection mode with an apparent reflectivity I/I_0 of 100%. DRS spectra recorded in non-invasive mode start with the blank at time 0 followed by a decrease in reflectivity with time that is due to the build-up of the oxide layer initially. The Beer Lambert law (Eq. 3.10) may be used in this domain, and the thickness of the oxide layer derived. Later the interferometric maxima (1st and 2nd later) appear in the recorded spectra (Fig. 3.23c). They may be used to estimate independently the oxide layer thickness using the relation:

$$\lambda_{\max(i)} = (2/i)\, n_{\mathrm{ox}}\, d \tag{3.20}$$

where n_{ox} denotes the refractive index of the material, d stands for the thickness of the corrosion layer and $\lambda_{\max(i)}$ the wavelength of the maximum of the interference peak i.

The experimental corrosion layer thicknesses are then plotted as a function of corrosion time and compared with model calculated data (Fig. 3.23d). The advantage of the non-invasive technique is to proceed to the investigation in situ without—as currently done when using gravimetry—taking the sample outside the autoclave or the loop for measurement.

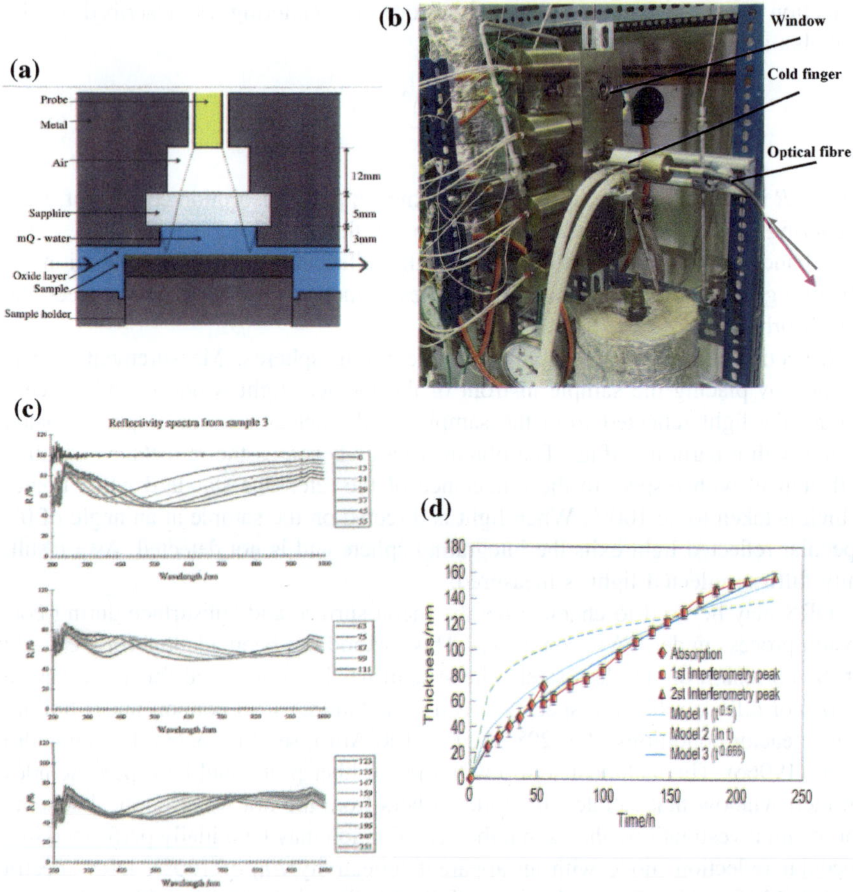

Fig. 3.23 In-line diffuse reflection spectroscopy study of the oxidation of stainless steel under boiling water reactor conditions. Degueldre et al. (1996c) with permission © Elsevier. **a** Optical scheme for in situ investigations, **b** Technical set-up with optical fiber probe and loop part equipped with inert window; **c** DRS spectra recorded in-line, in non-invasive mode; **d** Experimental corrosion layer *thickness* plot as a function of time in comparison with model data

The detection limit is found to be 5 nm for the stainless steel corrosion layer, and the lateral resolution could be reduced to 10 μm with microscopic optical fibers. The present technique is now used by various groups for nuclear application.

Similarly, in-line measurements of the oxide layer build-up on Zircaloy under Boiling Water Reactor conditions has been performed using DRS, Degueldre et al. (1998a). It was found that the build-up of the oxide layer follows a quadratic law from 2 to 200 nm.

3.2.3.4 Near Infrared—Visible—Ultraviolet Spectroscopy and Colorimetry

Near InfraRed-Visible-UltraViolet Spectroscopy (**NIR-VIS-UVS**) and **ColOrimetry (Col)** are used in analytical chemistry for the determination of transition metal, lanthanide and/or actinide ions, as well as highly conjugated organic ligands (Col). Spectroscopic analysis is done in solutions but gases and solids may also be investigated. Detailed information on UV-VIS spectroscopy and its applications may be found in the work of Perkampus (1992) and more recently the work on UV-VIS spectroscopy for nanomaterials characterization edited by Kumar (2013). The chapter of Monteit and Unsworth (2013) also provides specific information.

Solutions of transition metal ions absorb light when their d or f electrons are excited. The anions or ligands can affect their color. Specific ligands intensify the color of the solution and shift the wavelength of maximum absorption (λ_{max}). Specific organic compounds, for example with a high degree of conjugation, also absorb light in the UV or visible regions of the spectrum. The solvents can be water, or ethanol for organic-soluble compounds. The analysis can also be performed in the extraction phase. Solvent polarity and pH can affect the absorption spectrum of an organic compound.

The Beer-Lambert law (Eq. 3.9) states that the absorbance of a solution is directly proportional to the concentration of the absorbing species in the liquid sample and the path length. UV/Vis spectroscopy can be used to determine the concentration of the absorber in a solution. This can be taken from references, or more accurately, determined from a calibration curve. The method is often used in a quantitative way, measuring the absorbance A of an absorbing species in solution to determine its concentrations using the Beer-Lambert law:

$$A = \log \frac{I_0}{I} = \varepsilon c l \qquad (3.21)$$

where I_0/I is the ratio of the incident light intensity (e.g. at maximum absorption) over the transmitted intensity, l the light path length (e.g. analytical cell), and c the concentration of the absorbing species. For each species and wavelength, ε (molar absorptivity) is constant. This is a fundamental molecular property in a given solvent, at a particular temperature and pressure, and has units of M^{-1} cm^{-1} with or L mol^{-1} cm^{-1}. The light is analyzed in single beam or double beam mode using a prism or a grating as a monochromator.

The example on nuclear material analysis concerns colorimetry of uranium. This determination is important because of its use for uranium prospection but also to track contamination and associated health hazards after explosion of depleted uranium warhead.

Table 3.11 Determination of uranium by colorimetry

Reagent	Formula	λ_{max} (nm)	ε (M^{-1} cm^{-1})	C_{DL}* (M^{-})	References
PAR	4-pyridyl-2-azo resorcinol $(C_5H_4N)N{=}N(C_6H_3OHOH)$	530 Red	0.4– 3.2×10^4	3×10^{-8}	Ghosh et al. (1991)
DBM	Dibenzoylmethane $C_6H_4COCH_2COC_6H_4$	393 Yellow	–	#0.4×10^{-9}	Yoe et al. (1953)
Arsenazo III	1,8-dihydroxynaphtalene-3,6-disulphonic acid-2,7-bis [(azo-2)-phenylarsonic acid] $(HO)_2C_{10}H_2(SO_3H)_2(N{=}NC_6H_4AsO_3H_2)_2$	651-670* Bleu.green	0.5– $1.45 \times 10^{5*}$	8×10^{-8}	Khan et al. (2006), Savvin (1961)
TAC	2-(2-Thiazolylazo)-p-Cresol $(C_3H_3NS)N{=}N(C_6H_3CH_3OH)$	588 Bleu	1.31×10^4	1.0×10^{-7}	Gomes Teixeira et al. (1991)
Br-PADAP	2-(5-Bromo-2-pyridylazo)-5-(diethylamino)phenol $(C_2H_5)_2N(C_6H_3OH)N{=}N(C_5H_3NBr)$	578 Bleu	5.8×10^5	3.8×10^{-8}	Johnson and Florence (1971) Das et al. (2010)

Conditions: t = 25 °C, *l = 10–50 mm, # with solid-phase extraction of uranium(VI) ultratrace in natural waters using octadecyl silica membrane disks modified by tri-*n*-octylphosphine oxide (see Shamsipur et al. 1999).
*Compiled from Khan et al. (2006), Savvin (1961)

There are various methods of spectrophotometric determination of uranium (see Table 3.11). The most often used methods are the colorimetry with:

- 4-pyridiyl-2-azo resorcinol (**PAR**)
- 1,3-diphenyl-1,3-propanedione (**DBM**) and
- 1,8-dihydroxynaphtalene-3,6-disulphonic acid-2,7-bis[(azo-2)-phenylarsonic acid] (**Arsenazo III**).

PAR is a sensitive reagent for uranium, forming complexes in aqueous solutions suitable for the determination of uranium at concentrations between 2 and 400 μg L^{-1}, (see Ghosh et al. 1991). Determination of uranium(VI) with PAR and N-octylacetamide is carried out into benzene over pH 7.0–9.0. The molar absorptivity of the complex with 9 different amides is in the range of (0.40–3.2)· 10^4 L mol^{-1} cm^{-1} at the absorption maximum. Out of these, the most sensitive compound N-OctylAcetAmide (OAA) was chosen for detailed studies in the present investigation. The **uranium detection limit** of the method is 8 ng ml^{-1} (3×10^{-8} M). The system obeys Beer's law in the U concentration range of 0–5 μg ml^{-1} as described by Ghosh et al. (1991).

DBM is also suitable for the colorimetry of uranium at concentration around 50 μg L^{-1} (see Yoe et al. 1953).

Arsenazo III offers a rapid and sensitive method for quantitative colorimetry of U(VI) as reported by Fritz and Johnson-Richard (1959); masking agents such as tartaric acid, thioglycerol or EDTA make the analysis possible in presence of many metal ions. It was found that the 1:1 arsenazo III–uranyl complex has four protonation states which respectively dominate at pH 2.2, 5.0, 7.8 and >10.0, (see Hosten and Rohwer 1997). Anions have a significant effect on complexation by interacting with both the free uranyl and the arsenazo III–uranyl complex. Specific molar absorptivities of these complexes are known.

More recently, the determination of uranium using 2-(2-ThiAzolylazo)-p-Cresol (**TAC**) was investigated (see Gomes Teixeira et al. 1999). The reaction between TAC and uranium (VI) is instantaneous at pH 6.5 and the absorbance remains stable for over 3 h. N-cetyl-N,N,N-TrimethylAmonium Bromide (CTAB) and triton X-100 are used for increasing the sensitivity and solubility of the system respectively. The method allows the determination of uranium in the range from 0.30 to 12.0 μg mL^{-1} with a molar absorptivity of 1.31×10^4 L mol^{-1} cm^{-1} and features a **detection limit of 26 ng** $\mathbf{mL^{-1}}$ (10^{-7} M) at 588 nm.

Br-PADAP photospectrometry has been revisited by Johnson and Florence (1971) to upgrade the analytical performances. The sample solution is spiked with masking agent (EDTA/citrate) buffered at pH 8–12 spiked with solubilizing agent and Br-PADAP (2-(5-Bromo-2-Pyridylazo)-5-DiethylAminoPhenol: $C_{16}H_{17}N_4BrO$). The analysis is performed at 578 nm with a quantification limit of 30 ng mL^{-1} or a detection limit of 9 ng mL^{-1} (3.8×10^{-8} M).

The comparison of the complex spectrophotometric properties is reported Table 3.11.

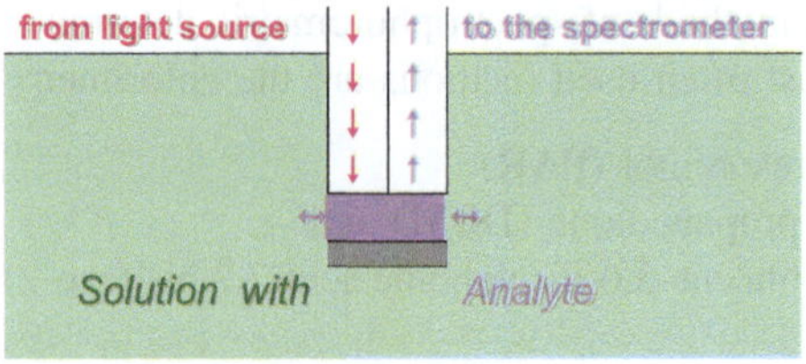

Fig. 3.24 Optode schema *down arrow* illumination with pink or white light, *filled square* colorimetric extractant gel, *up arrow* optical fiber collecting light back to the spectrometer

Optodes

An **optode** is a solid-phase extraction porous cylinder equipped by optical fibers to illuminate the sensing phase and to connect the system to a spectrometer as depicted in Fig. 3.24.

An optode consequently requires three components to operate: a colorimetric reagent that responds to the analyte, a porous matrix to immobilize the colorimetric chemical reagent, and the associated instrumentation (light source, optical fiber, the spectrometer). Optodes usually have the polymer matrix coated onto the tip of an optical fiber, but in the case of evanescent wave optodes the polymer is coated on a section of fiber that has been unsheathed, e.g. Chatni et al. (2009).

Optode for the analysis of uranium in natural waters was fabricated (see James et al. 2008). The sensing element was designed by the inclusion of uranyl ion imprinted polymer particles into polymethyl methacrylate followed by casting a thin film on a glass slide without any plasticizer. The ion imprinted polymer material was prepared via covalent immobilization of the newly synthesized ligand 4-vinyl phenylazo-2-naphthol by thermal polymerization. Operational parameters such as pH, response time and the amount of sensing material were optimized. The imprinted polymer inclusion optode responds linearly to uranium in the concentration range 0–1.0 μg mL^{-1} with a detection limit of 0.18 μg mL^{-1} (7.5×10^{-7} M), which is much better than the solution studies using 4-vinyl phenylazo-2-naphthol (1.5 μg mL^{-1}). Triplicate determinations of 100 μg of uranium(VI) present in 250 mL of solution gave a mean absorbance of 0.018 with a relative standard deviation of 8.33%. The superior sensitivity of imprinted polymer inclusion optode is exemplified by lower detection limits and broader dynamic range over non-imprinted polymer inclusion optode. The applicability for analyzing ground, lake and tap water samples collected in the vicinity of uranium deposits was successfully demonstrated.

Microscopy with visible and UV light

Microscopy carried out with both ocular and objective lenses allow morphological observation. Information on the basics and development of these techniques is provided in the review paper by Weisenburger and Sandoghdar (2015). Theoretically, the maximum resolution, *d*, that one can obtain with an optical

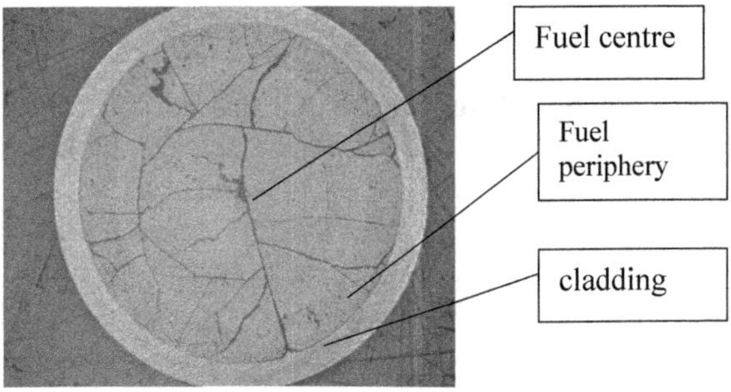

Fig. 3.25 Microscopic density changes across a fuel section due to burn-up variation over the pellet radius with increasing resolution optical microscopy of UO_2 fuel, burn-up 40 MW d kg^{-1}. Degueldre et al. (2016), with permission © Elsevier

microscope has been limited by the wavelength of the radiation that is being used to probe the sample, λ and the Numerical Aperture (NA) of the system, NA (corresponding to the angular aperture α):

$$d = \frac{\lambda}{2n \sin \alpha} \approx \frac{\lambda}{2NA} \tag{3.22}$$

with n the breaking index of the lens material. Early twentieth-century scientists theorized ways of getting round the limitations of the relatively large wavelength of visible light (wavelengths of 400–700 nm) by using electrons.

The nuclear material example is a ceramography of a fuel cross section. After cutting and polishing of the fuel pin, the observations can be done by optical microscopy: as shown in Fig. 3.25. The different gray levels also testify the occurrence of 3 radial zones, the grayest zone corresponding to the more porous one (less reflectivity): the periphery where the burn-up is larger than in the center of the fuel is because of the proximity of the moderator (see Degueldre et al. 2016).

Confocal microscope

Confocal Microscopy (CM) is an optical micrographic technique for increasing lateral resolution and contrast of an image by introducing a spatial pinhole at the confocal plane of the lens to eliminate out-of-focus light. Davidovits and Egger (1969) described the first confocal laser scanning microscope. It enables reconstruction of 3D structures from the recorded micrographies. This technique is successful in research and industry; typical applications are found in materials and life science.

An example of analysis of nuclear material with CM may be that of nuclear grade graphite from AGR. During reactor operation stresses develop in the graphite components, causing them to deform. It is important to understand how the microstructure of graphite affects the material's response to these stresses. Taylor et al. (2016) investigated the effects of stress on the pore structures of nuclear grade graphites using CM. They proposed that pore closure and pore reorientation were competing processes.

3.2.3.5 Single Particle Counting

Single Particle Counting (SPC) has been used for the analysis of particles in suspension in fluids, e.g. air, water. This technique is based on the phenomenon of **static light scattering**.

There are several methods used for detecting and measuring particle size or size distribution: light blocking (obscuration in transmission), light scattering, Coulter principle (see Sect. 3.3.4) and direct imaging e.g. Anderson et al. (2013) as well as advanced techniques Kohli (2012). The light-blocking optical particle counter method is typical useful for detecting and sizing particles greater than 1 μm in size. It is based on the amount of light a particle blocks when passing through the detection area of the particle counter. This type of technique allows high resolution and reliable measurement.

A single particle counter detects and counts particles according to their size one at a time. A high intensity light source (laser or halogen light) is used to illuminate the particle as it passes through the beam in the detection cell, e.g. see Fig. 3.26. The nature of particle counting is based upon light scattering, light obscuration, masking, or direct imaging. The light scattered when the particle passes through the

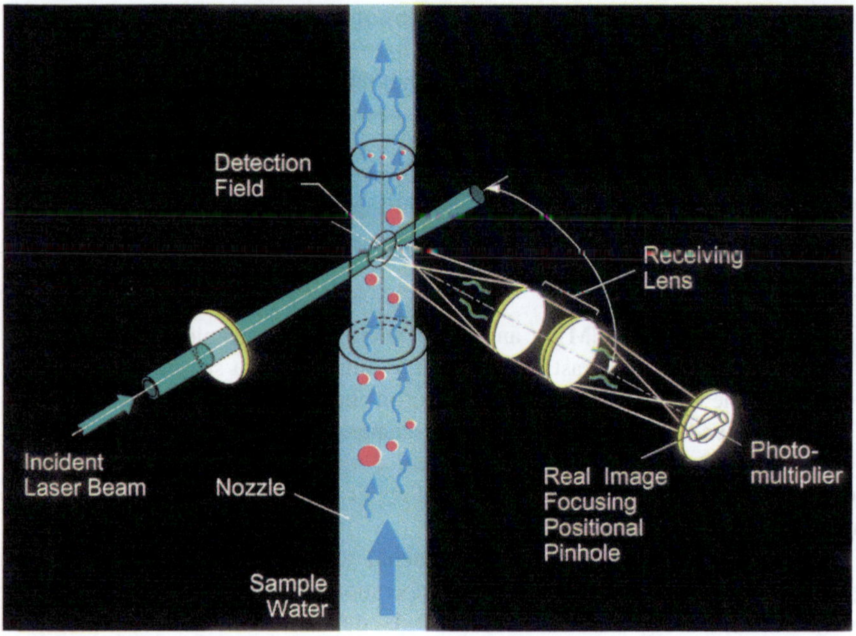

Fig. 3.26 Single particle counting (SPC) unit measurement principle in single particle light scattering set-up (1D). *Note* Particles injected in flow of pure water pass one at a time in the illuminated zone of the flow cell. Photo detector(s) set orthogonally to the incident light beam record(s) the scattering flashes of particles. Particles are counted and sorted out according to their intensity/size. With permission from Degueldre

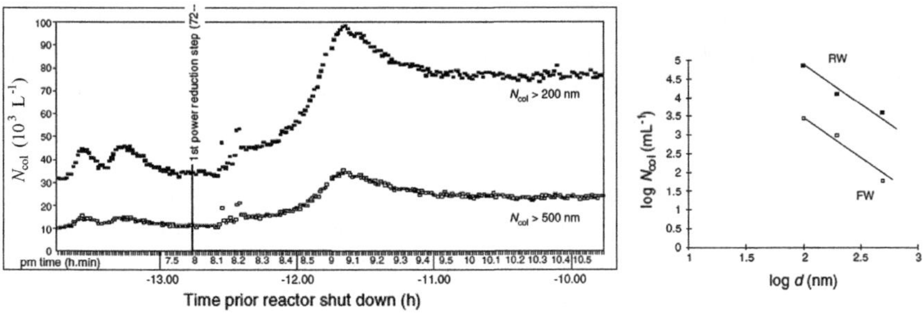

Fig. 3.27 Colloid concentration profile in reactor water—Above: before BWR power reduction and—Below size distribution during normal operation conditions for RW: primary reactor water, and FW feed water. Conditions: Horiba single particle counter unit used on-line at sampling point (RT). Time in hour before and after first power reduction (12 h before shutdown). Note the cumulative distribution slope is $1 - b$, with $b = 2.9$ (Eq. 3.23). Degueldre et al. (1996a) internal report

light beam is detected by a photo multiplier, for example. If direct imaging is used a high definition camera records the passing particles. Video records are subsequently analyzed by computer software to characterize the particle distributions (2D).

Applications of particle counters are separated into three primary categories:

- Aerosol particle counters (e.g. in contaminated environment)
- Liquid particle counters (e.g. after liquid extraction)
- Solid particle counters (e.g. directly in reactor water).

Solid particle counters are used to measure dry particles for various industrial applications. One such application could be for the detection of particle size coming from a rock crusher within a mining quarry. Sieves are usually the standard instruments utilized to measure dry particle size.

The light scattering method is capable of detecting smaller (<5000 nm) sizing particles. This technique is based upon the light intensity that is redirected by a particle passing through the detection area of the SPC unit. Typical detection sensitivity of the light scattering method is 50 nm. The model of size distribution commonly used to describe a continuous colloid size distribution is of a Pareto law (1896); for the colloid concentration N_{col} it reads:

$$\delta N_{col}/\delta d = \mathrm{A}\, d^{-\mathrm{b}} \tag{3.23}$$

with d the colloids size (diameter) and A and b empirical constants e.g. Degueldre et al. (2000). The technique covers orders of magnitude in size (e.g. 50–5000 nm) and in particle concentration such as 10 colloids per ml, of 100 nm size, to several ppm, depending on the size distribution, Silverman et al. (1971).

Single Particle Counting has been used on-line, at the sampling line of nuclear power plants for the analysis of particles in suspension in primary reactor water (see

Degueldre et al. 1996a). The test requires water flow control to feed the SPC unit with the required water flow rate for regular on-line monitoring. SPC results from a BWR nuclear reactor cooling water show that the clean-up (filtered water: FW) reduces the concentration of particles by a factor 20–50 compared to the reactor water (RW), see Fig. 3.27.

The colloid concentration was monitored before and shortly after reactor power reduction, before shutdown see Fig. 3.27. The monitoring was carried out for 200 and 500 nm colloids over several hours before power reduction before shutdown and about one hour after power reduction.

3.2.3.6 Photon Correlation Spectroscopy

Photon Correlation Spectroscopy (PCS), also called **dynamic light scattering**, is a technique that can be used to measure the size distribution of colloidal particles in suspension.

The scattered light produced by particles illuminated by a monochromatic polarized light beam may be analyzed through a second polarizer and recorded by a photon detector or on a screen as a speckle pattern. When light illuminates nano-particles, scattering (for particles smaller than the wavelength, e.g. below 250 nm, Rayleigh scattering) occurs in all directions, fluctuating over time. These fluctuations are due to colloids in suspension undergoing Brownian motion. These scattered light flashes yield constructive or destructive interferences. These intensity fluctuations allow quantification of the size of the scatterers in movement. Elimination of dust and artifacts from the solution is mandatory. Pike and Abbiss (1997) edited a work on light scattering and photon correlation spectroscopy that is a reference work for detailed information.

The intensity is measured as function of time. To model the flash intensity decay fitting methods are applied. They are based on assumed distributions. If the sample is monodisperse the decay is treated as a single exponential. The colloid diffusion coefficient D may be evaluated using Stocke-Einstein (1905a, b) equation:

$$D = \frac{kT}{6\pi R_{sph}\eta} \tag{3.24}$$

with R_{sph} the hydrodynamic radius of the colloid, η the viscosity of the fluid. The analysis is easier when effectively particles do not aggregate. Particle-particle collisions are reduced by dilution.

If the system is not widely dispersed in size, PCS can be utilized to measure the mean effective diameter of colloidal particles including polymers, micelles, nanobubbles, carbohydrates and nanoparticles. The quality of the measurement depends on the particle size, particle concentration, and the ions in the suspension. PCS monitoring of particles in suspension allows observation of the increase of the particle size over time. The technique covers size ranges from 5 to 1000 nm and in particle concentration from 100 ppb–100 ppm.

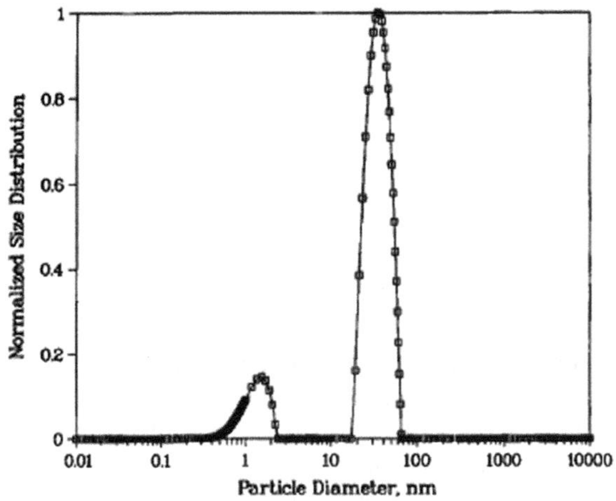

Fig. 3.28 Size distribution of Pu colloids obtained by auto-oxidation of Pu(III) in Pu(IV). Note that the concentration distribution is arbitrary normalized (1) for the larger maximum and that the area below the peak corresponds to the relative scattering intensity, distribution normalized (1) for the 33 nm size class. Data from Triay et al. (1991), with permission from © de Gruyter Verlag

The example concerning nuclear material sample deals with $PuO(OH)_n$ polymers (see Triay et al. 1991). A solution of Pu(III) 6×10^{-3} M was slowly auto-oxidized from pH 2.8–4 for 227 days. The colloidal phases developed were analyzed for their size distribution by PCS. The size distribution record is presented in Fig. 3.28. It shows the $PuO(OH)_n$ polymers which size are found around 1.5 nm. A colloid population of size around 33 nm is also observed (with stronger scattering). However, the Pu colloids placed in groundwater could not be analyzed using PCS. The formation of very large pseudocolloids in groundwater was identified as a potential interference for the failure of PCS data analysis.

3.2.3.7 Raman Spectroscopy

Raman Spectroscopy (RamS) is used in analytical chemistry to provide a fingerprint for identifying the molecules of a sample. It is based on inelastic scattering, also called Raman effect. The incident light (monochromatic in the near ultraviolet, visible, or near infrared) interacts with molecular vibrations in sample molecules, resulting in shifts of the photon energy. The shifts provide information about the vibrations in the molecules. Infrared spectroscopy provides complementary data. Detailed information on Raman spectroscopy theory and practice are available in the work edited by Szymanski (1970).

Typically, samples are illuminated with a laser beam. The Raman (1928) effect occurs when a photon interacts on a molecule with the bonds of the molecule in the solid, liquid or gaseous phase and its environment. The Raman effect concerns a photon that excites a molecule in either its lowest rotational and vibration energy level or an excited state. This yields in the molecule an energy state for a short period of time before a scattered photon is emitted. The photon can be either of lower (Stokes) or of higher (anti-Stokes) energy than the incident photon (see

Fig. 3.1). The difference in energy leads to a shift in frequency $\Delta\nu$ between the excitation ν_{ex} and the emitted ν_{em} photons:

$$\Delta E = \mathrm{h}\,\Delta\nu = \mathrm{h}\,\nu_{ex} - \mathrm{h}\,\nu_{em} \tag{3.25}$$

with h Plank's constant. Raman spectroscopy is used in chemistry, because vibrational information can be a fingerprint of molecules, specific to the molecular bonds allowing identification of the molecules. In solid-state physics, RAMS is used to characterize materials, find the crystallographic orientations and measure temperature of a sample.

Bore carbide is used as **neutron absorber** in the control rod of nuclear reactors. During utilization the material is irradiated and can undergo irradiation damage modifying the structure and integrity of the material. The reaction: $^{10}B(n,\alpha)^{7}Li$ induces damage due to recoil, change of B in Li and formation of He phase. It is consequently important to study the behavior of this safety-relevant material. This has been done by Raman spectroscopy before and after irradiation, and after healing by annealing under operational conditions. Study by Raman spectroscopy of boron carbide evolution under neutron irradiation has been reported by Simeone et al. (2000). Raman spectra comparison between neutron irradiated and unirradiated $B_{12}C_3$ and after annealing are shown in Fig. 3.29. The RAMS spectra testify that the healing of $B_{12}C_3$ repairs the damaged structure.

Another nuclear material example deals with Raman microscopy of irradiated carbon. The nature of vibration modes in graphite is analyzed in terms of the resonant excitation of π states and the long-range polarizability of π bonding. Raman spectroscopy is an appropriate technique to analyze defects in carbon-based materials (see Ammar et al. 2015), most often focused on the intensity ratio between vibration mode signals, which may be enhanced by various defects. They may appear in nuclear graphite during its manufacturing process, its operation in the nuclear reactor, or even during its preparation process such as polishing which is currently used before analysis.

Raman spectroscopy has also been used to monitor material in situ under irradiation during contact with fluid. This can be illustrated with the study of uranium dioxide alteration by water radiolysis (see Canizarès et al. 2012). The fuel integrity is highlighted by a comparative study of the time dependence of UO_2 surface modification induced, on one hand by contact with water under irradiation by 5 MeV He^{2+} particles, and on the other hand by pure chemical alteration, through contact with a hydrogen peroxide solution.

Raman microscopy

Raman Microscopy (**RAMM**) is a scattering technique for which specimens do not need to be cut and fixed. Raman spectra can be recorded from volume of <1 μm^3 in size; these spectra allow micro-identification of species present in that volume. Water does not interfere with Raman analysis. Consequently, RAMM is adaptable for examination of minerals, materials such as ceramics, and forensic trace evidence at the microscopic scale. A Raman microscope begins with an optical microscope

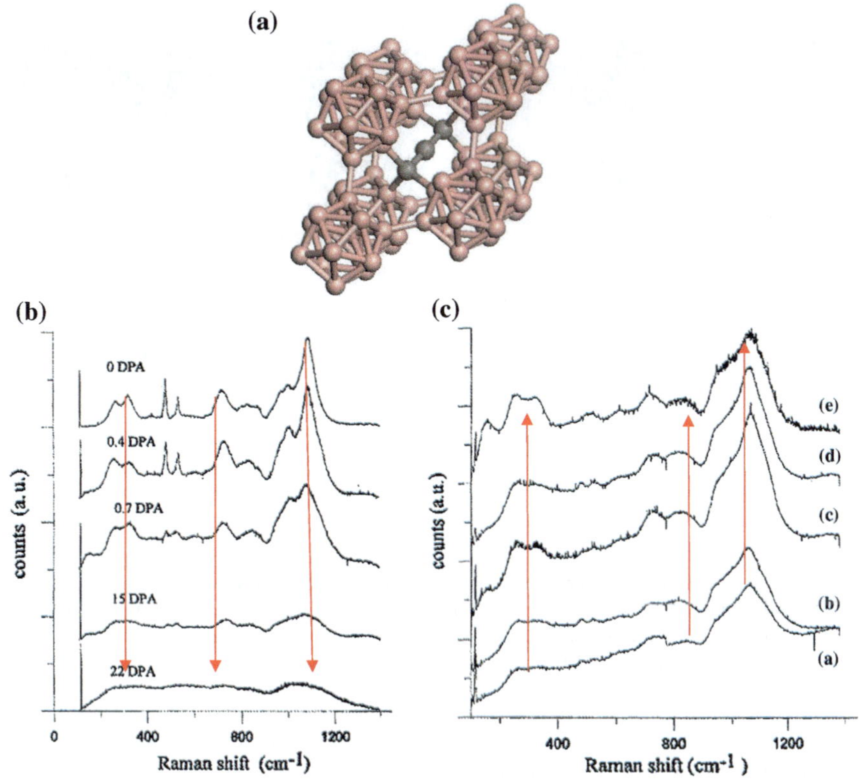

Fig. 3.29 **a** Molecular structure of boron carbide ($B_{12}C_3$); **b** Neutron irradiation effect on $B_{12}C_3$ for various damages level; **c** Annealing effect of $B_{12}C_3$ irradiated by neutrons: *a* unannealed neutrons irradiated at 2 dpa; *b* annealed 30 min at 373 K; *c* 30 min at 573 K; *d* annealed 30 min at 673 K; *e* 30 min at 373 K. Adapted from Simeone et al. (2000), with permission © Elsevier

(see Sect. 3.2.3.4), on which an excitation laser, a monochromator, and a sensitive detector are added. UV microscopes and UV enhanced optics must be used when a UV laser source is utilized. Information may be found in the work of Opilik et al. (2013).

The micro-Raman spectra of four oxides of uranium were recorded and discussed by Allen et al. (1987). Specific features led to the characterization of uranium oxide particles. Much lower laser powers than normally used for Raman measurements have proved to be possible because of the high collection yield of the microscope optics and the multi-scanning capability of the spectrometer. RamM data can be obtained for materials earlier thought to be too unstable to survive the long periods of continuous radiation necessary.

The nuclear material analysis example is an in situ Raman microscopic study to investigate the effect of hydrogen peroxide on uranium oxide films on stainless steel plates (see Wilbraham et al. 2015). Figure 3.30 gives optical and micro-Raman data

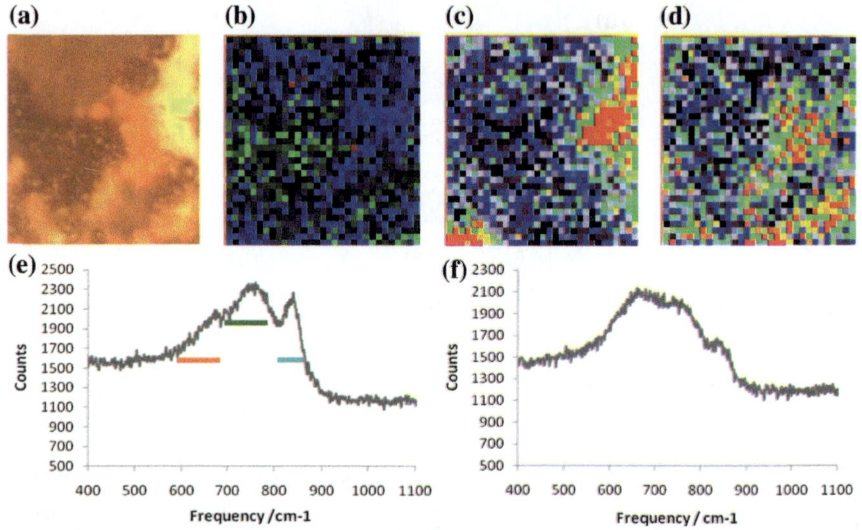

Fig. 3.30 Micro Raman of layer structure of untreated electrodeposited uranium oxide film. **a** White Light Image; **b** 582 to 703 cm^{-1} (*orange*); **c** 812 to 858 cm^{-1} (*blue*); **d** 703 to 801 cm^{-1} (*green*) (scale 1 pixel: 1 μm) **e** and **f** Raman Spectra of the cracked base layer and the crystalline middle layer. Wilbraham et al. (2015),with permission © Elsevier

of layer structure of untreated electrodeposited uranium oxide film. Basically, the white light image obtained by optical microscopy (morphological observation) is compared with the 582 to 703 cm^{-1}, the 812 to 858 cm^{-1} and the 703 to 801 cm^{-1} maps from the Raman Spectra of the cracked base layer and the crystalline middle layer.

3.2.3.8 Atomic Absorption Spectroscopy, Optical Emission Spectroscopy and Laser-Induced Breakdown Spectroscopy

Atomic Absorption Spectroscopy (**AAS**) makes use of a solution sample excitation tool which goes from flame to electrothermal oven, i.e. a graphite furnace. The sample is injected in a heated source (e.g. IR). Atoms in vapor phase are excited by irradiation with light and absorption for specific lines corresponding to the investigated elements. Information on analytical graphite furnace atomic absorption spectrometry is available in a laboratory guide monograph by Schlemmer (1999).

The nuclear material example deals with a method for the direct determination of Ag, Ca, K, Li, Mg, Na, Pb, Sn and Zn in U_3O_8 powder samples by electrothermal AAS as developed by Page et al. (1983). Nanogram and lower amounts of these elements have been determined with a relative standard deviation of 6–16% in mg amounts of sample.

Optical Emission Spectroscopy (OES) makes use of a solution sample excitation tool which goes from flame to electrothermal oven, i.e. a graphite furnace. During this excitation the light emitted by the heated phase (e.g. by IR) is analyzed by a spectrometer. The typical element lines for an element are emitted and analyzed by a spectrometer (see Sect. 3.2.3.4). The line waveline allows identification of the element and their intensity is proportional to the concentration of the element in solution. The excitation tools (Barnes 1974) for OES are arcs, sparks and plasmas (all ions and electrons) and lasers (photons). The latter are treated here.

Laser Induced Breakdown Spectroscopy (LIBS) is a type of atomic emission spectroscopy which uses a highly energetic laser pulse as the excitation source. The laser is focused to form plasma, which atomizes and excites samples. LIBS can analyze any matter regardless of its physical state. Because all elements emit light of characteristic frequencies when excited to sufficiently high temperatures, LIBS can detect all elements. The limits are dictated by the power of the laser and the sensitivity and wavelength range of the spectrometer. When the material constituents are known, LIBS may be used to estimate their fraction, or to monitor the presence of impurities. The detection limits are a function of:

- The plasma excitation temperature
- The light collection window
- The line strength of the viewed transition.

LIBS makes use of optical emission spectrometry and is to some extent similar to the arc or spark emission spectroscopy. For specific information. a work on the theory of laser-induced breakdown spectroscopy and its application has been edited by Musazzi and Perini (2014).

LIBS operates by focusing the laser onto a small area on the specimen; when the laser flash shoots the sample ablation of ng to pg of material generates a plasma plume of several 10,000 K. Data collection is carried out for colder plasma conditions. During early plasma, the ablated material dissociates, i.e. breaks down into excited ionic species. At this point the atomic emission lines of the elements can be observed.

The nuclear analysis example deals with monitoring of reactor tubes from heat exchangers. Remote material analysis of nuclear power station steam generator tubes by laser-induced breakdown spectroscopy has been reported by Whitehouse et al. (2001). A 75-m length umbilical fiber-optic LIBS system (see Fig. 3.31) was used for remotely determining the copper content of 316H austenitic stainless steel superheater tubing within the pressure vessels of an Advanced Gas-cooled Reactor (AGR). Copper content of the steel over the range $0.04\% < Cu < 0.60\%$ (by mass) was successfully determined. Measurement times per sample were typically less than 3 min with $\pm 25\%$ accuracy.

Finally, Laser Ablation Inductive Coupled Plasma Optical Emission Spectroscopy (LAICPOES) should be treated here. Its use is scarce, which contrasts with LAICPMS treated in Sect. 3.2.5.

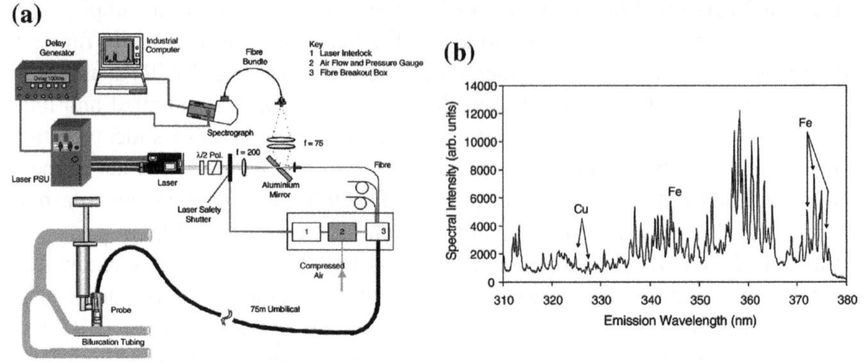

Fig. 3.31 **a** Schematic diagram of fiber-optic LIBS system. **b** Typical spectrum obtained using optimized experimental parameters, showing the copper and iron atomic emissions used to calculate the Cu/Fe spectral intensity ratio. Whitehouse et al. (2001) with permission © Elsevier

3.2.3.9 Ultraviolet Fluorescence Spectroscopy

Fluorescence Spectroscopy (FS) is a spectroscopy which analyzes fluorescence after illumination of a sample (see Fig. 3.1). It makes use of an ultraviolet light source that excites molecule electrons of specific sample compounds and causes them to emit photons, e.g. visible light that is measured by a fluorimeter. Fluorescence spectroscopy concerns electronic and vibrational states of the investigated molecules. At room temperature, the species are in a ground electronic state and may be excited in electronic states of higher energy. Within each of these electronic states are various vibrational states. Detailed information may be found in the work of Sharma and Schulman (1999).

In fluorescence spectroscopy, a molecule first absorbs a photon, exciting the molecule from its ground state to one of the excited electronic states. The excited molecules lose energy by collisions with other molecules until they reach the lowest vibrational state of the excited electronic energy level. The molecules then decay to one of the levels of the ground electronic state, emitting a photon. As molecules may drop down into any of the vibrational levels in the ground state, the emitted photons will have different energies that are characteristic of the illuminated molecules. A work on new trends in fluorescence spectroscopy and applications to chemistry was edited by Valeur and Brochon (2001).

The types of instruments are:

- Filter fluorimeters that use filters to isolate the incident light and fluorescent light
- Spectrofluorimeters that use diffraction grating monochromators to isolate both incident light and fluorescent light.

Both types use the following scheme: the light from an excitation source passes through a filter or monochromator and strikes the sample. A proportion of the incident light is absorbed by the sample, and some of the investigated material molecules fluoresce. The fluorescent light is emitted in all directions.

Various excitation sources may be utilized, including lasers, LED, or lamps; xenon arcs and mercury-vapor lamps in particular. Filters and/or monochromators may be used in fluorimeters. The monochromator can then be adjusted to select which wavelengths to transmit. To allow anisotropy measurements, the addition of two polarization instruments is required. The detector can either be single-channeled or multi-channeled. The different types of detectors have both advantages and disadvantages. Additionally, fluorescence spectroscopy can be adapted to the microscopic level using microfluorimetry.

Many analytical procedures involve the use of a fluorimeter, usually with a single exciting wavelength (or range) and single detection wavelength. Because of the sensitivity the method affords, fluorescent molecule concentrations as low as 1 ppt can be measured. Fluorescence in several wavelengths can be detected by an array detector, for example to detect compounds from High Performance Liquid Chromatography (HPLC) eluent flow. Also, thin layer chromatography plates can be visualized if the compounds or a coloring reagent is fluorescent.

The nuclear material example deals with the fluorimetric determination of uranium in natural waters. The UV spectroscopy analysis of U spiked in pellets of 2% LiF—98% NaF flux has been carried out by fluorescence (see Hues et al. 1977). The pellets are dried under heat lamps; then fused over special propane burners. The fused pellets are transferred to a fluorimeter, where they are excited with ultraviolet radiation and the fluorescence is measured. The uranium is calculated by comparing the measured fluorescence with that of other pellets, carried through the same procedure, which contain aliquots of standard uranium solutions. The sensitivity of the method is about 200 ppt of uranium, and the precision is approximately 15% in the 0.2–10 ppb uranium concentration range.

This method was found powerful. However, interferences such as absorption effect from other elements yield fluorescence signal reduction called quenching. At AIEA, Veselsky and Ratsimandresy (1979) investigated the quenching effects in the direct fluorimetric determination of uranium in minerals. A semi-logarithmic dependence of the reduction of uranium fluorescence intensity on the concentration of Fe(III), Cr(III) or Mn(II) in the flux was established. The quenching power was defined in terms of half-concentration, $C_{1/2}$, i.e. the quencher concentration which reduces uranium fluorescence to half its original value. The dependence of the results by the spike method on the degree of quenching was determined, and two improved analytical procedures were proposed for the determination of about 20–5000 ppm of uranium in minerals.

The interference study was completed by Veselsky and Degueldre (1986) with the quenching behavior of lanthanides. Ce, Pr, Nd and Tb showed different quenching behavior with half-concentrations lower than 300 ppm. This peculiar trend exhibited by these elements may be correlated with their chemical behavior during fusion in an ambient atmosphere yielding colored flux pellets owing to an at least partial formation of a higher oxidation state of the corresponding element.

Time-resolved laser induced fluorescence spectroscopy

Time-resolved spectroscopy is the study of dynamic processes after illumination of a material. The technique can be applied to any process that leads to a change in properties of a material. With the help of pulsed lasers, it is possible to study processes that occur on timescales as short as 10^{-16} s. **Time-Resolved Laser Induced Fluorescence Spectroscopy (TRLIFS)** is an extension of fluorescence spectroscopy. Here, the fluorescence of the material is recorded as a function of time after excitation by a flash of light. It also allows the quantification of the exited state lifetimes which can give insight on the excited state nature and the speciation of the luminescent ion at the molecular level. Specific information may be found in the work of Moulin et al. (1997) and in the chapter of Mauchin et al. (2014).

TRLIFS is a non-invasive in situ technique, allowing for a direct detailed probing of the local coordination environment of luminescent ions. The time resolution can be obtained in a number of ways, depending on the required sensitivity and time resolution. It can be done with:

- Fast-detection electronics (ns)
- Time-Correlated Single Photon Counting (ps)
- A streak camera (ps)
- Intensified cameras (200 ps)
- Optical gating (fs).

This technique uses convolution integral to calculate a lifetime from fluorescence decay.

The example of nuclear material investigation deals with analysis of luminescent ions such as Eu^{3+}, Cm^{3+}, Am^{3+} and U^{6+} in aqueous solutions, on the surface and within crystalline and amorphous solids. Phosphates are studied as potential waste form because of their low solubility.

TRLFS technique was used to analyze $LaPO_4$ monazite ceramic doped with 500 ppm Eu^{3+} or 50 ppm Cm^{3+}. Here, europium serves as an inactive surrogate for trivalent minor actinides (Am, Cm) and Pu. The use of curium as a luminescent probe allows for trace concentration measurements due to its low detection limit Bünzli (1989), and serves for a verification of surrogate studies with Eu^{3+}. The theoretical bases of the Eu^{3+} and Cm^{3+} TRLFS method and further experimental examples can be found e.g. in the reviews by Binnemans (2015) and Edelstein et al. (2006) respectively.

Highly crystalline Eu^{3+} and Cm^{3+} doped $LaPO_4$-monazite samples for the TRLFS-measurements were obtained by precipitation—e.g. Roncal-Herrero et al. (2011)—followed up by thermal treatment. The luminescence signal was plotted against excitation wavelength (Fig. 3.32a). In the case of europium, the transition is non-degenerated and results therefore in a single peak for a single Eu environment. The splitting in the excitation spectra of $LaPO_4$:Cm^{3+} is more complex due to the degenerate ground state. In both doped $LaPO_4$ one single incorporated species was found.

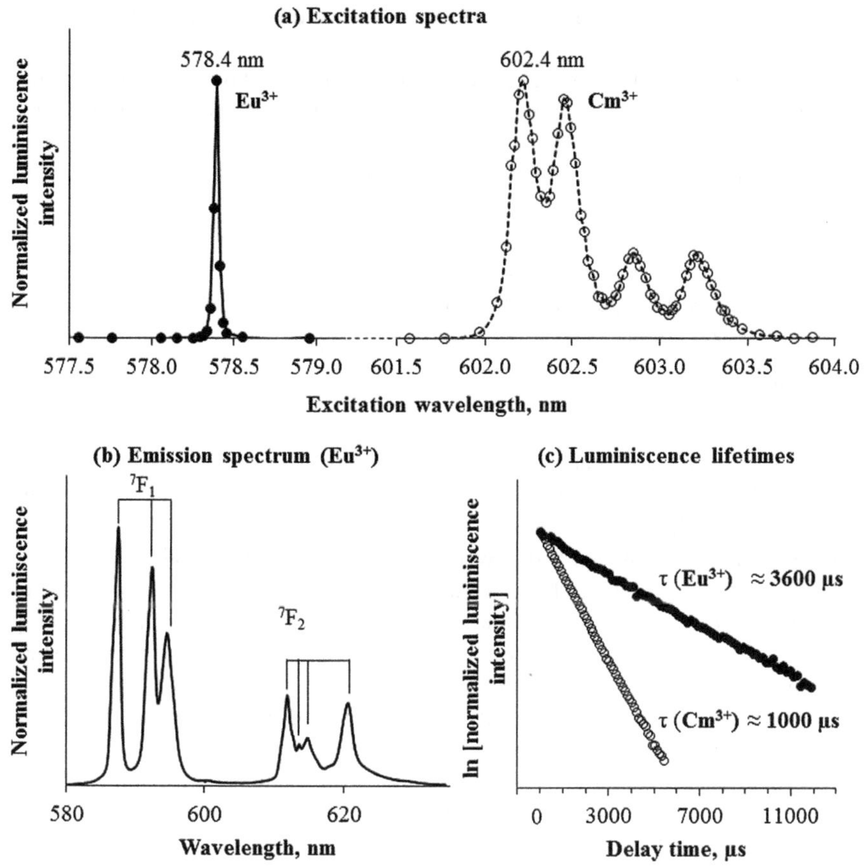

Fig. 3.32 TRLFS-measurements recorded for $LaPO_4$: Eu^{3+} and Cm^{3+}. **a** luminescence signal against excitation wavelength, **b** high resolution emission spectrum, **c** exponential slope luminescence decay curves. Arinicheva et al. (2016), with courtesy

By exciting Eu^{3+} and Cm^{3+} in $LaPO_4$ at the excitation peak maximum—i.e. at 578.40 nm for Eu^{3+} and e.g. 602.4 nm for Cm^{3+}—high resolution emission spectra were recorded. The high resolution emission spectrum is shown in Fig. 3.32b for $LaPO_4$:Eu^{3+3+} by way of example.

Maximal splitting of the ground state levels was characteristic for Eu incorporated into the monazite cation lattice site with a monoclinic low symmetry (see Holliday et al. 2012; Bünzli 1989). Lifetime measurements are obtained by monitoring the luminescence emission as a function of delay time between the laser pulse and the camera gating. The exponential (single slope) luminescence decay curves for both Eu^{3+}- and Cm^{3+}-doped $LaPO_4$ (Fig. 3.32c) confirm the presence of only one Eu^{3+} or Cm^{3+} species in the structure.

3.2.3.10 X-Ray Absorption and X-Ray Fluorescence Spectroscopy

X-ray absorption and fluorescence spectroscopy deal with photon energies ranging from about 0.002-200 keV corresponding to core electron excitation transitions. These absorptions correspond to edges which depend upon the core electron excitation: the principal quantum atomic numbers $n = 1, 2$, and 3 …, correspond to the K-, L-, and M- … edges, respectively. For instance, excitation of a 1s electron occurs at the K-edge, while excitation of a 2s or 2p electron occurs at an L-edge. Absorption and fluorescence are described by the transitions depicted in Fig. 3.13. Data on the X-ray absorption and emission are to be found in the *Handbook of X-ray Data* (see Zschornack 2007).

X-ray sources

The **X-ray sources** used for analysis are classed according to their properties. The X-ray tube, synchrotron beam lines and free electron lasers are examined below.

In the **X-ray tube** sources, electrons are accelerated onto an anode. The wavelength of the light produced is fixed by the anode material, e.g. Cu or Mo. The intensity is limited by the applied power and the cooling of the anode. In the X-ray tubes, electrons are emitted from a cathode and accelerated by a potential of $\sim$50 kV. The electrons collide on the metal atoms of the anode, emitting braking radiation and the defined lines due to the excitation of electrons from the inner-shell of the anode metal. The simplest and cheapest variety of sealed X-ray tube has a stationary anode and runs with $\sim$2 kW of electron beam power. More expensive varieties include rotating-anode type sources that run with $\sim$14 kW of e-beam power. X-rays are generally filtered to a single wavelength and collimated to a defined direction toward the sample. A good overview on the X-ray tubes appears in the article by Zink (1997).

Synchrotron radiation facilities are some of the brightest lights sources. They provide X-ray beams generated by acceleration of electrically charged particles, generally electrons, close to the speed of light, and confine them in a quasi circular loop using magnetic fields. The work edited by Jaeschke et al. (2016) provides general information on synchrotron light sources. These dedicated facilities generally host several beam lines where tests and analyses are carried out simultaneously without interruption. The main components of each synchrotron are its electron source and its storage ring. This ring is not circular, but a multi-side polygon. At each polygon corner magnets bend the electron beam path. The electrons deflection by magnetic fields is the source emitting X-ray photons. The other tools to generate X-rays are insertion devices like undulators. They are located in the storage ring as an integral part of its magnet structure. In such an undulator the electrons of the ring are forced to horizontal oscillations, thus yielding a high flux of X-rays. X-ray photons produced by bending magnets, undulators or wigglers may be used for specific applications in dedicated beam lines. Specific information may be found in the monograph by Marks (1995).

Over the last 40 years, **free electron lasers** have been developed for use of light (VIS, UV, X-ray) flashes in medical and biological sciences as well as in crystallography. The work edited by Jaeschke et al. (2016) also provides general information on the free electron laser light sources. These sources are the brightest sources available today, providing fs X-ray flashes; e.g. Feldhaus et al. (2005), Schlichting and Miao (2012). A beam of accelerated electrons passes through an undulator, a periodic arrangement of magnetic fields with alternating polarity across the beam. Along the direction of the incident beam the undulator produces photons. Since the electrons enter the undulator as a pulse, the photons also form a pulse (flash). After separation of the electrons from the photons, the flash of light can be used for specific applications. The source intensity allows diffraction patterns to be resolved at the atomic level for nanocrystals. This method, serial femtosecond crystallography, has been used in solving a number of crystal structures, sometimes revealing differences with equivalent structures gained from synchrotron sources.

Detectors

Detectors for X-ray spectroscopy have already been presented in Sect. 3.1.2. Proportional counters or various types of solid-state detectors (diode, Si(Li), Ge(Li), …) are currently used. **Si(Li) detectors** consist usually of a $\sim$ 0.4 cm thick silicon diodes with 1000 V across it. The lithium center part forms a non-conducting layer (see Szöghy et al. 1981). High-purity silicon wafers with low conductivity became available more recently. Cooled by the Peltier effect, they provide convenient detectors, although the liquid nitrogen-cooled Si(Li) detector still has the best resolution (i.e. ability to distinguish different photon energies). These detectors allow counting of the photon on an Energy-Dispersive Spectroscopy (EDS) basis.

In Wavelength Dispersive Spectroscopy (**WDS**) analysis, polychromatic photons are first analyzed through a monochromator; the resulting single-wavelength radiation is then detected by a photomultiplier, which counts individual photons. The signal is amplified and transformed into an accumulating digital count. These counts are then processed to obtain analytical data. Basic information on WDS may be found in the work of Goldstein et al. (1992).

Finally, for imaging, semiconductor **Charge-Coupled Devices** (**CCD**) cameras are extensively used for analysis. They provide initially 2D pictures that can be combined for 2D, 3D and 4D reconstructions. A CCD image sensor is an analog device formed of an array of microsensors integrated as a 2D pixel package. When photons strike the chip their energy is held as a small electrical charge in each photo sensor. The charges are converted to voltage one pixel at a time as they are read from the chip. Additional circuitry in the camera converts the voltage into digital information. General information on CCD cameras may be found in the monograph by Theuwissen (1995).

X-ray Absorption Spectroscopy (**XAS**) is based on the absorption of X-ray photons by the atoms (element) of the sample. All elements absorb X-rays to some extent. Each element has a characteristic absorption spectrum which consists of a succession of steps occurring at a wavelength close to an emission line of the element. Absorption (Eq. 3.9) also attenuates the secondary X-rays leaving the sample. Mass absorption coefficients are well known and can be calculated for a multi-element sample; however, the sample composition needs to be known. To derive accurately the mass absorption, data for the concentration of elements not measured by transmission or by fluorescence XRF may be needed, and various strategies are employed to estimate these absorption coefficients.

3.2.3.11 X-Ray Fluorescence

X-Ray Fluorescence (XRF) is the emission of specific fluorescent X-rays from atoms in material that has been excited after illumination by X-rays. Materials exposed to X-rays may undergo ionization of specific component atoms or ions. This ionization ejects one or more electrons from the atom or ion. This process occurs if the atom or ion reacts with a photon of energy greater than its ionization potential. X-rays can be energetic enough to expel electrons from the inner shells of atoms or ions. The removal of an electron in this way makes the electronic structure of the atom or ion unstable, and electrons in higher shells drop to the lower shell to fill the vacancy. The energy is released by emission of a photon of the energy which equals the energy difference between the two shells. The material subsequently emits radiations, which have a defined energy allowing identification of the sample atoms. The term "fluorescence" is used to define the reemission of radiation at lower energy than the absorbed one. Information on the practical aspect and on instrumentation and application are provided in the handbook edited by Beckhoff et al. (2006) and the monograph by Haschke (2014) respectively.

Each element has electronic shells at defined energies. Following removal of an inner shell electron by an energetic photon, an electron from an outer shell drops into its place. There is a limited number of possibilities for how this can happen. The higher energetic transitions have given names: an L $\rightarrow$ K transition is called K_α, an M $\rightarrow$ K transition is called K_β, an M $\rightarrow$ L transition is called L_α, and so on. Each of these transitions yields a fluorescent photon with a fixed energy. The wavelength of this fluorescent radiation can be calculated from Planck's law:

$$\lambda = \frac{hc}{E} \quad (3.26)$$

The fluorescent radiation can be analyzed either by sorting the energies of the photons by EDS or by separating the wavelengths of the radiation, i.e. WDS. The intensity of each characteristic radiation is directly related to the amount of each element in the material. This is the basis of a powerful technique in analytical chemistry. The wavelengths of the sharp element fluorescent spectral lines obtained in the wavelength-dispersive method are determined by the atomic numbers as expressed in Moseley's law (1913).

The XRF measurement may be applied in two different modes: WDS and EDS. The EDS is a multi-elemental technique that can determine the elements in a wide range of concentrations from 10^{-6} to 1 g g^{-1} (1–10^6 ppm). It is non-destructive with minimum spectral interference. The analysis is very fast and in many cases the sample preparation requires only grinding and pelletizing. Recently, use of detectors with better energy resolution down to 10 eV (super-conducting tunnel junction) makes this technique even more promising. X-ray fluorescence is also a technique that can be applied in the field with portable units. It may be applied for mineral exploration, mining, monitoring of polluted areas, and other environmental studies.

Nuclear material applications are developed below.

A nuclear material analysis illustrated for XRF concerns the Boiling Water Reactor (BWR) deposit product called CRUD (Chalk River Unidentified Deposit). In BWR, corrosion deposits on the fuel rods are produced by precipitation of the iron-based rust onto the fuel cladding, mostly at the place around the transition zone between the one phase (liquid) and the two phases (liquid-steam) of the coolant. CRUD phases on cladding samples were analyzed by μXRF (see Degueldre et al. 2009). CRUDs condense in the form of secondary phases where precipitation is dictated by solubility limits of oxides and binary oxides as well as of their solid solutions (as observed from element mapping); see Fig. 3.33. However, the addition of soluble Zn in the primary water proposed for reduction of ^{60}Co build-up in the corrosion layers induced unexpected problems including precipitation of magnetic-sensitive phases which falsified Eddy current measurements used to quantify the thickness of the Zircalloy corrosion layer. Initially, EPMA measurements were carried out to identify the element associated with the CRUD layer and it was found that Zn was associated with Fe and O. It was provisionally deduced that the material was made of a zinc ferritic spinel. Additional studies were needed to better understand the nature of the material. It was found by μXRF that the element distribution include Fe, Ni, Zn and Mn mainly together with traces of Cr, Co and their stoechiometry is buffered by Fe_2O_3, NiO and MnO in binary oxides (see Orlov et al. 2010). Figure 3.33 shows μXRF scans of Zn, Fe and Ni along and across a segment of the CRUD layer confirming occurrence of Zn, Ni and Fe rich deposits. The XRF scans were completed by combining XRD and XAS investigations as summarized in Sect. 4.2.

Aerosol particles collected on a filter create generally homogeneous and thin layers, which result in negligible matrix effects during XRF analysis. The most recent advances are the application of synchrotron-induced XRF. In water, suspended matter can be separated on a filter and analyzed by XRF. Actinides in samples like soils, sediments, rocks, minerals, fly ash, and solid waste … are often detected at the 10^{-6} g g^{-1} level when applying an XRF method. This technique has been applied for actinide analysis of particles from waters, sediments (Hunter and Bertsch 1998) and waste phases (e.g. Schoonover and Havrilla 1999), and may also be applied for determination in analysis of airborne particles (see Misaelides et al. 1993). Uranium can be determined in water by precipitation with a non-specific chelating reagent and collection on a filter, which is analyzed by XRF. The detection limit is 0.15 μg U (6×10^{-10} mol) or 10^{-9} mol for a 500 mL sample. It can be improved with a microbeam.

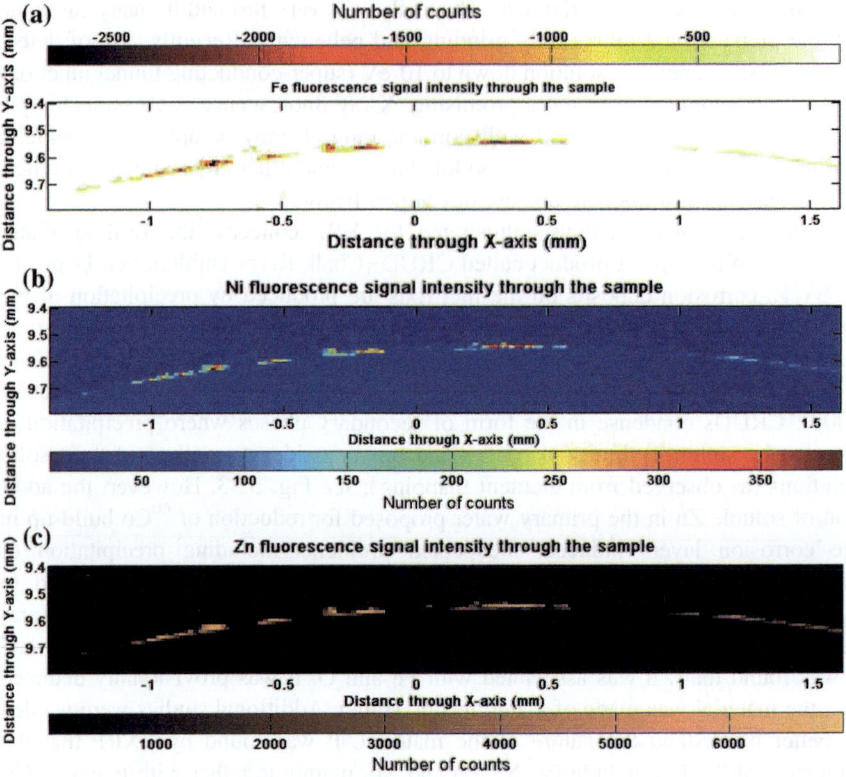

Fig. 3.33 X-ray fluorescence map of a BWR CRUD deposit on a BWR cladding. **a** Fe, **b** Ni, **c** Zn. Orlov et al. (2010) with permission © Elsevier

3.2.3.12 X-Ray Absorption Fine Structure Spectroscopy

X-ray Absorption Fine Structure (XAFS) spectroscopy is a widely used technique for determining the local atomic and/or electronic structure of matter. The experiment is usually performed at synchrotron radiation sources, which provide high flux and tunable X-ray beams. Samples can be in fluid form as: gas-phase or solution, or in condensed matter form as: crystalline or glass/amorphous phase. For a general overview, an introduction on X-Ray Spectroscopy is to be found in the work of Agarwal (1991).

XAFS data are obtained by tuning the photon energy using a monochromator to a range where core electrons can be excited (0.1–100 keV photon energy). The edge label depends upon the exited electron shell: the principal quantum numbers being $n = 1$, 2, and 3, corresponding to the K-, L-, and M-edges, respectively. Actually, a 1s electron excitation occurs at the K-edge, while 2s or 2p electron excitation occurs at an L-edge.

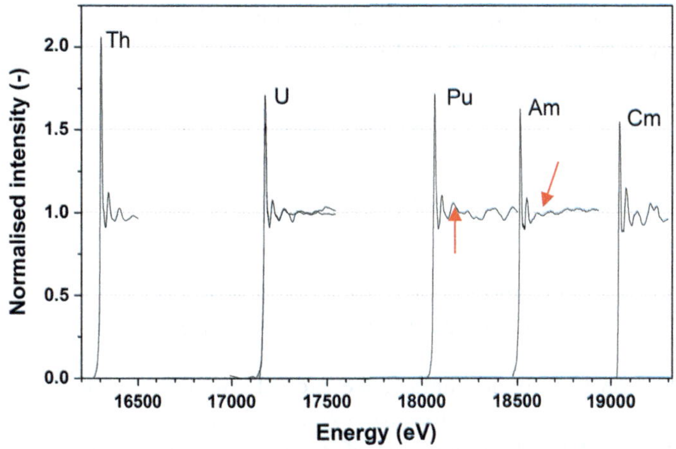

Fig. 3.34 Actinide XAFS spectra for spent fuel MOX (60 MW d kg^{-1}, sphere-pac) and redox states evaluation. The *arrows* indicate a decreasing wave when An^{4+} (Th, U, Pu) reduces to An^{3+} (Am, Cm); conditions: location in fuel periphery. Degueldre et al. (2011a, 2013b, c, 2016) with permission © Elsevier

There are three main regions found on a spectrum generated by XAFS data (e.g. Fig. 3.34):

- The absorption threshold including a pre-edge region (e.g. 0–10 eV) determined by the transition to the lowest unoccupied states
  - the states at the Fermi energy in metals giving a sharp edge
  - the bond core excitons in insulators
- The X-ray Absorption Near-Edge Structure (XANES), dominated by core transitions to quasi bond states for photoelectrons with kinetic energy in the range from 10 to 150 eV above the Fermi level
- In the high kinetic energy range (e.g. 50–500 eV) of the photoelectron the scattering cross-section with neighbor atoms is weak and the absorption spectra are dominated by Extended X-ray Absorption Fine Structure (EXAFS) where the scattering of the ejected photoelectron off neighboring atoms can be approximated by single scattering events.

XAS is a type of absorption spectroscopy from a core initial state with a well-defined symmetry, therefore the quantum mechanical selection rules determine the symmetry of the final states in the continuum, which are usually mixtures of multiple components.

XAFS may be applied in transmission or by fluorescence. An incident X-ray beam is absorbed in the material which component elements are excited. The intensity of the directly transmitted beam is reduced by absorption following the

Beer–Lambert law (Eq. 3.8), where the absorption coefficient, μ for X-rays, can be written derived from Moseley's (1913) law:

$$\mu \approx \frac{\rho Z^4}{A E^3} \tag{3.27}$$

with ρ the sample density, Z the atomic number, A the atomic mass and E the X-ray energyand

$$\begin{array}{ll} \text{in transmission} & \mu(E) = -[\ln(I/I_0)]/x \\ \text{in fluorescence} & \mu(E) \approx -[\ln(I_f/I_0)]/\Delta x \end{array}$$

Depending on the atomic number, the excited component element emits preferentially fluoresce light or electrons following inner-shell transitions (phenomenon treated in Sect. 3.2.4). The reemission of visible light may also be recorded. The absorption edge is followed by a series of oscillations due to the interference of the photoelectron on the surrounding atoms or ions.

X-ray absorption spectroscopy (XAS) today takes advantage of third-generation high-brilliance synchrotron radiation sources. This technique has revived interest in the field of hard X-rays in macro- and micro-probes. Recent developments are based on the use of Fresnel zone plates and containment techniques such as taped glass capillaries and plastic films, allowing microanalysis. Typical performance of synchrotron microprobes corresponds to a photon flux of 10^{10} s^{-1} μm^{-2} with minimum detection limits $\sim 10^{-15}$ g ($\sim 4 \times 10^{-18}$ mol).

The nuclear material example deals with actinide XAFS spectra recorded for a MOX fuel in the pellet periphery (see Fig. 3.34) after an irradiation in LWR corresponding to 60 MW d kg^{-1}. The redox states evaluation indicates that Th, U and Pu stay as An^{4+} and that Am and Cm reduces to An^{3+} Degueldre et al. (2016).

Extended X-ray absorption fine structure spectroscopy

Extended X-ray Absorption Fine Structure (EXAFS) spectroscopy can be used to determine the chemistry and coordination environment of the absorbing atoms. EXAFS basic principles and data analysis are to be found in the monograph by Teo (1986), and the theoretical approaches to X-ray absorption fine structure may be find in the work by Rehr and Albers (2000).

After recording of the EXAFS spectrum the data need to be handled to extract the relevant information. The operations to be performed are:

- Derivation of the raw μ spectrum to determine E_0
- Normalizing the spectrum to gain the χ spectrum
- Transformation of the XAFS spectrum from E space to k space
- Weighting the spectrum plotted in k space by the k^i factor
- Operate a Fourier transformation on the k^i-weighted spectrum
- testing reference atom cluster models to fit the k^i-weighted spectrum (backdoor approach).

Normalizing the raw spectrum data

The absorption coefficient is obtained by taking the log ratio of the incident X-ray intensity to the transmitted X-ray intensity.

$$\mu = \frac{-\ln(I_t/I_0)}{x} \tag{3.28}$$

EXAFS spectra are displayed as graphs of the absorption coefficient of a given material versus energy, typically in a 500–1000 eV range, beginning before an absorption edge of an element in the sample. The X-ray absorption coefficient plotted over the energy is normalized to unit absorption step height.

Determining E_0

The XAFS spectral data may be first reduced so that the relative absorption step for its asymptotic value may reach the arbitrary value of 1. The spectrum

$$\mu = \mathrm{f}(E)$$

may be displayed. The determination of E_0 is carried out by derivative

$$(\delta/\delta E)\,\mu = \mathrm{f}'(E) \quad \text{with} \quad \mathrm{f}'(E_0) = \text{Maximum}$$

followed by a second derivative

$$(\delta/\delta E)^2 \mu = \mathrm{f}''(E) \quad \text{with} \quad \mathrm{f}''(E_0) = 0.$$

Normalizing the absorption coefficient in the EXAFS region

To represent the EXAFS region from the whole spectrum, a function χ can be roughly defined in terms of the absorption coefficient, as a function of energy:

$$\chi(E) = \frac{\mu(E) - \mu_0(E)}{\Delta\mu_0} \tag{3.29}$$

Here, the subtracted term, μ_0, represents a spline function used to approach the atomic absorption background and the division by $\Delta\mu_o$ represents the normalization of the function. When interpreting data for the EXAFS, the use of the photoelectron wave vector, k, given by $k = 2\pi/\lambda$ is suggested as an independent variable.

Transformation in k space

The wave vector k may be solved by first assuming that the photon energy E will be greater than E_0. Since energy is conserved, excess energy given by $E - E_0$ is converted into the kinetic energy of the photoelectron wave. Since wavelengths are dependent on kinetic energies, the photoelectron wave propagates with a velocity of v which gives the relation, $(E - E_0) = m_e v^2/2$. The photoelectron wavelength is inversely proportional to their momentum; $(m_e v)$: $\lambda = h/m_e v$ consequently k becomes:

$$k = \frac{\sqrt{2(E - E_0)m_e}}{\hbar} \tag{3.30}$$

From *E* space to *k* space

On the basis of an expression for k, the EXAFS equation may be developed referring to the probability that the photoelectron is found as a function of radius R (distance from the absorbing atom).

This wave function describes the constructive/destructive nature of the wave interactions, and varies depending on the phase difference of the waves interacting. This phase difference can be easily expressed in terms of our photoelectron wave number and R, the distance to the neighbor atom shell from the absorbing central atom, as $2k/R$. Another characteristic of this wave interaction is the amplitude of wave backscattering to the central atom. This amplitude can provide coordination number since it is directly proportional to the number of scatters. The final function for $\chi(k)$ is then the EXAFS equation:

$$\chi(k) = \sum_j \frac{N_j f_j(k) \exp\left[-2k^2\sigma_j^2\right] \exp\left[-2R_j/\lambda\right]}{k R_j^2} \sin\left[2k R_j + \delta_j(k)\right] \tag{3.31}$$

where $f(k)$ is the amplitude and $\delta(k)$ the phase shift. These parameters can be calculated for fixed k value; radius R, coordination number (N), and the measure of disorder in neighboring atoms (σ) remaining unknown. These unknown values are the information to be extracted from Eq. 3.31.

The factors of Eq. 3.31 that contribute to the EXAFS spectrum are namely: the sine factor that fixes the shape, with a phase shift modulating the oscillations. This oscillation depends on the energy difference with the edge value $(E - E_0)$ and on the radial distance. The Debye-Waller factor explains the decay with increasing energy, as well as increasing disorder. Actually, EXAFS does not work over long distances (e.g. >5 Å) because of the $\exp(R_j)$ factor that decrease exponentially over large values of R_j, limiting the potential of EXAFS.

Fourier transformation of the k^n weighted plotted spectrum

To amplify the oscillations graphically, χ may be plotted as $k^n \chi$.

The transformation of the k^n weighted normalized absorption in R space applying Fournier transformation (FT) for a frequency window allows production of a quasi pair distribution function that represents the presence of next neighbor atoms at a certain distance of the absorber. The distance between atoms (or ions) may be used to build a 3D model of the atomic structure.

Testing reference atom cluster models with k^i weighted k-plotted spectrum (backdoor approach)

The backdoor approach consists of starting with a 3D atomic cluster. Calculate what would be the resulting distribution, then the normalized spectrum, and compare the theoretically derived normalized spectrum with the one obtained from experiment.

The specific nuclear analysis example deals with speciation of actinides in water. EXAFS technique is quasi non-destructive, element-specific (L edges for actinides) and is very attractive for actinide speciation, for the definition of their atomic environment and redox state. It makes use of their L edges using mono-energetic beams in the range up to ~20 keV, focusing a very intense beam over a small area. The micro- X-ray absorption beam may be used to focus on local details, for example to better understand the sorption of uranium in soil sediments (see Bertsch et al. 1994), comparing the fine structure for various valence state of uranium. EXAFS of Np(IV) and Np(V) showing the fine structure of the redox species (see Denecke et al. 2004) gives evidence for Np(V) reduction in humic solutions. The Np L3 XANES is unclear, the white line position at 17.6201 keV for Np(V) and 17.6209 keV for the Np(IV) are difficult to differentiate (see Fig. 3.35). The EXAFS data however unequivocally reveal Np(IV) formation in the solutions (Denecke et al. 2004). Actually, the EXAFS FT magnitude plots versus R (see Fig. 3.35) reveal the neptunyl oxygen's (a) while they disappear as expected after reduction to neptunium(IV) state (c).

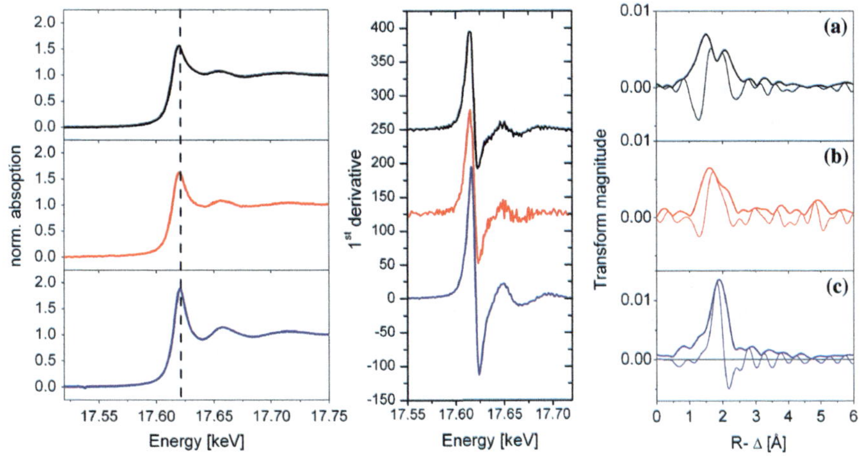

Fig. 3.35 XANES, differential XAFS data and EXAFS of Np(IV) and Np(V) showing the fine structure of the redox species in humic solutions. Evidence for Np(V) reduction (*black* (0%) → *red* (50%) → *blue* (100%)): the Np L3 XANES is unclear (white line 17.6201 keV for Np(V) and 17.6209 keV for the Np(IV)). Only the Fourier transformation allows discrimination between both redox species. Denecke et al. (2004) with permission © Elsevier

Variable Incidence Angle X-ray Absorption Fine Structure (VIAXAFS) spectroscopy has been used for its non-destructive ability to probe and investigate nano-structures in material films. The theory of X-ray absorption-detection in fluorescence mode was revisited and applied to the variable incidence angles of the photon beam for studying microscopic films (see Degueldre et al. 2007). The expression derived for the depth profile was used to evaluate nano-structures and atom environments in corrosion film. This technique was applied, spanning angles from a grazing toward normal incidence on a sample layer obtained by zirconium alloy corrosion. XAFS analysis on the Zr K edge suggests a lower number of next neighbor Zr atoms in the corroded samples. This may be due to the relative density for given average atom distances suggesting a high density of dislocations and defects. The technique reveals nano-pores, dislocations, vacancies or defect features. The presence of tetragonal zirconia is not observed for zirconium alloy corrosion layers.

X-ray absorption near edge spectroscopy

X-ray Absorption Near Edge Spectroscopy (XANES) is a type of absorption spectroscopy due to the photoabsorption for electronic transitions from an atomic core level to final states in the energy region up to 50–100 eV above the selected atomic core level ionization energy. XANES analyses are performed to map elemental distribution and to determine the oxidation states.

In the XANES region, starting about 5 eV beyond the absorption threshold, because of the low kinetic energy range (5–150 eV) the photoelectron backscattering amplitude by neighbor atoms is very large so that multiple scattering events become dominant in the XANES spectra.

The different energy range between XANES and EXAFS can be explained by the comparison between the photoelectron wavelength λ and the interatomic distance of the photoabsorber-backscatterer pair. The photoelectron kinetic energy is connected with the wavelength λ by the following relation:

$$E_{\text{kin}} = h\nu - E_{\text{bin}} = \frac{h^2 k^2}{(2m)} = \frac{(2\pi)^2 h^2}{(2m\lambda^2)} \tag{3.32}$$

which means that for high energy the wavelength is shorter than interatomic distances and hence the EXAFS region corresponds to a single scattering regime; while for lower E, λ is larger than interatomic distances and the XANES region is associated with a multiple scattering regime.

The great power of XANES derives from its elemental specificity. Because XANES can also determine the chemical state of elements, it has found widespread use in environmental chemistry and geochemistry. Much chemical information can be extracted from the XANES region: formal valence (very difficult to experimentally determine in a non-destructive way); coordination environment (e.g., octahedral, tetrahedral coordination) and subtle geometrical distortions of it.

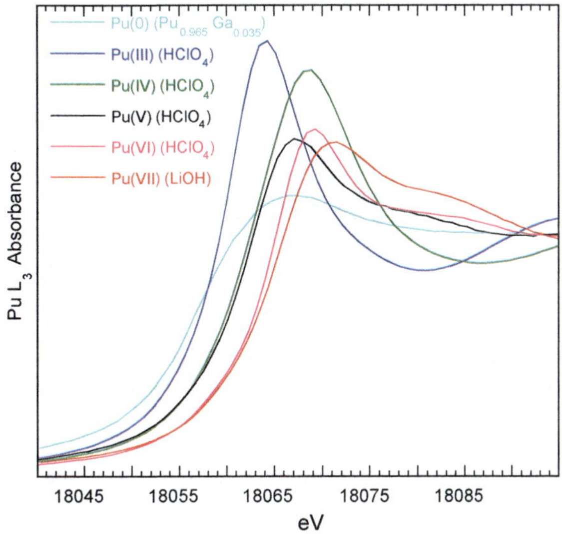

Fig. 3.36 Comparison of the plutonium XANES spectra recorded for 6 redox states of Pu. Conradson (1998) with permission © Springer

The nuclear material examples deal with the XANES of actinides. Several authors (Nitsche 1995; Hess et al. 1997; Antonio et al. 2001) suggested in the mid-1990s the use of X-rays provided by synchrotron beams in absorption spectroscopy (XAS) as a tool for actinide speciation in solids (soils) and liquid (waters) as well as at their interface. This quasi non-destructive technique is element-specific (L edges for actinides) and very attractive for actinide speciation and for the definition of their atomic environment, e.g. Conradson (1998), Degueldre et al. (2004) and redox state (see Figs. 3.34 and 3.36).

The example of such a use is the determination of the oxidation state of plutonium (see Fig. 3.36). It shows a systematic shift of the E_0 going from Pu metal through Pu(III), Pu(IV), Pu(V), Pu(VI) in acidic to Pu(VII) in basic solution. Their E_0 passes from ~18,055 eV for Pu(0) to ~18,062 eV for Pu(VI) and Pu(VII). This shift is also visible on the maximum of the white line that also reduces in intensity when the redox state increases. This may be due to the contraction of the Pu ion from valence state III to VII and change of the unoccupied density of state accessible for the excited 2p electron.

Scanning transmission X-ray microscopy

Scanning Transmission X-ray Microscopy (**STXM**) is a type of X-ray microscopy in which an X-ray beam is focused by a zone plate array onto a small spot on a sample. The sample is scanned in the focal plane of the zone plate and the transmitted X-ray intensity is recorded as a function of the sample position. X-ray microscopes work by exposing a film or CCD camera to detect X-rays that are transmitted through the specimen. The image formed is of a thin section of specimen. Newer X-ray microscopes also use X-ray absorption spectroscopy to heterogeneous materials at high spatial resolution. The technique is a combination

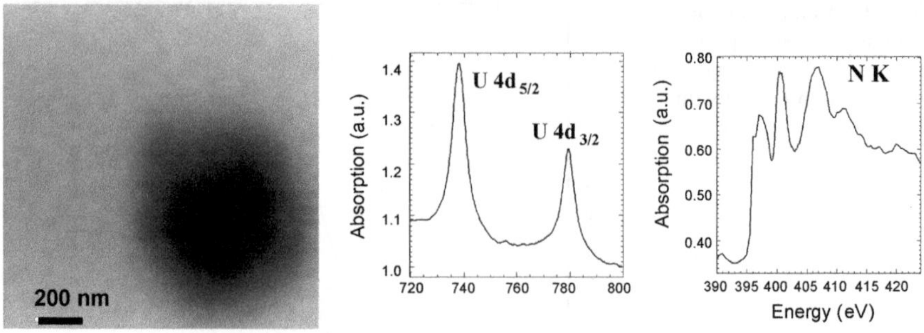

Fig. 3.37 Soft X-ray Scanning Transmission X-ray Microscope (STXM) data collection from a mixed phase uranium nitride ($UN_{1.2–1.8}$). Glans et al. (2010), courtesy of Seaborg Institute

of spectromicroscopy, imaging with spectral sensitivity, and microspectroscopy, recording spectra from very small spots. Detailed information may be found in the Status Report from the Fifth International Conference X-Ray Microscopy and Spectromicroscopy, edited by Thieme et al. (1998). The detection limit for a given element is similar to that of XAFS spectroscopy and the lateral resolution may be assimilated to the pixel size, e.g. 20 nm (depending on the zone plate geometry).

The nuclear material example deals with advanced fuel investigations (see Fig. 3.37). Among the advanced nuclear fuels, actinide nitrides are of interest for some of the Generation IV reactors, e.g. LMR. A particle of uranium nitride ($UN_{1.2–1.8}$) was recently characterized by STXM (see Fig. 3.37) for its absorption edges of U (absorption edge N_5 and N_4 respectively around 740 and 780 eV) and of N (absorption edge K around 395 eV with maxima at 397, 400.5 and 407 eV); see Glans et al. (2010). These data could be used for identifying at the nano-scale the chemical properties of the actinide nitride, working similarly to the process on another scale by Ward et al. (2015) for the uranyl compounds with the oxygen K edge.

3.2.3.13 X-Ray Diffraction, Small Angle X-Ray Scattering and Laue Diffraction

The principles of X-ray diffraction and scattering are dictated by wave diffraction on the lattice planes. These techniques are performed in quasi-transmission or in back diffraction mode using **monochromatic** X-rays. When the grains are smaller than the X-ray beam, width rings are recorded in quasi-transmission and backscattering mode by application of Bragg (1908) law. This is the classical domain of powder X-ray diffraction. When the grains are larger than the X-ray beam, the reflex spots are recorded in a so-called micro- X-ray diffraction mode. In contrast, the von Laue (1913) diffraction method was the first to be utilized in 1912, using **white** or pink X-rays (i.e. continuous spectrum).

X-Ray Diffraction (XRD) is a technique used for characterizing the atomic and molecular structure of a crystal. Basically, the atom or ion array diffracts an incident X-ray beam in specific directions. A 3D picture of electron density in the crystal can be reconstructed by measuring angles and intensities of the diffracted photons. From this measured electron density, the atom or ion positions in the crystal can be determined, as well as their distances and their chemical bonds. Information on X-ray diffraction crystallography is to be found in the work of Waseda et al. (2011).

For a single-crystal XRD, a crystal is mounted on a goniometer. The goniometer is used to fix the crystal position during the measurements. The crystal is illuminated with a focused monochromatic X-ray beam forming a diffraction pattern with defined orientations (see Fig. 3.38). The 2D images taken for different orientations are converted in a 3D model of electron density in the crystal applying Fourier transformation and taking into account the chemical data known for the studied sample. For XRD, the scattering is elastic, the scattered X-rays keep the same wavelength as the incident X-rays. Inelastic X-ray scattering methods are used for investigating excitations of samples, instead of atom distributions.

Atoms or ions from the crystal scatter X-ray waves through the atoms' electrons by elastic scattering. The electrons are also called scatterers. A regular scatterer array produces a regular spherical wave array. Some of these waves cancel one another out through destructive interference. They add constructively in specific directions defined by Bragg's law as sketched in Fig. 3.38.

In the diffraction scheme, equally spaced atomic sheets through the crystal are contributing to that given reflection. Their orientation is fixed by their Miller indices (h, k, l), and their spacing d. For Bragg, the incident X-rays are reflected by each plane; acting with constructive interference for the angle θ between the plane and the resulting X-ray, the path-length difference being a multiple n of λ. The spacing d between planes is consequently deduced from:

$$2d \sin \theta = n\lambda \tag{3.33}$$

The intensity of the diffracted X-ray is dictated by the density of electrons in the scattering atoms. Since the X-ray photon energy is much larger than the binding

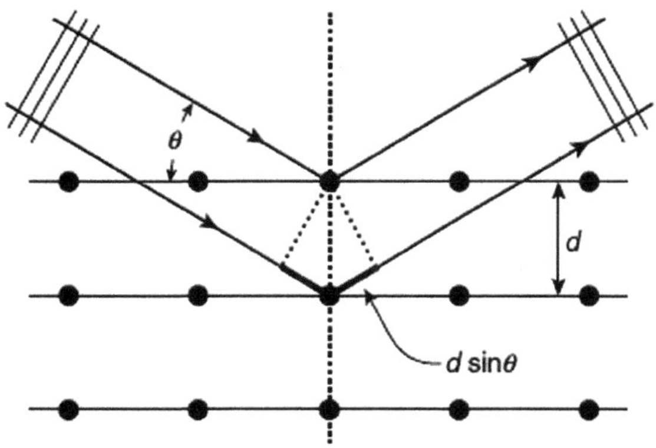

Fig. 3.38 Bragg diffraction scheme guiding diffraction rule. The incident beam causes each scatterer in ordered groups of atoms to reflect a fraction of its intensity as a spherical wave

energy of the valence electrons (see Table 1.1), their interaction may be that of a photon with a free electron, e.g. a Thomson scattering. Thomson (1923) deduced that the photon scattering on one particle of mass m and charge q is described by the differential cross-section σ for a photon scattered at angle 2θ per unit solid angle Ω as:

$$\frac{\mathrm{d}\sigma}{\mathrm{d}\Omega} = \left(\frac{q^4}{m^2 c^4}\right) \frac{1 + \cos^2 2\theta}{2} \tag{3.34}$$

For the electron, the q^4/m^2c^4 factor equals 7.94×10^{-26} cm^2 sr^{-1}. Now the atom nuclei, which are heavier than one electron, do not contribute significantly to the scattering of the X-rays.

The recording of XRD reflections and their analysis require a single crystal or polycrystalline sample and an intense X-ray beam. The exposed sample diffracts X-rays into a reflection pattern of rings or spots that are usually observed on a screen behind the sample. The reflected feature intensities may be revealed on photographic film, or recorded on an area detector or with a CCD camera. Low-resolution data are found at small angles, whereas with lower intensities high-resolution data are recorded at high angles, limiting the data quality. The mosaicity, polygonalization and disorder in the crystal can be observed in the peak width and form. Specific crystal properties are consequently detected. The reflex relative intensities allow determination of the molecular structure. In non-crystalline matter, e.g. liquid, amorphous sample, glass or nano powder, the random orientations may provide specific information.

The nuclear material example treated for XRD deals with the miscibility gap within the UO_2–PuO_2–Pu_2O_3 system in a plutonium rich MOX (see Bellin et al. 2015). The UO_2–PuO_2–Pu_2O_3 phase diagram is known to exhibit a miscibility gap at low temperature. MOX fuels with a composition in this region could form two fluorite phases and thus exhibit heterogeneities. Experimental data in this domain are scarce; however, Bellin et al. (2015) reported in situ XRD experiment from room temperature to 1750 K. Phase equilibria of UO_2–PuO_2–Pu_2O_3 are known to be a $(U_{1-y},Pu_y)O_{2-x}$ solid solution for low plutonium fractions. At higher Pu fraction the system becomes more complex, a second phase may be found due to the existence of a miscibility gap depending on the temperature. This miscibility gap may cause cracks or density changes that affect pellet integrity.

Samples were prepared by co-milling of UO_2 and PuO_2 powders. The humidity was adjusted to prepare stoichiometric materials to obtain monophasic $(U_{1-y}Pu_y)O_2$ solid solution as a starting material. Stoichiometric $(U_{1-y}Pu_y)O_2$ solid solutions with $y = 0.14$, 0.24, 0.35 and 0.46 were prepared, and reduced at 1750 K under He —5%H_2 followed by in situ high temperature XRD analysis. XRD patterns were recorded back to 300 K steps. Data were refined using the Rietveld method (1969).

XRD measurements were performed at ambient pressure as a function of temperature with an advanced X-ray diffractometer using copper radiation from a conventional tube source (Kα1 + Kα2 radiation, $\lambda = 1.5406$ and 1.5444 Å). The

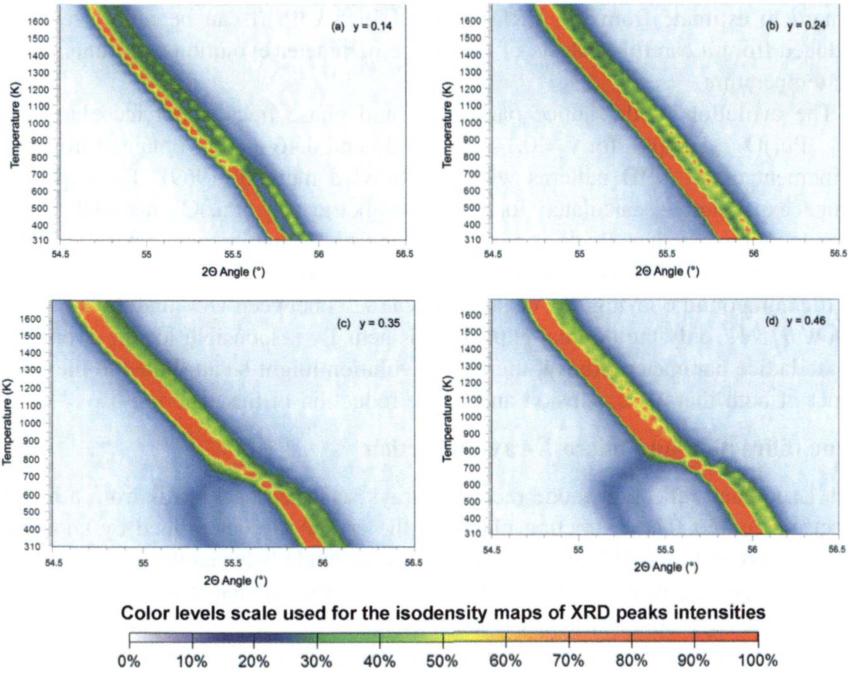

Fig. 3.39 Isodensity maps of XRD intensity of the (311) peak of the *fcc* fluorite structure of $(U_{1-y}Pu_y)O_{2-x}$ samples at $y = 0.14$, 0.24, 0.35 and 0.46 between 54.5° and 56.5° 2θ, as a function of temperature, Conditions XRD from copper source ($K\alpha1 + K\alpha2$ radiation, $\lambda = 1.5406$ and 1.5444 Å) operating at 40 kV and 40 mA, fast counting detector, opening angle $2\theta = 3°$. Adapted from Bellin et al. (2015) with permission © Elsevier

lattice parameters uncertainty was estimated to be ±0.001 Å corresponding to data point size in Fig. 3.39.

Figure 3.39 shows the isodensity map of XRD peak intensity of the (311) diffraction peak of the fluorite structure for the compounds $(U_{1-y}Pu_y)O_2$ at $y = 0.14$, 0.24, 0.35 and 0.46 as a function of temperature during cooling. At $y = 0.14$ (Fig. 3.39a), no phase separation is observed as only one reflection corresponding to a single fluorite phase is found at all temperatures. This is expected since this plutonium content range is located outside the miscibility gap. However, a break in the slope of the lattice parameter evolution is observed at ~700 to 800 K. At $y = 0.24$ (Fig. 3.39b), only one contribution is observed above ~573 K, demonstrating the presence of an *fcc* solid solution. Below this temperature, although hardly visible in Fig. 3.39b, diffraction peaks are slightly broader and require the use of two *fcc* structures to obtain a suitable refinement of the data (see later in Fig. 3.39). This suggests existence of the phase separation region as reported in the literature at such Pu content. At $y = 0.35$ and 0.46, two diffraction peaks with different intensities are clearly observed below a given temperature and the phase separation is now obvious. However, the temperature of phase separation is hard to estimate from the XRD maps of Fig. 3.39. It can be more precisely deduced from a careful analysis of the lattice parameter evolution as a function of the temperature.

The evolution of the lattice parameters and phase fractions of fcc phases in $(U_{1-y}Pu_y)O_{2-x}$ samples for $y = 0.14, 0.24, 0.35$ and 0.46 may be obtained from the refinement of all XRD patterns with the Rietveld method (1969). The expected lattice expansion is calculated for a stoichiometric $(U_{1-y}Pu_y)O_2$ material at the corresponding y value. Back at room temperature, the lattice parameter reaches a value of 5.459(1) Å, which perfectly matches that of a $(U_{0.86}Pu_{0.14})O_2$ stoichiometric compound obeying the Vegard's law (1921) between UO_2 and PuO_2. Thus, below 715 K, only the thermal expansion is actually responsible for the evolution of the lattice parameter. Above, the faster evolution might be attributed to the joint effect of both thermal expansion and of the reduction of the material.

Laue diffraction and micro-X-ray diffraction

The Laue back reflection mode records X-rays scattered backwards from a broad spectrum source. The diffracting planes in the crystal are determined by knowing that the normal to the diffracting plane bisects the angle between the incident beam and the diffracted beam. von Laue (1913) diffraction method was first used with X-ray white light sources; it is less used today.

Methods with unprecedented signal-to-noise, spatial and time resolution have been developed on synchrotron X-ray sources using microbeam, timing and signal-resolved methods for studying nuclear materials down to ultra-small sample volumes (Ice and Specht 2012).

The nuclear material example is that recorded for fuel analysis. First identified in Degueldre et al. (2011a) and further investigated as reported in Mieszczynski et al. (2012), the μXRD diffraction effects recorded on fuel grains may be precious for characterizing the fuel (see Fig. 3.40).

Specimens of a 4.7 wt% plutonium MOX fuel, irradiated for a period of 6 years up to 60 MW d kg^{-1} were investigated using synchrotron-based μXRD (Mieszczynski et al. 2012). The μXRD measurements were carried out with a beam spot size of about 2×2 μm on the center and periphery region on the cross-section of the irradiated fuel, as well as on non-irradiated reference MOX material (sample preparation see Fig. 2.3). The profiles of characteristic (111), (200), (220), (311) and (222) diffraction features of the irradiated samples were analyzed in detail and compared with the data of the non-irradiated MOX material, revealing the lattice distortion and dilatation in the irradiated fuel matrix. Broadening of reflexes can be observed due to the increasing stress finally yielding strain up to fuel grain fracture. This phenomenon is quantified thanks to the analysis of the reflexes' arcs fine structure. The sub-grain number per grain volume is analyzed as a function of fuel location and history.

The use of μXRD provides information on the grain structure and its changes due to irradiation (Mieszczynski et al. 2012). When performing measurements in

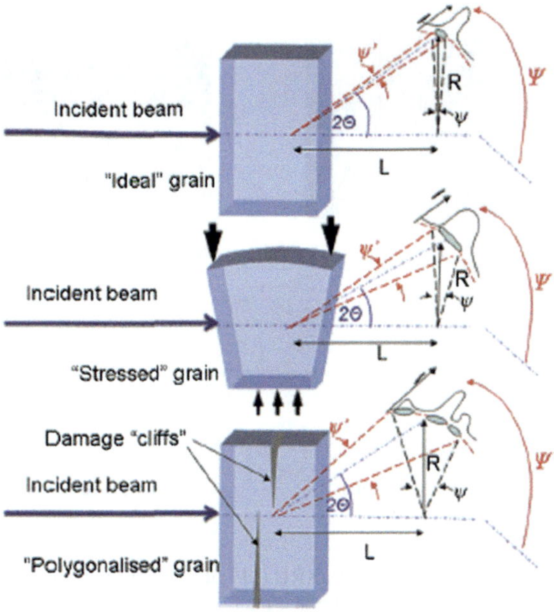

Fig. 3.40 Schematic view of μXRD principle in transmission mode. The X-rays are monochromatic and the beam size is smaller than the sample single crystal size. *Top* ideal crystal grain without stress. *Middle* bent crystal deformed by stress, broadened peak due to the diffraction pattern. *Bottom*: crystal grain with sub-grains, peak splitting on the diffraction pattern. Ψ reflex circular position, Ψ reflex azimuthal width. Mieszczynski et al. (2012) with permission © Elsevier

un-irradiated fuel and in the center and rim of an irradiated fuel pellet, changes in the 2θ position of the diffraction and the azimuthal spread and intensity modulations can be observed (Fig. 3.41). The position and diffraction spot width in 2θ direction delivers information about lattice plane distances and their distribution, respectively, and are an indicator for the local strain and the strain distribution or for a changed fuel composition e.g. due to fission products. The azimuthal elongation is a hint for grain deformation; the duplication and multiplication of reflex peaks are a sign for sub-grain occurrence (polygonization), the sub-grains exhibiting low angle boundaries (Fig. 3.41b, c). In the concrete case: the higher the local burn-up the more the diffraction spot is elongated (pellet rim); at higher temperature during irradiation this effect is less pronounced (pellet center) due to partial healing. Actually, the number of reduplicated peaks in the arc shaped reflex indicates the apparent number of sub-grains.

Small angle X-ray scattering

Small Angle X-ray Scattering (**SAXS**) is an elastic X-ray interaction technique ($\lambda \approx 0.1$–0.2 nm) where photons scatter over small-angles (in contrast with WAXS) on nano- inhomogeneities in a sample. These heterogeneities have size distributions ranging in the nm-domain and are consequently recorded at very low angles e.g. 0.1°–10° versus the incident beam. This record contains information on the size and shape of the nano-heterogeneities, distances between partially ordered objects, precipitates or pore sizes and other data. SAXS is also able to provide structural information about molecules between 5 and 25 nm, and about pores in partially ordered systems of up to 150 nm. Information on SAXS may be found in

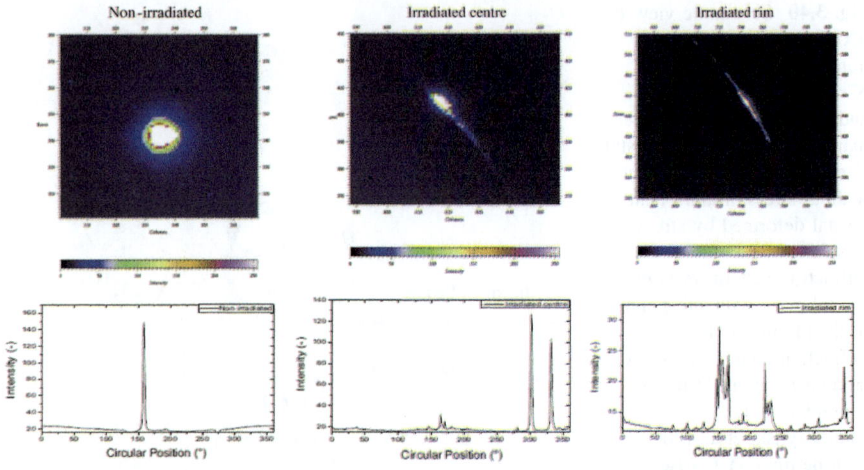

(a) Non-irradiated no subgrains

(b) Irradiated center subgrains per grain ~6

(c) Irradiated rim subgrains per grain ~30

Fig. 3.41 Comparison of μXRD (111) diffraction spots. **a** non-irradiated, **b** irradiated pellet center and **c** irradiated rim. Conditions: Fuel grain particle of the order $\sim 30 \times 20 \times 10\ \mu m^3$, beam size $\sim 2 \times 2\ \mu m^2$, photon energy set at 18.6 keV, average burn-up 40 MW d kg^{-1}, used software FIT2D and PeakFit (see Mieszczynski et al. 2012), with permission © Elsevier

the work edited by Glatter and Kratky (1982). Ultra-SAXS can even resolve larger dimensions. SAXS patterns are typically represented as scattered intensity as a function of q the scattering vector

$$q = \frac{4\pi \sin\theta}{\lambda} \tag{3.35}$$

Here 2θ is the angle between the incident X-ray beam and the detector measuring the scattered intensity, and λ is the wavelength of the X-rays. In the case of a two-phase sample, e.g. small particles in liquid suspension, the only contrast leading to scattering in the typical range of resolution of the SAXS is simply $\Delta\rho$, the difference in *average* electron density between the particle and the surrounding liquid, because variations in ρ due to the atomic structure only become visible at higher angles in the WAXS regime. This means that the total integrated intensity of the SAXS pattern (in 3D) is an invariant quantity proportional to the square $\Delta\rho^2$. In 1-dimensional projection, as usually recorded for an isotropic pattern, this invariant quantity becomes

$$\int I(q) q^2 \, \mathrm{d}x \tag{3.36}$$

where the integral runs from $q = 0$ to where the SAXS pattern is assumed to end and the WAXS pattern starts. It is also assumed that the density does not vary in the liquid or inside the particles, i.e. there is binary contrast. At wave numbers that are relatively large on the scale of SAXS, but still small when compared to wide-angle Bragg diffraction, local interface intercorrelations are probed, whereas correlations between opposite interface segments are averaged out. For smooth interfaces, Porod's law may be applied:

$$I(q) \approx Sq^{-4} \tag{3.37}$$

The nuclear material example considered here is that of a material foreseen for Generation IV reactors. Park et al. (2015) applied high-energy synchrotron X-ray techniques for studying irradiated Oxide Dispersion-Strengthened (ODS) steel samples. These specimens were measured by **SAXS** and WAXS. The samples investigated are nanostructured ferritic alloys. Non-destructive in situ experimental techniques utilizing high-energy X-rays from synchrotron sources were found to be attractive to investigate the structure–processing–property relationships systematically at smaller length scales and help build better material models.

3.2.3.14 X-Ray and γ-Ray Tomography

The principle of **transmission Tomography** (Tom) is quite simple. Photons from a source penetrate the sample and are subsequently detected on the other side of the sample. Information on the transparent phases in the sample may be gained. Transmission tomography, using the absorption of γ-rays or X-rays crossing a sample, is able to reveal the morphology inside the object. In the specific case of application to nuclear fuels, transmission Tomography can be utilized to investigate fuel density and to determine the 3D distribution maps of linear attenuation coefficients (e.g. to use in emission tomography). The Beer–Lambert law (Eq. 3.9) is applied. To obtain information regarding the nature and position of materials inside a body, tomography may be performed in single or double beam mode.

Single beam tomography

In this mode, a single γ-ray or X-ray beam is used in a system including a γ source, a shielding with an entrance collimator, with the object and with an exit collimator and a detector (see Sect. 3.1.1) along the photon propagation axis. Once the counting has been performed, the 2D or 3D reconstruction follows in two main steps: first, an analytical method, filtered back-projection, may be used to generate a first-guess image. Then, an iterative statistical method, the penalized maximum-likelihood (PML) expectation maximization for transmission tomography (Fessler and Hero 1994), may be used to iterate on the first-guess noisy image and to generate the final linear attenuation coefficient matrix. This technique may be integrated with a paraboloidal-surrogates coordinate ascent (PSCA) algorithm (Erdoğan and Fessler 1999) to accelerate the iterative convergence process.

The nuclear example deals with a nuclear fuel rod tested on a tomographic cell station designed, built, and used to perform transmission and emission tomography

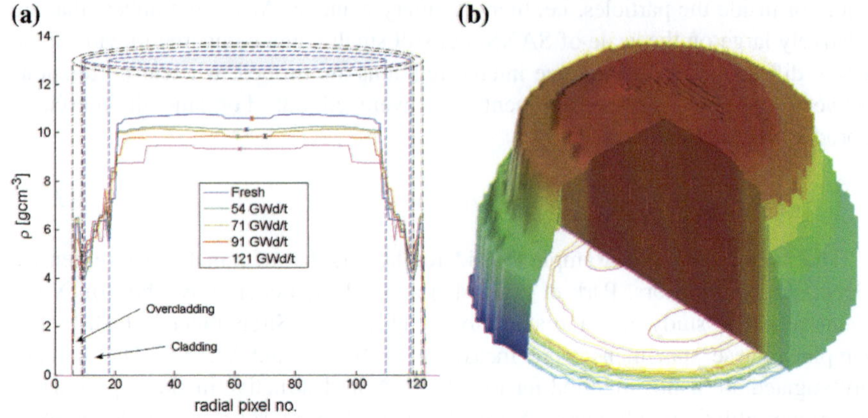

Fig. 3.42 **γ-tomography** of UO_2 fuel. **a** comparison of density radial profile from fresh to irradiated fuel at various burn-up and **b** density 3D data for burn up of 121 MW d kg^{-1}. Conditions: incident beam source ^{60}Co, slit 20 mm × 1 mm. Caruso et al. (2014) with permission © Elsevier

(see Caruso et al. 2014). It mainly consists of a lead shielding (200 mm thick), a tungsten collimator (150 mm length with an internal slit of 0.2 × 20 mm), a remote handling system, a high-purity germanium (HPGe) detector and a system for the injection of an external gamma source (a 2.4 GBq ^{60}Co source). Transmission tomography was employed by measuring the relative intensity of the 1.332 MeV γ-ray peak (see Fig. 3.42).

The samples are a fresh 3.5% UO_2 and four burnt PWR UO_2 fuel rod segments of 400 mm length, two with very high (52 and 71 MW d kg^{-1}) and two with ultra-high (91 and 121 MW d kg^{-1}) burn-up values have been investigated (see Caruso et al. 2014). The burnt samples have about 10 years' cooling time in the first three cases and 4 years' cooling time in the last case. All the rod segments were encapsulated in Zircaloy over-cladding (300–400 μm thickness) to prevent their being damaged and to guard against contamination, the void between the cladding and over-cladding being filled with helium at 200 kPa.

The γ-tomography of UO_2 fuel allows comparison of density radial profiles from fresh to irradiated fuel at various burn-up (Fig. 3.42). On the 121 MW d kg^{-1} fuel a lower density is observed in the periphery due to the very high burn-up structure. Detection of density variations is based on absorption differences of the γ-photons in the materials. The detection limit may here be based on the smallest density difference one can measure between 2 pixels. For the resolution, the pixel is about a 1/1000 of the sample cross-section diameter.

Dual beam tomography

In this mode an X-ray beam is used in a system including an X-ray source with dual energy photons, a shielding with entrance collimator, the object and the exit collimator and a detector.

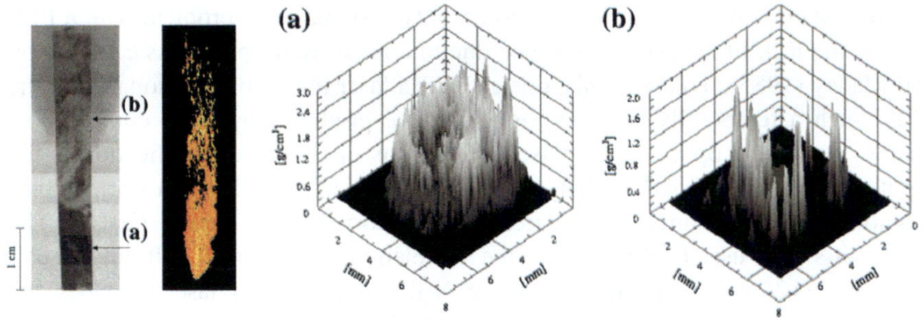

Fig. 3.43 Radiographies and tomograms of the U phase in a sample from the Okelobondo site. Conditions: sample size: 1 mm × 1 mm × 5 mm, incident beam with dual light energy 114.6 and 116.6 keV. Baechler et al. (2001) with permission © Springer

The nuclear example deals with geological samples from the Oklo region. X-ray tomography investigations have been carried out to identify and locate the spatial extension of phases. When used with dichromatic or dual-energy mode for the element discrimination, the irradiation is first performed below the element edge and then above the edge. This mode of tomography allows analysis of a specific element by comparing the intensity of X-ray absorption below and above the edge. Since the actinide atomic numbers are high, the energy of the X-ray absorption K-edges are rather high too ($\sim$100 keV). The sample is placed on a rotation table; the beam size is several mm with photon intensities of the order of 10^8 mm^{-2} s^{-1}. The detector is a CCD camera that records the image obtained on an X-ray converter screen. Images collected are translated into a 3D tomogram of the element considered, visualizing the element densities. Figure 3.43 presents typical radiograms and tomograms obtained for a sample from Okelobondo, Oklo, Gabon (Baechler et al. 2001).

3.2.3.15 Analysis Using a Free-Electron Laser

A **Free-Electron Laser** (**FEL**) is a type of laser whose lasing medium consists of very-high-speed electrons moving freely through an undulator magnetic structure. The free-electron laser is tunable and has the widest frequency range of any laser type, currently ranging in wavelength from microwaves, through far infrared and near infrared, to the visible spectrum, ultraviolet, and X-ray. Information on ultraviolet and soft X-ray free-electron lasers' physical principles, experimental results and technological challenges, is to be found in the work of Schmüser et al. (2009).

The term "free-electron laser" was suggested by Madey (1971) at Stanford University. The work emanates from research done by Hans Motz and his co-workers, who built an undulator at Stanford in 1953, using the wiggler magnetic configuration which is the heart of a free electron laser. Madey used a 43-MeV electron beam and 5 m long wiggler to amplify a signal. One problem with a FEL may be the lack of temporal coherence due to a noisy start-up process e.g. Benson and Madey (1985). To avoid this, one can use a laser tuned to the resonance of the FEL. Such a temporally coherent signal can be produced by more conventional means, such as an optical laser pulse. This results in coherent amplification of the input signal. In late 2010, in Italy, the FEL source FERMI@Elettra started commissioning at the Sincrotrone Trieste Laboratory, e.g. Couprie (2014). The single-pass FEL user-facility coverts the wavelengths range from 100 nm (12 eV) to 10 nm (124 eV). The advent of the X-ray free-electron laser provides the potential for ultra-short (some fs) pulses and high transverse coherence, e.g. Rudenko and Rolles (2015).

Froideval et al. (2011) evaluated possible opportunities in nuclear materials science and technology to investigate samples with an X-ray free-electron laser. Spectroscopy and imaging of condensed matter have benefited greatly from the availability of intense X-ray beams from synchrotron sources, both in terms of spatial resolution and of elemental specificity. The advent of the X-ray FEL provides the additional features of ultra-short pulses and high transverse coherence, which greatly expand possibilities to study dynamic processes and to image non-crystalline materials. In nuclear materials science and technology, benefits from X-ray FEL sources are foreseen. X-ray FEL-based experiments are proposed on nuclear energy-related materials using single-shot X-ray spectroscopy, coherent X-ray scattering and/or X-ray photon correlation spectroscopy in order to address relevant scientific questions such as the evolution in time of the irradiation-induced damage processes, the deformation processes in nuclear materials, the boiling heat transfer in nuclear reactors, as well as the structural characterization of graphite dust in advanced nuclear reactors, the ion diffusion processes in the barrier systems of geological repositories, and clay colloid aggregates in the groundwater near a radioactive waste repository.

3.2.3.16 Mössbauer Absorption

Mössbauer Absorption Spectroscopy (MoAS) is a γ photon absorption technique that can provide structural and chemical information from daughter nuclides labeled molecules irradiated by a source of father radionuclides. The emission mode is described in Sect. 3.1.2.4; *this* section focuses mainly on absorption. For information, the Mössbauer effect methodology has been reported, for example, in a chapter of the book by Weyer (1976) and in the work "Mössbauer Spectroscopy" by Nasu (2012).

Conservation of momentum requires a nuclide to recoil during absorption of a photonγ. If a nucleus ^{A}X at rest absorbs aγ, to yield $^{A}X^{*}$, the γ energy must be slightly larger than the natural energy, because energy is lost to recoil. Consequently, the emission and absorption of the same γ photon by identical nuclei, called nuclear resonance, is unobservable with free nuclei. This is due to the shift in energy that is too large and the emission and absorption spectra have no significant overlap. The reactions read:

$$^{A}X + \gamma \rightarrow {}^{A}X^{*} \rightarrow {}^{A}X + \gamma \qquad \{3.13\}$$

with $^{A}X^{*}$, ^{A}X in exited state. However, nuclei in a solid crystal are not free to recoil because they are bond and located in given positions in the crystal lattice. When a nucleus in a solid absorbs a γ photon, some energy can be lost by recoil. Mössbauer found that a fraction of emitted photons are recoil-free (see Mössbauer 1958). This makes Mössbauer spectroscopy possible; it means that γ photons emitted by one $^{A}X^{*}$ nucleus can be resonantly absorbed by a sample containing nuclei of the same isotope ^{A}X, and that this absorption can be measured in the absorber sample using a set-up as described in Fig. 3.44.

In the most common form of MoAS, a solid sample (absorber) is exposed to a beam of gamma radiation, and a detector measures the intensity of the beam transmitted through the sample (see Fig. 3.44).

If the emitting and absorbing nuclei are in the same chemical environments, the nuclear transition energies should be exactly equal and resonant absorption would be observed with both materials at rest. The difference in chemical environments, however, causes the nuclear energy levels to shift in a few different ways. Although these energy shifts are tiny, often less than one MeV, the extremely narrow spectral linewidths of gamma rays for some radionuclides make the small energy shifts correspond to large changes in absorbance. To bring the two nuclei back into resonance it is necessary to change slightly the energy of the γ-ray, and in practice this is always done using the Doppler effect.

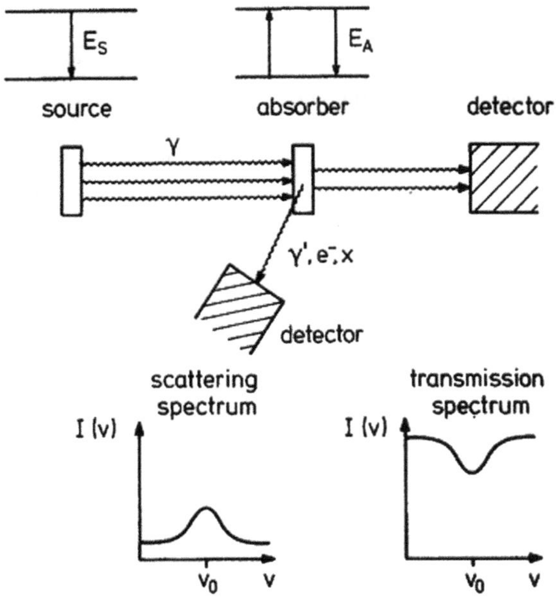

Fig. 3.44 Illustration of transmission and scattering MoAS. Adapted from Weyer (1976), with permission © Springer. Note here the sample is the absorber

During the Mössbauer absorption test, the source is accelerated through a range of velocities using a linear motor to produce a Doppler effect and scan the gamma ray energy through a given range.

In the resulting spectra, gamma ray intensity is plotted as a function of the source velocity. At velocities corresponding to the resonant energy levels of the sample, a fraction of the gamma rays are absorbed, resulting in a drop in the measured intensity and a corresponding dip in the spectrum. The number, positions, and intensities of the dips (also called peaks; dips in transmitted intensity are peaks in absorbance) provide information about the chemical environment of the absorbing nuclei and can be used to characterize the sample. A typical range of velocities for ^{57}Fe, for example, may be ±11 mm s^{-1} (1 mm s^{-1} = 48.075 neV).

Suitable gamma-ray sources consist of a radioactive parent (emitter) that decays to the desired isotope (absorber). Ideally the parent isotope will have a convenient half-life. Also, the γ energy should be relatively low, otherwise the system will have a low recoil-free fraction resulting in a poor signal-to-noise ratio and requiring long collection times. Table 3.12 indicates isotopes suitable for Mössbauer spectroscopy. Among these, ^{57}Fe is by far the most common element studied using the technique, although fission product isotopes ^{129}I, ^{119}Sn, and ^{121}Sb are also frequently studied (see Table 3.12).

Among the long-life fission products ^{126}Sn is safety relevant. Other tin isotopes may be used as analogues. Mössbauer emission spectra of ^{119}Sn have been measured in metals, oxides, and chalcogenides of antimony and tellurium labeled with ^{119}Sb or ^{119m}Te. Sources of ^{119}Sb and ^{119m}Te metals gave an emission line in the Sn (0) region (see Ambe et al. 1974). In the emission spectra of $^{119}Sb_2O_3$ the species attributable to the Sn(II) state were dominant, while only the Sn(IV) species was

Table 3.12 Isotope of interest for nuclear material analysis suitable for Mössbauer spectroscopy

Isotope	**Half-life (h∩d∩a)**	**γ energy (keV)**	**Species**	**Refs.**
^{57}Co → **^{57}Fe**	271.8 d	121, 136	Fe(II, III)	Fitch and Rees (1981)
^{99}Rh → **^{99}Ru**	16 d	354	Ru(III, VIII)	Ohkubo et al. (1995)
^{119}Sb → **^{119}Sn**	38.5 h	24	Sb(II, IV)	Ambe et al. (1974)
^{121m}Sn → **^{121}Sb**	50 a	6	Sb(III, V)	Ambe and Ambe (1975)
^{129}Te → **^{129}I→**	33.6 d	27.8	I(−I, 0, I,)	Boolchand et al. (1981)
^{231}Th → **^{231}Pa→**	25.5 h	84.2	Pa(IV, V)	Croft et al. (1968)
^{242}Pu → **^{238}U→**	3.75×10^5 a	44.91	U(IV, V, VI)	Tsutsui et al. (1998)
^{241}Am → **^{237}Np→**	4.32×10^2 a	59.54	Np(III, IV, V, VI)	Smith et al. (2015)
^{247}Bk → **^{243}Am→**	1.38×10^3 a	84	Am(III, IV, V, VI)	Smith et al. (2015)

Emitter → **absorber (as sample)** and their characterization potential found in a $^{119m}TeO_2$ samples. The $^{119}Sb_2S_3$ and $^{119}Sb_2Se_3$ gave two emission lines attributable to the Sn(II) and Sn(IV) states. The $^{119}Sb_2Te_3$ showed only an emission line in the Sn(II) region. These observations were discussed in terms of the after-effect of the electron capture decays and the properties of the matrices. Mössbauer spectroscopic verifications of two different states of impurity ^{119}Sn atoms in Sb_2Te_3, have been carried out by Ambe and Ambe (1973).

MoAS may be applied to provide information on several actinide isotopes: ^{231}Pa, ^{232}Th, ^{238}U, ^{237}Np and ^{243}Am (see Table 3.12). Preparation details of source and sealed absorption holders for ^{237}Np and ^{238}U Mössbauer measurements were reported (Nakada et al. 1998). However, little has been done for actinide speciation of environmental samples, perhaps because this technique requires macro quantities. However, the study of the interaction of Pu(IV) and Np(IV,V,VI) with iron hydroxides has been performed by Fe Mössbauer spectroscopy in order to predict the behavior of actinides in environmental media (Grigoriev et al. 2001).

The Mössbauer effect was observed with the 84.2 keV gamma ray of ^{231}Pa, following beta decay of 25 · 5 h ^{231}Th. In experiments at 4.2 K using ThO_2 sources, resonance absorption was detected with absorbers of both $^{231}Pa_2O_5$ and $^{231}PaO_2$ (see Croft et al. 1968). In each case, the resonance was broad and unresolved, with the maximum absorption dip 0·1% (uncorrected for background). The spectra show no discernible isomer shift between the valence states of protactinium.

The ^{237}Np MoAS measurements were carried out in transmission, using a ^{241}Am metal source with a sinusoidal driving mode. The effect was measured with photon energy of 59.54 keV. The powder samples, in 3 concentric aluminum containers, were probed in the temperature range 4.2–20 K, while the source was kept at a constant temperature of 4.2 K inside an independent chamber in the stainless steel cryostat. The velocity scale was moreover calibrated with respect to $NpAl_2$. Recently, Smith et al. (2015) performed structural investigation of Na_3NpO_4 and Na_3PuO_4 using ^{237}Np Mössbauer spectroscopy and compared the Mössbauer shift of these compounds with several others (see Fig. 3.45). The comparison shows an interesting shift from low valence Np compounds to oxidized forms.

The structures of Na_3NpO_4 and Na_3PuO_4 compounds were also shown to exhibit a completely disordered NaCl type of structure at low temperatures ($T < 873$ K), corresponding to a metastable m phase, that transforms to the equilibrium ordered α phase when heated at higher temperatures ($T > 1093$ K).

Bogé et al. (1986), studied ^{237}Np Mössbauer spectroscopy on neptunium doped borosilicate glasses. Different melting and annealing times, in various crucibles, have been investigated. ^{237}Np valency states of neptunium ions in glasses were characterized using the 59.5 keV Mössbauer resonance. Roughly, absorption spectra show two sites easily differenciated above 50 K by two lines. The neptunium isomer shifts were respectively: -10.0 ± 1.0 and $+32.5 \pm 0.5$ mm s^{-1} relative to $NpAl_2$. These lines can be assigned, the first at Np^{4+}, the second at Np^{3+} species. The intensity ratio Np^{4+}/Np^{3+} was found to vary with the glass production conditions (Fig. 3.45).

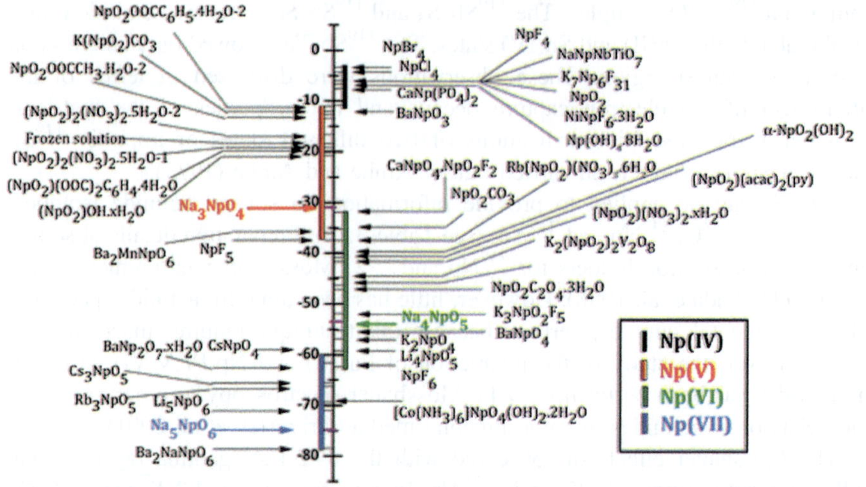

Fig. 3.45 Isomer shifts versus $NpAl_2$ of Np(IV), Np(V), Np(VI), and Np(VII) compounds and for Na_3NpO_4 is shown in *red* together with Na_4NpO_5 (*green*) and Na_5NpO_6 (*blue*) reported in Smith et al. (2015), with permission © RSC

3.2.3.17 Photoactivation

Photoactivation (PhotA) consists of γ absorption by nuclides followed by the decay of their product nuclides. This is a form of inelastic light scattering on the nuclides and is driven by specific reaction such as

$$^{A}_{Z}X^{*} + \gamma \rightarrow {}^{A}_{Z}X^{*} + \gamma' \qquad \{3.14\}$$

Photoactivation cross-sections for short-lived ($\tau < 20$ s) isomeric states in ^{167}Er, ^{79}Br, ^{191}Ir, ^{183}W, ^{197}Au, ^{77}Se, ^{137}Ba and ^{89}Y were determined by Anderson et al. (1989) using bremsstrahlung from a 6 MeV medical therapy linac. A portable NaI (Tl) spectrometer and pneumatic transfer system were designed to avoid impacting normal use of the linac. Comparison with data obtained at a flash X-ray source (1.5 MeV endpoint energy) indicates that cross-sections for populating these isomers through gateway states in the range 1.5–6 MeV are more than three orders of magnitude greater than corresponding cross-sections below 1.5 MeV. The results are significant for the development of photoactivation techniques to measure short (<100 ns) bremsstrahlung pulses.

A compact X-ray source based on inverse-Compton scattering was designed with a laser-driven electron beam. This source produces a beam of high-energy

X-rays in a narrow cone angle (some mrad), at a photon emission rate of 10^8 s^{-1} (see Banerjee et al. 2015). Tunable operation of the source over a large energy range, with energy spread of $\sim$50%, has also been demonstrated. Photon energies >10 MeV have been obtained. The narrowband nature of the source is advantageous for radiography with low dose, low noise, and minimal shielding. A compact high-energy X-ray source generating photons with MeV energy makes radiography of dense objects possible for actinides photonuclear detection. Small source size enables high-resolution imaging. Radiography of a shielded uranium target was experimentally demonstrated e.g. for radiography of cargo containers.

Photoexcitation of radioactive ^{99}Tc was studied by irradiation with 15–50 MeV bremsstrahlung (see Sekine et al. 1991). The integral cross-sections of the ^{99}Tc(γ, γ')^{99m}Tc reaction were found to be nearly constant as $(5.8 \pm 1.0) \times 10^{-27}$ cm^2 MeV in the given energy range.

The ideal photon source for active interrogation of fissile materials would use monoenergetic photons to minimize radiation dose to surroundings (see Morse et al. 2007). The photon energy would be high enough to produce relatively large photofission signals, but below the photoneutron threshold for common cargo materials in order to reduce background levels. To develop such a source, use of low-energy, proton-induced nuclear reactions have been investigated to generate monochromatic, MeV-energy γ-rays. Of particular interest are the nuclear resonances at 163 keV for the ^{11}B(p,γ)^{12}C reaction producing 11.7 MeV gamma-rays, 340 keV for the ^{19}F(p,$\alpha\gamma$)^{16}O reaction producing 6.13 MeV photons, and 441 keV for the ^{7}Li(p,γ)^{8}Be reaction producing 14.8 and 17.7 MeV photons. A 700 keV Van de Graff ion accelerator was used to test several potential (p,γ) materials and the γ-ray yields from these targets were measured with a 5″ × 5″ NaI detector. A pulsed proton beam from the accelerator was used to induce prompt neutron and delayed neutron and gamma-ray photofission signals in uranium which were measured with ^{3}He and NaI detectors.

3.2.3.18 Detection Limit for the Interactive Photon-Photon

The photon-photon interactive analytical techniques compared include VIS, CoL, XRF and TRLIFS. Comparison of VIS, CoL, XRF and TRLIFS detection limits applied for actinide elements or species and for their characterization are given in Table 3.13. A comparison of detection limits is reported in Sect. 4.1.

Table 3.13 Comparison of detection limits for photon-photon interactive analytical techniques for actinide elements or species characterization

Detection	Goal	Sample	^{A}An(*Y*)	C_{DL} and N_{DL}	Remarks
Transmitted photon: IRS, IRFT, NIR-VIS, PCS CoL, AAS, UVS, XAS, ToM, MoAS	Identification and determination of species	Solid or liquid bulk, or interphases	Ac(III) Th(IV) Pa(V) U(VI) Np(V) Pu(IV-VI) Am(III-VI) Cm(III)	– 3×10^{-8} M CoL – 4×10^{-8} M CoL 1×10^{-7} M VIS – – –	VIS or CoL Keil (1981) Keil (1979) Gauthier et al. (1983)
Reflected scattered photon: DRS, RAMS, PHOS, UVF, XRF	Determination of species	Solid bulk or liquid bulk	Ac(III) Th(IV) Pa(IV,V) U(IV-VI) Np(IV-VI) Pu(IV-VI) Am(III-V) Cm(III)	– – – 6×10^{-10} mol 2×10^{-9} mol 1×10^{-9} mol 1×10^{-9} mol –	XRF Akopov et al. (1988)
Delayed photon: LAICPOES, TRLIFS, PhotA	Determination of elements species or isotopes	Liquid bulk or solid bulk	Ac(III) $^{227-232}$Th(IV) $^{2331-234}$Pa(V) $^{234-238}$U(VI) ^{237,239}Np(IV-VI) $^{238-244}$Pu(IV-VI) ^{241,243}Am(III) $^{242-250}$Cm(III)	– – – 1×10^{-12} M 1×10^{-9} M 4×10^{-8} M 4×10^{-9} M 4×10^{-11} M	TRLIFS Moulin et al. (1991) Moulin et al. (1995) Beitz et al. (1980, 1988)

3.2.4 *Techniques with Photon Interaction and Lepton Detection*

3.2.4.1 Ultraviolet and X-Ray Photoelectron Spectroscopy

Ultraviolet Photoelectron Spectroscopy (UPS) and **X-ray Photoelectron Spectroscopy (XPS)** apply to spectroscopy of electrons excited by UV or X-ray photons from samples under vacuum. When applying Einstein's (1905a) photoelectric law to a free molecule, the kinetic energy (E_K) of an emitted photoelectron is given by:

$$E_K = \mathrm{h}\,v - E_\mathrm{I} \tag{3.38}$$

where *h* is Planck's constant, *v* is the frequency of the ionizing light, and E_I is an ionization energy for the formation of a singly charged ion in either the ground state or an excited state. According to Koopman (1934), each of the ionization energies may be identified with the energy of an occupied molecular orbital. The ion in its ground-state is formed by removal of an electron from the highest occupied molecular orbital. Information on XPS in Materials Science is available in the monograph by Hofmann (2013).

- UPS was applied to thin layer actinide research only (Gouder 1998).
- XPS was applied by Allen et al. (1982) to study UO_2 under low pressure of O_2.

XPS techniques have also occasionally been used to characterize actinides associated with environmental samples. They help to understand how uranium is absorbed onto phosphate (Drot et al. 1998) or calcite (Geipel et al. 1997) surfaces, as well as how actinides behave in contaminated soil (Dodge et al. 1995). In all cases, XPS allows redox and spatial speciation of the actinides by use of energy shifts. For XPS, the detection limit may be estimated as 4×10^{-11} mol for a sample of 1 cm^2 with a 10 nm thickness and a density of 1 g cm^{-3} and 1% actinide mass. This detection limit can be upgraded for microbeam XPS units.

3.2.4.2 Surface Extended X-Ray Absorption Fine Structure

Surface Extended X-ray Absorption Fine Structure (**SEXAFS**) spectroscopy is related to the analysis a layer slimmer than 10 nm recording the electron yielded by ionization above the absorption edge e.g. Landman and Adams (1976).

The experimental set-up needed for SEXAFS investigations is similar to that required for EXAFS (see Sect. 3.2.3.8). In SEXAFS an electrode is used to collect the electron generated on and over the sample, e.g. at the mm over the sample and under atmospheric pressure (see Long et al. 1989). However, an electron detector and a high-vacuum chamber are required to detect the calculated electron yields instead of the intensity of the transmitted X-ray waves.

The example that could be cited here is that of Citrin et al. (1979) on the absorption of iodine on a single crystal of silver and copper that may be utilized in the frame of radioiodine decontamination.

3.2.5 Techniques with Photon Interaction and Ion or Neutron Detection

3.2.5.1 Photon-Induced Mass Spectroscopy

Thermal Ionization Mass Spectrometry (**TIMS**) is a very sensitive mass spectrometry technique (see Gross 2004). TIMS exploits the thermal ionization (InfraRed, to simplify) effect, in which a purified sample is heated to ionize the atoms. The ions are focused electromagnetically into an incident beam and separated into individual sub-beams for fixed mass/charge ratio of the ions. TIMS may require isotope dilution calibration to quantify isotopic abundances with 50–500 ng filament loads and chemical separation to remove interferences. Latest generation of TIMS instruments have 3 × or higher precision on the U isotope ratios than former instrument generation. The relative abundances of different isotopes are then used to describe the chemical fractionation of different isotopes, and to travel in different systems of non-radiogenic isotopes. The technique is used in isotope geochemistry, cosmochemistry and in nuclear material science, e.g. Aggarwal (2016a, b).

The nuclear material example is the uranium isotope analysis (see Mennecart et al. 2016).

TIMS analyses were conducted with an intact as well as a declad spent fuel pellet six months in contact with a reference carbonate solution. The initial enrichment in ^{235}U of 4.25 wt% was irradiated for two cycles, each of 18 months, with an average linear power of 321 W cm^{-1} and a final average rod burn-up of about 50 MW d kg^{-1}, based on the reactor power history and the calculated power mapping across the core. These results were compared to the total inventory measured with radiochemical burn-up analyses, as well as with the calculated inventory profile along the pellet radius. The comparison gives information about the origin of the dissolved uranium. The release into solution of the uranium isotopes ^{233}U, ^{234}U, ^{236}U and ^{238}U, after a period of six months leaching (see Table 3.14) gives the abundances of ^{234}U to ^{238}U for:

- the inventory before irradiation
- the initial inventory (before leaching) of the spent fuel obtained from destructive radiochemical analysis (column 2a) and the average initial inventory by calculation of the radial distribution of the uranium isotopes (column 2b)
- the leaching solution in contact with the clad segment (column 3a) and the declad segment (column 3b) after 6 months, measured by means of TIMS.

TIMS is also used for analyzing environmental samples. Chen et al. (2015) applied this method to measure the uranium isotopic ratio in uranium bearing particles.

3.2.5.2 Laser Ablation Inductively Coupled Plasma Mass Spectrometry

In **Laser Ablation Inductively Coupled Plasma Mass Spectrometry (LAICPMS)** the sample is directly analyzed by ablating with a pulsed laser beam. The created aerosols are transported by argon flow into the Inductively Coupled Plasma (ICP, see details in Sect. 3.2.14.1), which temperature reaches about 8000 °C. The plasma in ICP is used to generate ions that are then introduced to the mass spectrometer where these ions are separated and detected according to their mass to charge ratios. The constituents of an unknown sample can be identified and measured. ICPMS offers very high sensitivity to detect a wide range of isotopes.

Laser ablation can be applied to all type of solid sample; there are no sample preparation procedures and no size requirements. Isotope analysis using laser ablation requires very small amount of sample, actually a smaller amount of sample (micrograms) than for solution nebulization (milligrams). In addition, a focused laser beam permits spatial characterization of heterogeneity in solid samples, with typically the micron in terms of lateral and depth resolution. The potential of LAICPMS is:

Table 3.14 Isotopic uranium abundance (weight %) from fuel pellets and leachates

Isotopic abundance (wt%) isotope.	**Inventory before irradiation (1)**	**Initial inventory**		**Leachate inventory (TIMS)**	
		TIMS (2a)	**Calculation (2b)**	**Intact (3a)**	**Declad (3b)**
^{234}U	0.038	0.022062 ± 0.000098	0.02217	0.02519 ± 0.00011	0.02231 ± 0.00016
^{235}U	4.254	0.63214 ± 0.00098	0.61295	0.59767 ± 0.00093	0.59551 ± 0.00094
^{236}U	0.001	0.63637 ± 0.00066	0.63479	0.64335 ± 0.00067	0.64307 ± 0.00068
^{238}U	95.708	98.7095 ± 0.0017	98.73009	98.7338 ± 0.0016	98.7391 ± 0. 0017
Total	100.001	100.000072	100.000000	100.00001	99.99999

- major, minor and trace level compositional analysis of conductive, semi-conductive and nonconductive materials
- contamination of plastics, pharmaceuticals, organics or biological materials
- failure, contamination and inclusion analysis
- forensics analysis
- environmental and mineral sample analysis
- isotopic and elemental distribution analysis.

The beam diameter can vary from ~10 to 300 μm. The method allows a depth profile analysis in the range from about 10 μm to several mm. The detection limit for a 10 μm^3 crater and a 1 ppm limit as C_{DL} gives an N_{DL} that would be 10^{-19} mol. The ICPMS technique is described and illustrated in Sect. 3.2.18.4.

LAICPMS has been applied for the in situ determination of thorium and uranium in silicate rocks and soils (Perkins et al. 1993). Powdered geological materials have been prepared both as pressed powder disks and fused glasses. Uranium was also analyzed successfully by LAICPMS in various geological samples such as in Neo-Proterozoic (1.0–0.5 Ga ago) or older zircons or baddeleyite with U contents ≥65–270 ppm as reported by Horn et al. (2000). These samples could be used as natural analogues of waste forms.

Laser ablation ICPMS may be used for the analysis of Kr and Xe in irradiated fuel at the 20 μm level. Analyses of inclusions in solids have been a major interest in various fields and have been carried out to determine the concentration of Xe in nuclear fuel (Horvath et al. 2017). The calibration strategy was based on the direct injection of a known quantity of a reference gas into the LAICPMS carrier gas system. Further, the ablation of a 'matrix-matched' standard of a non-irradiated UO_2 sample, implanted with a known amount of ^{129}Xe, was also applied. Quantitative LAICPMS measurements were performed on high burn-up nuclear fuel. It was shown that direct gas injection is most suitable for the quantification of fission gas in micron-sized inclusions. The direct gas addition is simple and linear calibration curves were obtained. Good reproducibility was obtained and matrix effects were within the uncertainty of the measurements. For the quantification of fission gases in nuclear fuel, aerosol particles were filtered before entering the ICP to remove interferences on the Xe isotopes from the solid fission product matrix (see Fig. 3.46). The first quantitative determinations of the amount of gas in nuclear fuel using the direct injection method for calibration led to sample pressure calculations which were in good agreement with pressures estimated from computer simulations.

3.2.5.3 Resonance Ionization Mass Spectroscopy

In **Resonance Ionization Spectroscopy** (RIS) atoms or molecules are excited stepwise from the ground state to highly excited states by resonant absorption of photons followed by an ionization process e.g. Trautmann et al. (1986). An additional mass separation step completes the method of **Resonance Ionization Mass**

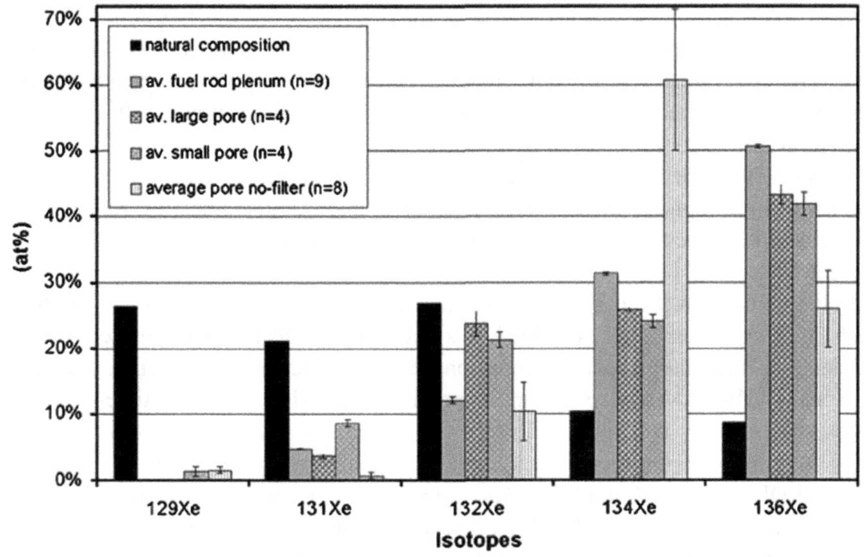

Fig. 3.46 Isotopic composition of Xe in nuclear fuel, rod plenum, in comparison to the natural Xe composition. Conditions: unfiltered measurements, small and large signals/pores from single laser shots. Horvath et al. (2017) with permission © RSC

Spectroscopy (RIMS). Spectroscopic study of thorium using continuous-wave RIMS with ultraviolet ionization provided sufficient signal for isotopic analysis of volcanic-like samples containing as little as 1–5 ng of thorium. The ability to determine accurately and precisely the $^{230}Th/^{232}Th$ isotopic ratios for 1 ng samples represented an improvement over TIMS (Johnson and Feary 1993). Detection and speciation of traces amounts of neptunium and plutonium is possible by RIMS, with a 3-step photo-ionization in combination with time of flight mass spectrometry. The detection efficiency allows analysis of 10^7 atoms (2×10^{-17} mol) of an actinide. The analysis at very low concentrations is consequently feasible. RIMS has been applied to plutonium and transplutonium elements and should allow detection down to 10^6 atoms (3×10^{-18} mol) (Trautmann et al. 1986; Erdmann et al. 1998).

3.2.5.4 Laser microprobe mass analysis (LAMMA)

A **Laser Microprobe Mass Spectrometer** (LMMS), also **LAser Microprobe Mass Analyzer** (LAMMA), **Laser Ionization Mass Spectrometer** (LIMS), or **Laser Ionization Mass Analyzer** (LIMA) are units that make use of a local molecular ionization by a pulsed laser and subsequent mass analysis of the generated (molecular) ions. Information on these mass spectroscopic techniques can be found in the work edited by Cai and Liu (2013).

In LAMMA, a highly focused laser beam is pulsed on a micro sample usually with a volume of approximately 1 μl. The resulting ions generated by this laser are then analyzed with time-of-flight mass spectrometry to give composition, and concentration, of matrix and dopant molecular fragments or atoms with potential structural information.

Unlike other methods of microprobe analysis which involve ions or electrons, the LMMS microproble fires an ultraviolet pulse in order to create ions. As a result, this method is much better at detecting qualitatively rather than quantitatively.

LAMMA could be used to study particles on filters or directly on solid samples, but very little has been reported on nuclear materials investigations with this technique. As an example, LAMMA has been applied to fingerprinting coal constituents in bituminous coal. Uranium phases were detected (Lyons et al. 1987).

Finally, **Matrix-Assisted Laser Desorption Ionization** (MALDI) time-of-flight mass spectrometry is a diagnostic tool that combines laser desorption ionization mode and mass spectrometry for the analysis of microscopic features.

MALDI analysis of uranium (VI) leads to the formation of uranium oxides clusters (Soto-Guerrero et al. 2001). The formation of clusters can be eliminated using selected matrices, and a more sensitive uranium determination with a detection limit down to 10^{-12} M is possible. The instrument finds its place in nuclear forensics. This study has been completed recently by Ivanova and Spiteller (2014) work on uranyl water containing complexes (e.g. $UO_2(OH)_2$, or $UO_3{\cdot}xH_2O$ and $UO_4{\cdot}2H_2O$) showing how sensitive the method is for speciation.

3.2.5.5 Photoactivation

PhotoActivation (PHOTA) consists of γ-ray absorption by nuclides followed by neutron, proton or alpha particle production. This is a form of photon interaction on the nuclides is driven by specific reaction such as:

$$ {}_Z^AX + \gamma \rightarrow {}_Z^{A-1}X + {}_0^1n \qquad \{3.15\} $$

$$ {}_Z^AX + \gamma \rightarrow {}_{Z-1}^{A-1}X + {}_1^1p \qquad \{3.16\} $$

$$ {}_Z^AX + \gamma \rightarrow {}_{Z-2}^{A-4}X + {}_2^4\alpha \qquad \{3.17\} $$

The ideal photon source for interactive analysis of fissile materials would be monoenergetic to minimize radiation dose to surroundings (see Morse et al. 2007). The photon energy would be high enough to produce relatively large photo-fission signals, but below the photo-neutron threshold for common cargo materials in order to reduce background levels. To develop such a source, use of low-energy, proton-induced nuclear reactions have been investigated to generate monochromatic, MeV-energy gamma-rays, e.g. 340 keV from the $^{19}F(p,\alpha\gamma)^{16}O$ reaction.

3.2.6 *Techniques with Lepton Interaction and Phonon Detection*

Several techniques using leptons, e.g. electrons, as incident particles are also applied to characterize nuclear materials. These are electron-phonon techniques such as scanning electron acoustic microscopy. The sources of electrons are mainly electrodes and electron guns.

3.2.6.1 Techniques with Electron Interaction and Phonons Detection

Scanning Electron Acoustic Microscopy (SEAM) is a technique that does not see the object, such as optical microscopy, but hears where it is. To overcome the optical microscope resolution limit, a variety of acoustic microscopes has been developed. The principle is based on an acoustic wave production in a tiny area just in the vicinity of surface through different interaction mechanisms and by detecting this acoustic wave. The acoustic properties of materials can be gained at a high resolution which is not dependent on wavelength. The **Scanning Electron Acoustic Microscope (SEAM)**, which has been developed from the commercial Scanning Electron Microscope (SEM), is a typical microscope of this kind with highest resolution. Information on the Scanning Probe Microscopy Electrical and Electromechanical Phenomena may be found in the work of Kalinin and Gruverman (2007). Figure 3.47 describes the main structure of SEAM:

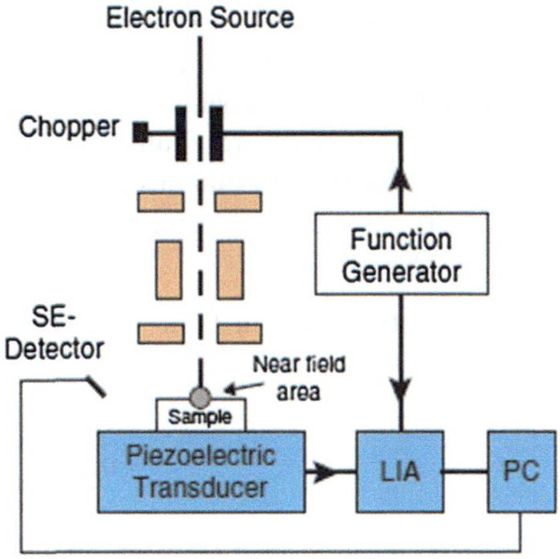

Fig. 3.47 Principle of operation of a Scanning Electron Acoustic Microscope (SEAM)

3.2.7 *Techniques with Lepton Interaction and Photon Detection*

This section concerns electron-photon and positron-photon techniques. Electron-photon methods, such as energy or wavelength dispersive spectroscopy and electron microprobe analysis (EDS, WDS, EMPA) are treated here. Techniques including positrons are also discussed and illustrated in this section. This section concerns microscopy analyses utilizing the scanning electron microscope or the scanning positron microscope (SPM).

3.2.7.1 Electron Probe Microanalysis

Electron Probe Microanalysis (EPMA) works by bombarding a micro-volume of a sample with a focused electron beam typically 5–30 keV and collecting the emitted X-rays by the various element of the specimen. Because the wavelengths of these X-rays are characteristic of the emitting atoms of the sample, its elemental composition can be identified by recording Wavelength Dispersive Spectroscopy (WDS) spectra using a similar approach to XRF (see Sect. 3.2.3.9). Information on quantitative EMPA can be found in the monograph by Theisen (1965).

- EPMA is a qualitative and quantitative method of **non-destructive elemental analysis of micron-sized volumes** at the surface of materials, with sensitivity at the level of ppm. Routine quantification to 1% reproducibility is obtained over several days.
- EPMA is the most precise and accurate micro-analysis technique available and **all elements from B to U and above** can be analyzed.
- EPMA is compatible with routine and repetitive analysis, with easy and direct interpretation of the results.
- EPMA instruments are equipped with a complete kit of built-in microscopy tools that allow **simultaneous X-ray (WDS and EDS)**, SEM and BSE imaging. They provide flexible sample inspection with image magnification ranging from 40 to 400,000.

The detection limit in term of concentration C_{DL} (wt%) is given by Ziebold (1967) formula:

$$C_{DL} = \frac{3.29\,\alpha}{(n\,t\,p^2/b)} \tag{3.39}$$

with n the number of measurement, t the time per measurement, p the element peak counting rate (s^{-1}), b the blank count rate (s^{-1}) and α the matrix correction (α factor or ZAF: atomic, absorption and fluorescence corrections).

Major applications are found in geochemistry, mineralogy, geochronology, physical metallurgy, nuclear metallurgy, materials science including glass (e.g.

waste form), ceramics (e.g. fuel), superconductors (e.g. accelerator magnet), cements (e.g. waste form).

The nuclear material example presented for EPMA concerns an inert matrix fuel produced for the disposition of plutonium excesses (see Restani et al. 2009). The samples in the form of pellets have been fabricated by internal gelation of zirconium yttrium erbium plutonium oxides and **irradiated** 3 years in a test reactor since 2000. The fuel has been examined by EPMA after irradiation and compared with data gained for the **unirradiated** material. The examined pellet cross-section was estimated to have an equivalent burn-up of 22 MW d kg^{-1}. EPMA measurements demonstrate that the burn-up was rather flat over more than the half pellet radius. A Pu consumption of about 2.5 wt% has been measured with a higher rate in the fuel border zone. The high fuel temperature is responsible for a certain homogenization of the mineral phases in the fuel center region whereas the border zone has remained instead with an as-fabricated phase distribution. The central part is characterized by a dense porosity distribution. Element mapping of irradiated stabilized zirconia-based inert matrix fuel (IMF) at pellet center. The material is heterogeneous as shown in Fig. 3.48 and the content range (dark blue–red) is the following: Zr: 41–53 wt% (21–27 at.%); Pu: 8–19 wt% (1.5–3.8 at.%); Er: 4–6 wt% (1.1–1.7 at.%); Y: $\sim$10 wt%.

Energy dispersive spectroscopy

Energy Dispersive Spectroscopy (**EDS**) is an analytical technique used on the electron microscope for the elemental analysis or chemical characterization of a sample. It relies on an interaction of some source of X-ray excitation and a sample. Its characterization capabilities are due in large part to the fundamental principle that each element has a unique atomic structure allowing a unique set of peaks on its X-ray emission spectrum.

Electron beam excitation is used in electron microscopes, scanning electron microscopes (SEM) and scanning transmission electron microscopes (STEM). X-ray beam excitation is used as X-ray fluorescence (XRF) spectrometer. A detector is used to convert X-ray energy into voltage signals; this information is sent to a pulse processor, which measures the signals and passes them onto an analyzer for data display and analysis. The most common detector now is Si(Li) detector cooled to cryogenic temperatures with liquid nitrogen; however, newer systems are often equipped with silicon drift detectors (SDD) with Peltier cooling systems. See Sect. 3.1.2 for more details.

The application on a nuclear material deals with a thoria doped inert matrix fuel: a calcia stabilized zirconia doped with urania-thoria partially in solid solution. The fuel is made of these two ceramic components with 40% zirconia based fluorite and 15% thoria with urania. The former exhibits cellular structure with Th and U rich phases surrounding dense grains of stabilized zirconia of at least 10 μm in size, see Fig. 3.49 (see Ronchi et al. 2003). The material contains about 18% of isolated globular particle of urania. This heterogeneous material, somewhat analogous to an inert matrix thorium MOX, has been successfully irradiated in the Halden reactor.

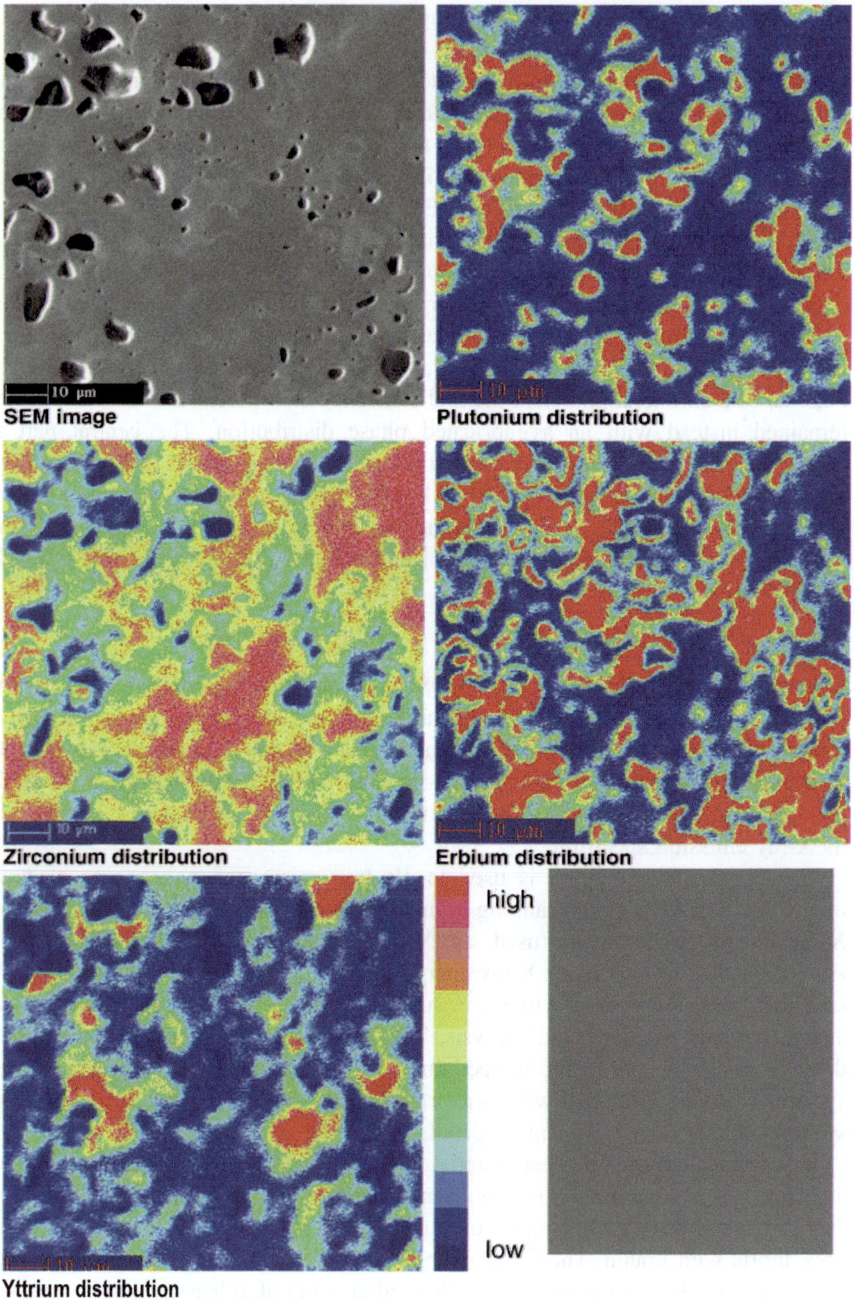

Fig. 3.48 Element mapping of irradiated YSZ–IMF at pellet center. The content range (*dark blue–red*): Zr: 41–53 wt% (21–27 at.%); Pu: 8–19 wt% (1.5–3.8 at.%); Er: 4–6 wt% (1.1–1.7 at.%). Y: ~10 wt%, C_{DL} ~ 0.5–0.2 wt%. Restani et al. (2009) with permission © Elsevier

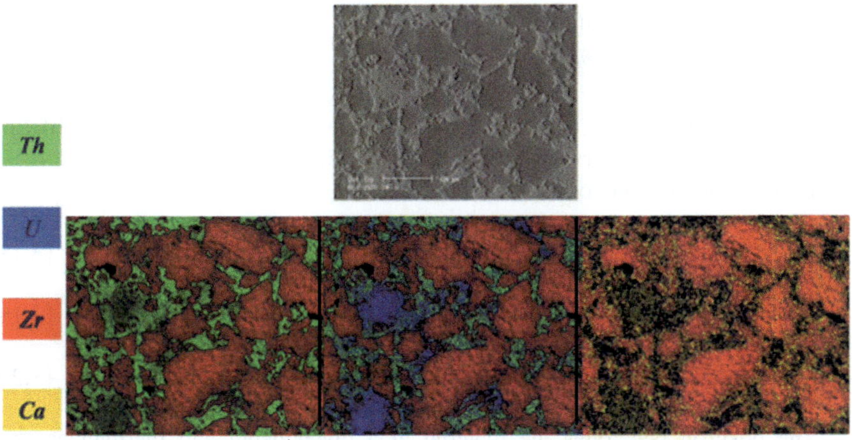

Fig. 3.49 SEM picture and EDS mapping for structural analysis of components in a CaO–ZrO_2–ThO_2–UO_2 composite fuel. Conditions: scale bare 100 μm Ronchi et al. (2003) with permission © Elsevier

3.2.7.2 Positron Annihilation Lifetime Spectroscopy

Anti-electron or positron annihilation spectroscopy also called **Positron Annihilation Lifetime Spectroscopy** (**PALS**) is a non-destructive spectroscopy technique to study voids and defects in solids. The technique is based on the principle that a positron, e^+, annihilation results from its interaction with an electron which yields two γ-ray photons in coincidence. The time between emission of positrons from a radioactive source and detection of the two γ-rays due to annihilation corresponds to the lifetime of the positron. Details on PALS can be found in the monograph by Mogensen (2004).

A positron is detected at time zero by its father nuclide γ emission. In solids containing free electrons (metals or semiconductors), the implanted positrons annihilate rapidly unless voids such as vacancy defects are present. If voids are available, positrons reside in them and annihilate less rapidly than in the bulk of the material. This happens on timescales up to ~ 1 ns. For insulators, implanted positrons may interact with electrons in the material to form positronium: a metastable hydrogen-like atom of an electron and a positron. Para-positronium, with e^+ and e^- spins anti-parallel is a state with a self-annihilation lifetime of 125 ps in vacuum, e.g. Mogensen (2004). Ortho-positronium is a state for which e^+ and e^- spins are parallel with self-annihilation lifetime of 142 ns in vacuum. In molecular materials, the lifetime is environment-dependent and it delivers information pertaining to the size of the void in which it resides. The spectrum can consequently be analyzed as a sum of exponential decay functions (Kansy 1996):

$$-\frac{\delta n(t)}{\delta t} = \sum_{i}^{n} I_i \tau_i \exp(-\tau_i t) \quad \sum_{i}^{n} I_i = 1 \tag{3.40}$$

where $n(t)$ is the number of positron alive at time t after its birth, τ_i is the lifetime associated to a decay curve with its initial intensity I_i.

Positron production source reactions are:

$$^{22}\mathrm{Na} \rightarrow {}^{22}\mathrm{Ne} + \beta^{+} + \nu_{e} + \gamma\ (1.27\ \mathrm{MeV}) \quad \{3.18\}$$

$$^{44}\mathrm{Ti} \rightarrow {}^{44}\mathrm{Sc} + \beta^{+} + \nu_{e} + \gamma\ (1.15\ \mathrm{MeV}) \quad \{3.19\}$$

$$^{64}\mathrm{Cu} \rightarrow {}^{64}\mathrm{Ni} + \beta^{+} + \nu_{e} + \gamma\ (1.35\ \mathrm{MeV}) \quad \{3.20\}$$

The positron penetrates the material at a given depth and thermalizes on a timescale of some 0.1 ns before annihilation. The annihilation takes place faster in an electron rich zone than in a non-rich zone (e.g. vacancy cluster). The larger the vacancy the longer lifetime expected due to the low e^- density. Detection of the 2 antiparallel 511 keV γ-rays is carried out in coincidence as described in Sect. 3.1.2. Resolution in depth is dictated by the energy of the positron. To reduce the e^+ energy an absorber may be used or the e^+ in a form of a beam may be accelerated or decelerated in an adapted electric field. The DL in term of defect concentration is given as 10^{14} cm^{-3} (see Chen and Milnes 1980); analysis may be carried out up to densities of 10^{19} cm^{-3}.

Beam-based PALS measurements, often using "soft" positrons, can be carried out using a ^{22}Na source moderated by solid Ne to produce a continuous beam of low-energy positrons. This is then trapped using a buffer gas trap and released as a pulsed beam. A customized analysis software package uses the annihilation in a reference sample to establish the instrument function of the positron pulse. This is then used in a fitting routine which convolutes the instrument function with one or more lifetimes to find the best fit to the sample lifetime data. For a laboratory arrangement, a ~30 μCi (1110 kBq) source encapsulated in an 8 μm Kapton foil may be used in standard sandwich geometry. Scintillators coupled to photomultiplier tubes can acquire the lifetime spectra. Typical spectra (counts vs. channel or time, see Fig. 3.50) from either laboratory or beam instruments contain over 1 million counts and errors are given through the least squares fitting process employed in the data analysis.

Examples from waste form material studies are given in Fig. 3.50.

Nickel metal (as well as other metals e.g. Cu, Pb …) may be used as a PALS standard. In recrystallized nickel the positron lifetime is rather short (of the order of 0.2 ns), see Fig. 3.50a. The positron annihilation spectrum in Ni metal may be

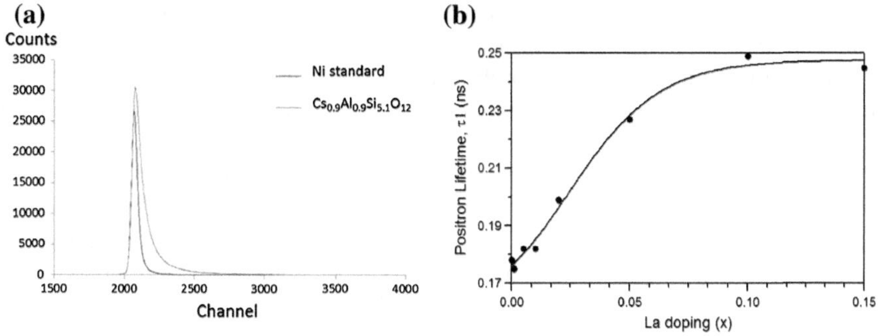

Fig. 3.50 **a** PALS spectra of Ni (small vacancies) and of $Cs_{0.9}Al_{0.9}Si_{5.1}O_{12}$ (larger vacancies), **b** lifetime increase with doping for $Ca_{(1-x)}La_xTiO_3$. Hadley et al. (2005)

recorded before testing PALS on a sample, e.g. $Cs_{0.9}Al_{0.9}Si_{5.1}O_{12}$ (Courtesy: Vance from ANSTO). In this material, partial substitution of Al and Cs generates locally rather large vacancies increasing the lifetime of the trapped positrons in this material as observed in Fig. 3.50a. Equation (3.39) could be used to derive the specific lifetimes (τ_i).

Perovskites may also be used as waste forms. It was shown that the major lifetime increases with doping level for $Ca_{(1-x)}La_xTiO_3$ perovskite in which the charge compensators for the substitution of La^{3+} (simulating An^{3+}) on the Ca^{2+} site are cation vacancies (see Hadley et al. 2005). The doping modifies the density of vacancies and their size distribution, which is observable with the increase of positron lifetime (Fig. 3.50b).

The relationship lifetime-vacancy size has been intensively investigated. For a given temperature, the size of the voids/vacancies (i) is approximately given by:

$$\tau_i^{-1} = 2\left[\frac{1 - R_i}{R_i + \Delta R_i} + \frac{1}{2\pi}\left(\frac{2\pi R_i}{R_i + \Delta R_i}\right)\right] \tag{3.41}$$

where τ_i (ns) is the lifetime component, R_i (nm) the vacancy radius and ΔR_i is $\sim$0.16 nm (see Tao 1972; Eldrup et al. 1981).

Other nuclear material applications are for example positron annihilation of electron-irradiated (Fe–Cu) and (Fe–Cu–C) alloys, as reported by Hori et al. (1993), and the application of positron annihilation analysis to fatigue damage for nuclear plant materials (316 SS and SA508 steel), as investigated by Maeda et al. (1996).

Scanning positron microscopy

A **Scanning Positron Microscope (SPM)** is a similar unit to a scanning electron microscope (SEM) where the electron gun is substituted by a positron production source.

The SPM delivers a micrograph of defect distributions. A SEM micrograph contains less information on surface defects than that of the SPM (David et al. 2001). Actually, it is possible to probe one e^- with one e^+. Results of a microscope that uses positrons to create an image of a surface show that positrons can see defects. Because of their positive charge, positrons are drawn to vacancies, locations within a crystal–or on its surface–where a positively charged nucleus is missing. Without the nucleus, there are fewer electrons in the neighborhood, consequently a positron can survive longer before colliding with and annihilating its electron counterpart. By measuring a positron's survival time inside a section of material, defect distributions can be evaluated.

Analysis begins by scanning the surface with electrons to create an image of the pattern, a procedure similar to that for a standard SEM. The most interesting regions of the surface can be selected to probe with positrons. The positrons, from a ^{22}Na source, filtered through a series of electric fields that homogenized their energies, are bunched into short pulses. They subsequently pass through another series of fields that accelerate the positron pulse to a fixed energy before focusing it into a 2 μm spot on the sample surface. By measuring the time between the pulse and the photon flash of electron-positron annihilation the defect distributions can be mapped. Positrons lodged inside defects were found to avoid annihilation for twice as long as those in others regions. By scanning the beam across the surface, an image of the defect distributions in the sample can be recorded.

A positron survives for different lengths of time inside different kinds of defects. The analysis of a spectrum of positron lifetimes can provide information on the types of defects present in a material. The lifetime measurements with a Scanning Positron Microscope also gives unprecedented sensitivity, about defects at the subnanometer level (David et al. 2001), while electrons are sensitive to only about a micron. The SPM device is consequently promising for nuclear material testing, since the unit can carefully identify and locate defects, allowing diagnosis and later production of better materials.

Copper is foreseen as the material for canister containing nuclear waste package in Scandinavian countries. The integrity of this material must be assured. Scanning Positron Microscopy allows investigations of fatigue cracks in copper. Maps of mean positron lifetime (ps scale) at 16 keV positron implantation energy (Egger et al. 2002) permit identification of the type of defect as well as their distribution (see Fig. 3.51).

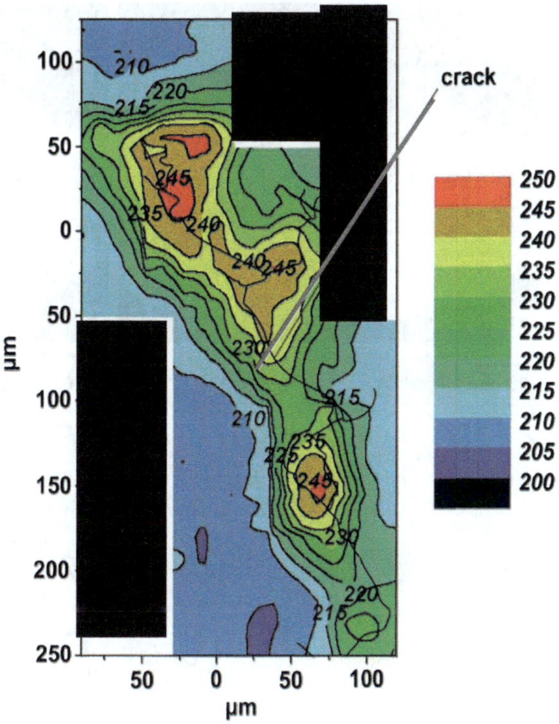

Fig. 3.51 Scanning Positron Microscopy. Fatigue crack in copper and map of mean positron lifetime (ps) at 16 keV positron implantation energy, adapted from Egger et al. (2002) with permission © Elsevier

3.2.8 Techniques with Lepton Interaction and Lepton Detection

This section deals with regular electrons and heavy electrons: e.g. muons.

3.2.8.1 Scanning Electron Microscopy

Scanning Electron Microscopy (**SEM**) provides a description of the electron-specimen interations. The different imaging and analytical modes using secondary and backscattered electrons, electron-beam-induced currents, X-rays, electron channeling effects, and cathodoluminescence reveal specific contrasts and quantitative information. Detailed information may be found in the monograph by Reimer (1998).

The example involving nuclear material deals in Fig. 3.52 with the observation of a natural analogue of a zirconia inert matrix fuel. SEM allows excellent morphological analysis, which can be completed by backscattered electron analysis to

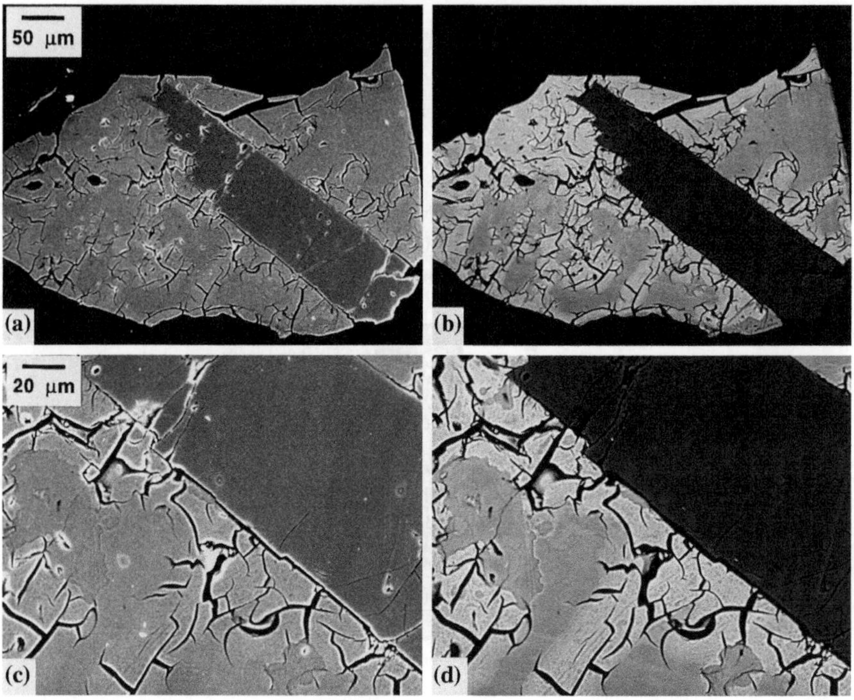

Fig. 3.52 SEM Comparison of secondary (**a**, **c**) and backscattered (**b**, **d**) electron images. Conditions: pairs of intergrowth between uranpyrochlore and baddeleyite from Jacupiranga carbonatite complex, Brazil. Images taken at low magnification (**a**, **b**) show the general features, including microfracturing and alteration in the uranpyrochlore (lighter gray, $(U, Ca, Ce)_2(Nb, Ta)_2O_6(OH, F)$ as waste form analogue) and the platy habit of the baddeleyite (darker gray, ZrO_2) as inert matrix fuel analogue. Lumpkin et al. (1999) with permission © Elsevier

complement the phase study, as in the investigation of an actinide doped phase (uranpyrochlore) adjacent to an actinide depleted phase (baddeleyite) as presented in Fig. 3.52. Here cumulative α, β, γ fluences over Ma's have restricted effect on the stability of zirconia (low crack density) even in the vicinity of the actinide doped phase.

3.2.8.2 Transmission Electron Microscopy and Associated Techniques

Transmission Electron Microscopy (**TEM**) is a technique in which the electrons forming a nano-beam are transmitted through an ultra-thin specimen. The electron beam interacts with the specimen creating by transmission an image; the image is magnified and focused onto a detector: a fluorescent screen, a layer of photographic film, or a CCD camera (see Sect. 3.1). Detailed information on TEM is available in reference monograph such as that provided by Fultz and Howe (2013).

A TEM is capable of imaging at a higher resolution than a light microscope, because the wavelength of electrons is shorter than that of visible light. This enables the TEM to examine details of the specimen at the atom level. TEM finds application in scientific fields, including among other physical and material sciences.

At small magnifications the contrast of TEM images results from the electron absorption due to the sample thickness and composition. At high magnifications wave interactions modulate the intensity of the micrograph. Analysis of modulations due to chemical composition, crystal orientation and electronic structure allows imaging these features.

Theoretically, the maximum resolution, *d*, that one can obtain with a microscope is limited by the wavelength of the incident electrons, λ_e and the numerical aperture of the microscope Eq. 3.22. As predicted by de Broglie (1923), electrons have wave and particle properties. The wavelength of electrons is related to their kinetic energy. However, since the electron's velocity approaches the speed of light, *c*, in a TEM, a relativistic factor must be added.

$$\lambda_e = \frac{h}{\sqrt{2m_0E\left(1 + \frac{E}{2m_0c^2}\right)}} \tag{3.42}$$

where, h is Planck's constant, m_0 the electron rest mass and E is the incident electron energy with the relativistic correction: $1 + E/(2m_0c^2)$.

Electrons are usually generated in an electron microscope by thermionic emission from a filament, usually tungsten or alternatively by field electron emission. The electrons are then accelerated by an electric potential. After collimation, the incident electron beam penetrates the sample lamella. The specimen preparation includes mechanical thinning, chemical etching, FIBing or replication as described in Sect. 2. The resolution is limited by the properties of the electron beam as well as by the sample thickness, porosity, nature, density. With a 200 keV TEM the lateral resolution can reach 50 pm. TEM has been extensively used to characterize nuclear materials (e.g. Linsmeier et al. 2013 for Gen V materials; Cameron et al. 2017 for actinide bearing oxides; and Gunther-Leopold et al. 2012 for CRUD studies). However, the combination of TEM with EDS, ED or EELS is more efficient for improving characterization as treated below.

Energy dispersive spectroscopy (**EDS**) is currently being used coupled with the TEM unit. EDS has been described in Sect. 3.2.4.2.

Electron diffraction

Electron Diffraction (**ED**) refers to the wave nature of **electrons**.

Electron diffraction of solid specimens may be performed in a **TEM** where the electrons pass through the sample thin film. Selected area (electron) diffraction is a crystallographic experimental technique that is usually performed with the TEM set-up as sketched in Fig. 3.53.

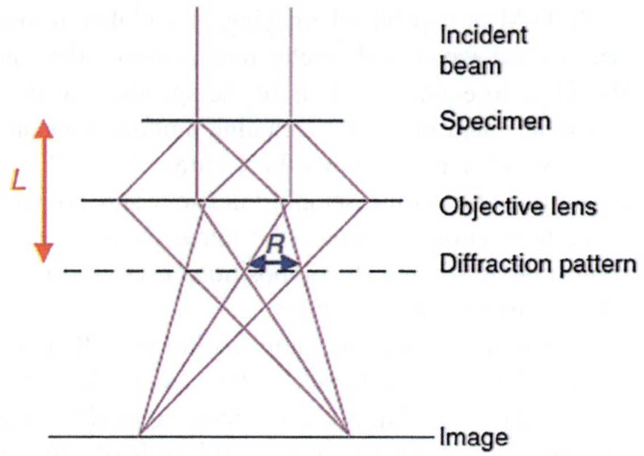

Fig. 3.53 Sketch of the electron beam-path in a TEM form morphologic, electron diffraction (ED); *R* distance pattern center—reflex and *L* distance specimen—screen for diffraction pattern

Electron diffraction is a collective scattering (nearly elastically) of electrons by atoms in a regular atom array. Most **ED** is performed with high-energy electrons with wavelengths, given by Eq. 3.50, smaller than the interplanar spacing's. The wavelength of the electrons in a 10 kV SEM is then 12.2 pm while in a 200 kV TEM the wavelength is 2.5 pm. Solid TEM-ED is carried out by the electrons passing through a thin specimen and observation of the resulting diffraction pattern on a fluorescent screen, on photographic film, on imaging plates or on a CCD camera.

Figure 3.53 is a simple sketch of the path of a parallel beam of electrons in a TEM from just above the sample and down the column to the fluorescent screen. After the electrons have left the sample they pass through the electromagnetic objective lens. This lens acts to collect all electrons scattered from one point of the sample in one point on the fluorescent screen, causing an image of the sample to be formed. At the dashed line in the figure, electrons scattered in the same direction by the sample are collected into a single point. This is the back focal plane of the microscope, and is where the diffraction pattern is formed. The lattice parameter *d* may be deduced from an equation derived from Eq. 3.33:

$$d = \lambda_e (L/R) \tag{3.43}$$

with *R* is the distance between the pattern center to the reflex and *L* is the distance between specimen and screen.

By manipulating the magnetic lenses of the microscope, the diffraction pattern may be observed by projecting it onto the screen instead of the image.

Electron diffraction in TEM is subject to some limitations. First, the sample to be studied must be 'electron transparent', meaning the sample thickness must be of the order of 100 nm or less. Many samples are affected by radiation damage caused by the electron beam. The study of magnetic materials is affected by the electrons deflection in magnetic fields. However, this can be exploited to investigate the magnetic domains of samples. But the main limitation of electron diffraction in TEM remains the degradation of the sample by heat, by defect induction and by chemical reduction as discussed by Degueldre and Alekseev (2015) for a specific specimen of UO_3.

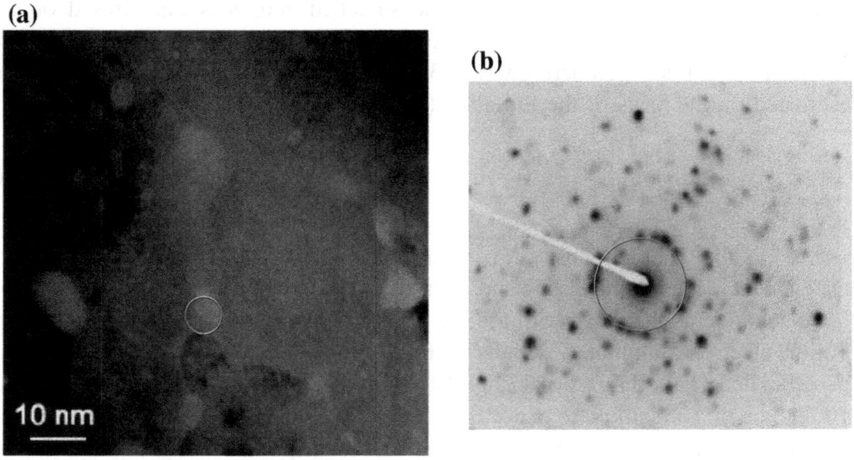

Fig. 3.54 **a** DF TEM image of nano-sized oxide particles located inside the grains and at the grain boundaries and **b** corresponding selected area electron diffraction pattern, adapted from Veleva et al. (2011) with permission © Elsevier

The example treated here deals with material proposed for divertor in a fusion reactor. The ODS including yttria dispersion in tungsten matrix (W-2Y) has been produced and studied by TEM and ED to reveal its structure at the nanometric scale. Dark field (DF) TEM images reveal nano-sized yttrium oxide (Y_2O_3) particles located inside the grains and at the grain boundaries. These particles are observable as light spots in the tungsten gray background (W), see Fig. 3.54. Zones corresponding to the selected area are analyzed by ED yield electron diffraction pattern where the ED reflexes suggest the nano-crystallinity of the particles (see Veleva et al. 2011).

Electron energy loss spectroscopy

In **Electron Energy Loss Spectroscopy** (**EELS**) electrons with fixed incident energy are transmitted through a specimen. Some electrons undergo inelastic scattering, losing their energy. The amount of energy loss is measured using an electron spectrometer and interpreted. Inelastic interactions include phonon excitations, band transitions, plasmon excitations, inner shell ionizations and Cherenkov radiation. The inner-shell ionizations are useful for detecting the elements of the specimen. From the energy loss spectrum, the types of atoms, and the numbers of atoms of each type, traversed by the beam can be determined. The scattering angle can also be measured, providing information about the material excitation by inelastic scattering. Details on Electron Energy-Loss Spectroscopy in the Electron Microscope are presented in a monograph by Egerton (2011).

Detection is carried out by an electron spectrometer with an energy resolution better than 1 eV e.g. 0.1 eV for example. The lateral resolution is dictated by the size of the electron beam which is below the nm. The detection limit is also dictated by the beam size but also by the sample thickness. This can be determined by EELS comparing the whole spectrum with the so-called zero loss line. Based on the average Z of the material and the energy of the electrons a thickness may be deduced using Beer–Lambert law (Eq. 3.8) and (Eq. 3.10). The thickness x may be estimated on the basis of the Equation (Lenz 1954; Pozsgai 2007):

$$\log_{10}\frac{I_0}{I} = \frac{\rho N_{Av}}{A}(\sigma_e + \sigma_i)x = \left(\frac{kZ^{\xi}\log_{10}e}{A}\right)\rho x \qquad (3.44)$$

where ρ is the material density N_{Av} the Avogadro constant, A the atomic weight of the element σ_{e} and σ_{i} the elastic and inelastic atomic scattering cross-sections, k and ξ are coefficients that vary with the incident energy and the collection semi angle. For the electron illuminated cylinder in the specimen the total number of atoms may be estimated. This figure can be used to derive a DL and for a 100 nm lamella, 100 atoms or 10^{-22} mol are expected to be detectable. Potential interferences such as plasmon peak and interfering core electron transitions from other elements may however affect the detection limit.

The analysis of UO_2 by TEM and EELS requires production of a 10–200 nm thick lamella. These sub-samples are provided by cutting and production using a FIB followed by its attachment of the microscopic piece of UO_2 on a TEM sample carrier. The piece is further FIBed by polishing down to a lamella with thickness going from 10 to 200 nm over sizes of 1–2 μm. The FIBing is carried out with Ga^+ ions of 30 keV for cutting and 3 keV for polishing (see Sect. 2.2.2). The TEM observation (Fig. 3.55a) and EELS records (Fig. 3.55b) are made within a flux of 200 keV electrons and an intensity of I = 30 pA cm^{-2}. The EELS spectra recorded from 0 to 300 eV show the zero energy loss peak followed by the plasmon (below P_3 and after) and core (P_3, P_2, P_1, O_5, O_4) peaks.

The EELS spectra can be used to estimate the local thickness of samples (see Fig. 2.2). The effect of the thickness on the spectrum is plotted in Fig. 3.55c: the zero energy loss peak follows Beer–Lambert law (Eq. 3.8) and the core peaks increase for very slim lamella toward a plateau followed by a decrease for thickness larger than 120 nm.

The speciation may be derived from the M4/M5 edges position and structure. Low levels of transuranics may be characterized with a lateral resolution of the order of 1 nm and concentrations of <200 ppm (Buck and Fortner 1997). This was applied to redox studies in glasses for plutonium immobilization (Fortner et al. 1997), in uranium pyrochlore and natural analogues of Pu- and U-bearing waste forms (Xu and Wang 1999).

Use of the high-energy $M_{4,5}$ edges to detect transuranium elements is the only method available to positively identify them because the weaker lower-energy $N_{4,5}$ edges overlap with the more intense $M_{4,5}$ edges of the trace levels of lanthanides that may be present in the same alteration phases. The position and intensity of the $M_{4,5}$ absorption edges of the transuranium elements were confirmed with samples of contaminated soils and data from the literature. The M_4/M_5 ratio for the actinide absorption edges was used, in combination with crystal chemical considerations to determine chemical state.

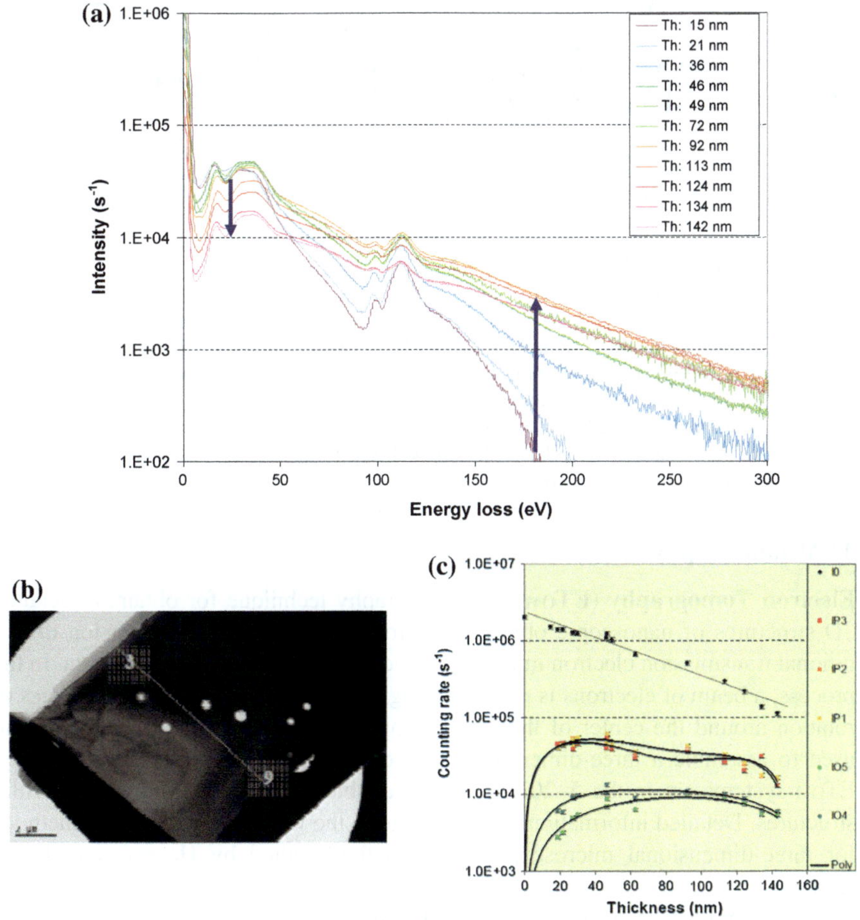

Fig. 3.55 TEM **a** and EELS analysis (**b** and **c**) of a FIBed UO_2 lamella. Conditions fuel sample thickness (Th) from ~15 to 150 nm, 200 kV incident electrons, *I:* 30 pA cm^{-2}. Degueldre et al. (2013a) with permission © Elsevier

In the environment of nuclear bomb test at Johnston Island, the detection of a low level of actinides has been reported by Buck and Fortner (1997) using EELS, see Fig. 3.56. The EELS spectra were recorded with an energy resolution of 1.0 eV per channel, an integration of time 5 s, for 9 readings and background subtracted. The normal EELS of a sample of contaminated soil present the EELS signal for U and Pu. The EELS spectrum of a site particle containing RE elements, U and Np was recorded for lower counting rate for the Np edge with a factor 20× relative to original spectrum. The energy dispersion was also 1.0 eV per channel but the acquisition time 3 s the reading was repeated 50 times. Because of its excellent lateral resolution (2 × 2 × 10) nm and high sensitivity (1.0–0.1%), EELS has a demonstrated superior detection limit

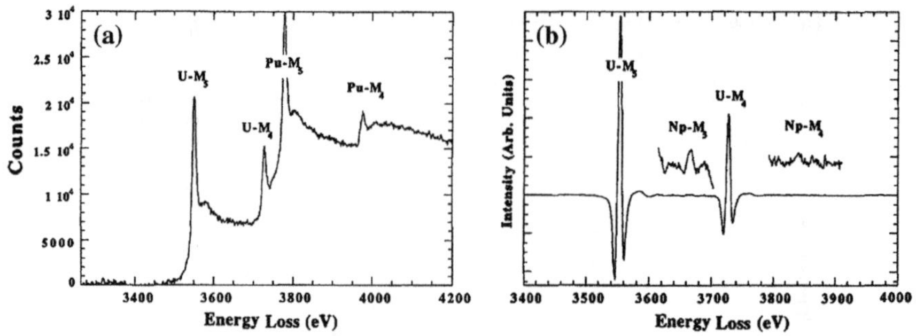

Fig. 3.56 Detecting low level of actinides with EELS, Buck and Fortner (1997) with permission © Springer. **a** Normal EELS of Pu-U mixed oxide from a sample of contaminated soil from Johnston Island. Conditions: resolution = 1.0 eV per channel, integration time 5 s, 9 readings, background subtracted. **b** EELS of a week site particle containing RE elements, U and Np. Np edge feature multiplied 20×, relative to original spectrum, energy dispersion = 1.0 eV per channel, integration time 3 s, 50 readings

TEM tomography

Electron Tomography (ETOM) is a tomography technique for obtaining detailed 3D structures of nanoscopic objects. Electron tomography is an extension of traditional transmission electron microscopy and uses a TEM to collect the data. In the process, a beam of electrons is passed through the sample at incremental degrees of rotation around the center of the target sample. This information is collected and used to assemble a three-dimensional image of the target. Current resolutions of ETOM systems are in the 5–20 nm range, suitable for examining supra-molecular structures. Detailed information may be found in the work 'Computational methods for three-dimensional microscopy reconstruction' edited by Herman and Frank (2014).

This technique has recently been used to directly visualize the atomic structure of screw dislocations in nanoparticles. Bidimentional pictures recorded for different tilting angle may be used to rebuild the 3D picture. Presently, the highest electron tomography resolution is around 2.4 Å as demonstrated by Scott et al. (2012) using a gold nanoparticle.

Promotion of the technique for the analysis at the atomic level of materials from Generation IV and V reactors has been carried out by Marquis et al. (2009).

3.2.8.3 Techniques with Muon Interaction and Muon Detection

Muon microscopy

Scanning muon Microscopy (SMUM) and **Transmission muon Microscopy** (TMUM) are the two muon variants of SEM and TEM. If electron microscopy provides a greater image resolution than optical microscopy, this is because electrons have a smaller wavelength than photons, which corresponds to the greater momentum of the electrons, since their wavelength is inversely proportional to the momentum (Eq. 3.42). Consequently, heavier particles than either electrons or photons provide a greater image resolution. Muons are all heavier than the electron and their potential is real for applications such as TMuM or SMuM (e.g. Doyama et al. 2009).

A muon beam intensity distribution perpendicular to the beam may be obtained using imaging plates employed in X-ray science, and transmission electron microscopy. The positional resolution is 50 μm × 50 μm. When the beam is one kind, because a coincidence measurement is not necessary, the muon distribution perpendicular to the beam is taken in a very short time, and the spatial resolution is 50 μm × 50 μm. Such a resolution cannot be obtained by the usual muon measuring method. This is quite convenient to adjust muon beams to reach the resolution of 10 μm × 10 μm. Such an intensity resolution of muon beams could not be imagined without using imaging plates. This microscopy technique can be used as a non-destructive test (see Doyama et al. 2009).

Muon Radiography (muRAD) and muon Tomography (muTOM) are techniques that use muons and more precisely cosmic ray muons to generate 2D or 3D images of volumes using information contained in the scattering of muons. Since cosmic muons (average energy 4 GeV (λ = 0.31 pm) and flux 1 cm^{-2} min^{-1} at sea level) are much more deeply penetrating than X-rays of comparable energy, muTOM can provide radiography through thicker objects than X-ray. Since its development in the 1950s (George 1955), muon tomography has had many applications, the most current being muon transmission radiography and muon scattering tomography. Muon tomography is under development, for non-proliferation purposes, for detecting nuclear material in road vehicles and cargo containers.

Muon Tomography uses multiple scattering radiographies in addition to energy loss and stopping cosmic muons undergo scattering. The angular distribution results from many single scatters that provide a way for obtaining information from charged particle beams. When the thickness of the object increases, multiple scattering can result from the number of interactions N, and the angular dispersion $dN/d\theta$ can be modeled as a Gaussian:

$$\frac{\mathrm{d}N}{\mathrm{d}\theta} = \frac{1}{2\pi\sigma^2} \mathrm{e}^{-\frac{\theta^2}{2\theta_0^2}} \tag{3.45}$$

where θ is the polar angle, θ_0 is the multiple scattering angle and σ the standard deviation. The Image can then be reconstructed for example by use of a specific code e.g. GEANT4 [see Agostinelli et al. (2003)]. The incident flux projected to the core location is used to normalize transmission radiography. From here the calculations are normalized for the zenith angle of the flux.

New imaging techniques based on muon scattering tomography (MuTOM) have been and are being developed, e.g. Parker and Joyce (2015). With muon scattering tomography, both incoming and outgoing trajectories for each particle are reconstructed. This technique has been shown to be useful to find materials with high atomic number in a surrounding of low-Z material such as uranium in material with a low atomic number. Companies have started to use it for several purposes, most notably for detecting nuclear cargo entering ports and crossing over borders. Teams have built portable muon trackers. These trackers are constructed from sealed aluminum drift tubes, which are grouped into 24 1.2 m squared planes. The drift tubes measure particle coordinates in X and Y with a typical accuracy of several 0.1 mm. A tracker can be moved via a pallet jack on a fork lift. If a nuclear material has been detected it is important to be able to measure details of its construction in order to correctly evaluate the threat.

Non-proliferation

Muon Tomography can be used for treaty verification, e.g. NPT. It is safe for humans and will not apply an artificial radiological dose to the controlled fissile material. Warheads can be imaged in a container behind significant shielding. Exposure times depend on the object and detector configuration ($\sim$few minutes if optimized). While sensitive nuclear material detection can be reliably confirmed, and discrete sensitive nuclear objects can be counted and localized, the system can be designed to not reveal potentially sensitive details of the object design and composition. The nuclear material example is that of muon tomography imaging for nuclear threat detection inside large volume containers with the muon portal detector (see Riggi et al. 2013). Muon tomographic visualization techniques reconstruct a 3D image (Fig. 3.57) as close as possible to the real localization of the objects being probed. Relative merits of all such methods are discussed, with reference to full GEANT4 simulations of different scenarios, incorporating medium and high-Z objects inside a container. Muon event of energy 1 GeV with hits produced in the four detector planes. The color scale of the plots represents the energy deposited. Acquisition time: 3 months from bottom panels. In the panels the XY tomography view for a fixed z depth ($z = 5$ cm) is shown.

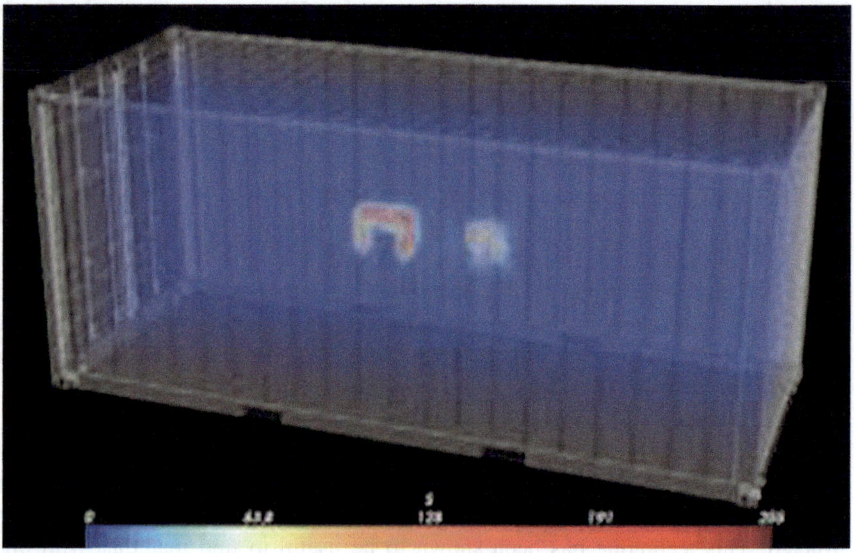

Fig. 3.57 Muon tomography imaging algorithms for nuclear threat detection inside large volume containers with the muon portal detector (bottom panels). Riggi et al. (2013) with permission © Elsevier

Reactor

Non-destructive inspection method by cosmic-ray muons can be used to observe the reactor internal from outside the reactor pressure vessel (RPV) and containment vessel (CV). An observation test on the high-temperature test reactor (HTTR) has been conducted to visualize reactor internal (Takamatsu et al. 2015; see Fig. 3.58).

The analytical evaluation of the resolution of the existing muon telescopes has been assessed for their suitability to investigate the high-temperature test reactor. Detection of the major structures of the test reactor is feasible based on the distribution of the surface densities calculated from the coincidences measured by the telescopes. It was found that the existing muon telescopes could be used for muon observation of the internal reactor from outside the reactor pressure vessel.

The analysis was based on point of closest approach, where the track pairs were projected to the mid-plane of the target, and the scattered angle was plotted at the intersection. This test object was successfully imaged, even though it was significantly smaller than expected at Fukushima Daiichi. One of the critical problems that have arisen from the accident at Fukushima nuclear power plant is the location of fuel debris. For solving this problem, internal reactor examinations have been planned to identify the fuel debris.

Nuclear waste imaging

Tomographic techniques can be effective for non-invasive nuclear waste characterization and for nuclear material accountancy of spent fuel inside dry storage containers. Cosmic muons can improve the accuracy of data on nuclear waste and Dry Storage Containers. This imaging exceeds the IAEA detection target for nuclear material accountancy. In Canada, spent nuclear fuel is stored in large pools (wet storage) for a nominal period of 10 years to allow for sufficient radioactive cooling. The waste packages need however characterization.

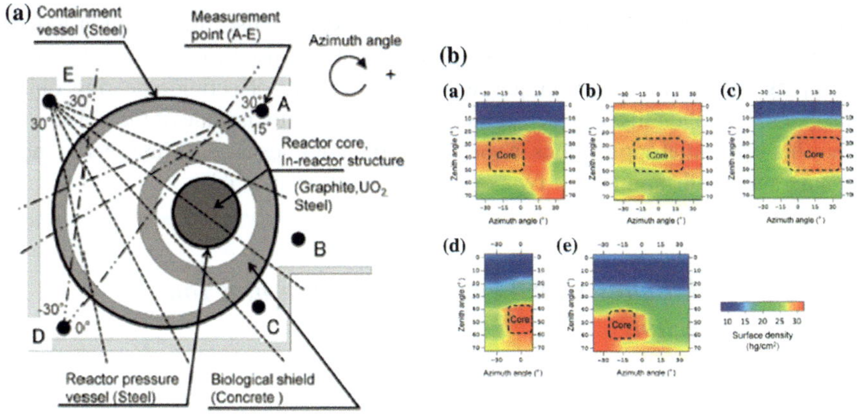

Fig. 3.58 **a** Measurement points and azimuthal angles for single channel telescope; **b** Surface density recorded from each point. Takamatsu et al. (2015) with permission © Elsevier

Challenges and issues for nuclear waste characterization are covered at great length, summarized below:

- Historical waste. Non-traceable waste stream poses a challenge for characterization. Different types of waste can be distinguished: tanks with liquids, fabrication facilities to be decontaminated before decommissioning, interim waste storage sites, etc.
- Some waste forms (i.e. encapsulated alpha/beta emitters, heavily shielded waste) may be difficult or impossible to measure and characterize.
- Direct measurements, i.e. destructive assay, are not possible in many cases and Non-Destructive Assay (NDA) techniques are required, which often do not provide conclusive characterization.
- Homogeneity of the waste needs characterization (i.e. sludge in tanks, in-homogeneities in cemented waste, etc.).
- Diagnostic of the waste and waste package: breach of containment, corrosion, voids, etc.

Accounting for all of these issues can take a great deal of time and effort. Muon Tomography can be useful to assess the characterization of waste, radiation cooling, and condition of the waste container (see Wanga et al. 2009; Jonkmans et al. 2013).

3.2.9 Techniques with Lepton Interaction and Neutron Detection

A novel technique based on cosmic-ray **muon interaction and neutron detection** has been proposed by Guardincerri et al. (2015) utilizing the penetrating ability of cosmic ray muons to probe for imaging actinides in containers. Experimental results from this new technique use neutrons generated by cosmic-ray muons to identify the presence of nuclear material. Neutrons emitted from actinides are used to tag muon-induced fission events in actinide materials. Laminography is used to form images of the stopping material. This technique allows the imaging of actinide-bearing objects tagged using muon tracking detectors located above or to the side of the objects, and may have potential applications in warhead verification scenarios. The experiment did not attempt to distinguish the type or grade of nuclear material.

3.2.10 Techniques with Lepton Interaction and Ion Detection

3.2.10.1 Electro-Spray Ionization Mass Spectroscopy

Electro-Spray Ionization Mass Spectroscopy (ESIMS). is a recently developed analytical technique that seems promising for speciation studies. This technique applies a spray of electrons on a material surface followed by mass spectrometry of the cluster ions formed in "soft conditions". First developed for biological samples, it is now used for direct speciation studies in solution. It is now possible to directly couple a liquid at atmospheric pressure to mass detection at reduced pressure, even if the mechanisms taking place are rather complex. ESIMS investigations were adapted for the speciation of Th(IV) hydrolysis species (Moulin et al. 2001) and U (VI) in neutral solutions (Moulin et al. 2000). Figure 3.59 shows a typical spectrum obtained for a uranyl solution at pH 3.5. Peaks can be interpreted for the species $[(UO_2)_i(OH)_j(H_2O)_n]^{(2i-j)+}$.

The detection limit is 10^{-15} mol for 1 ml aliquot (see Ho et al. 2003).

Increasing the energy of the electronic stimuli permits the technique to be extended from molecular to atomic analysis.

3.2.10.2 Spark Source Mass Spectroscopy

Spark Source Mass Spectroscopy (SSMS) is similar to the former technique but the source of electrons is driven at higher potential. This allows direct quantitative analysis of elements and isotopes of the material. SSMS was used to determine trace elements in U_3O_8 standard (Li et al. 1991) as well as other actinide oxides. In a complementary way, SSMS may be used to analyze uranium from environmental samples. The detection limit is $10^{-15} - 10^{-18}$ mol for 10^{-7} g AnO_2 sample (see Becker 2007).

3.2.11 Techniques with Neutron Interaction and Photon Detection

The neutron sources currently used are reactors, spallation sources (α,n) sources, and spontaneous fission sources such as ^{252}Cf. If **Neutron Activation Analysis** (NAA) is conducted directly on irradiated samples the term "**Instrumental Neutron Activation Analysis**" (INAA) is used. In some cases irradiated samples are subjected to chemical separation to remove interfering species or to concentrate the radioisotope of interest; this technique is known as **Radiochemical Neutron Activation Analysis** (RNAA).

3.2.11.1 Instrumental Neutron Activation Analysis

Instrumental Neutron Activation Analysis (INAA) is a nuclear process used for determining the concentrations of elements in a vast amount of materials. INAA allows discrete sampling of elements as it disregards the chemical form of a sample, and focuses solely on its nucleus. The method is based on neutron activation according to the reaction:

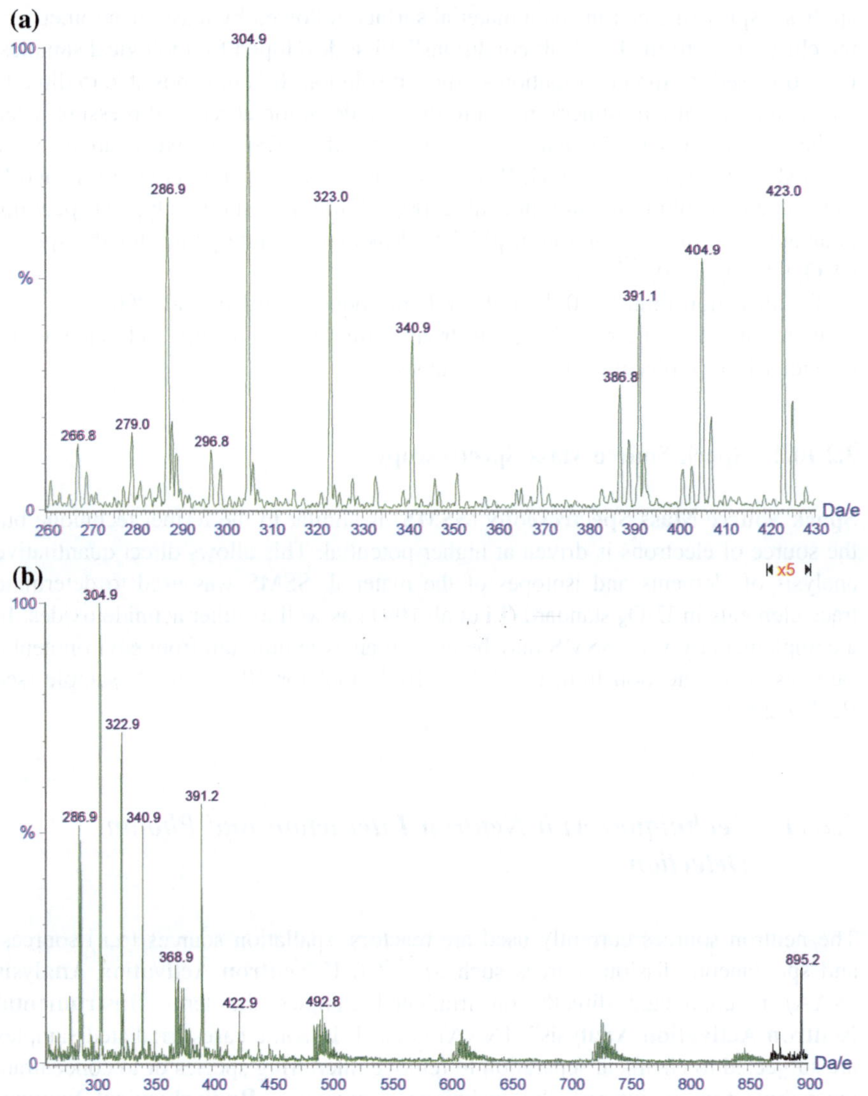

Fig. 3.59 Electro-spray ionization mass spectrum of uranium at 10 mg l^{-1} at pH 3.5 and at pH 6.5. Note the presence of species such as **a** $[UO_2OH(H_2O)_n]^+$ with masses 286.9 + n · 18 at pH 3.5 and **b** at pH 6.5 where $[(UO_2)_3(OH)_5]^+$ with mass 895.2 is also observed. Moulin et al. (2000) with permission ©Springer

$$ {}_{Z}^{A}X + {}_{0}^{i}n \rightarrow {}_{Z}^{A+1}X + \gamma \qquad \{3.21\} $$

The sample of ^{A}X is bombarded with neutrons, causing the elements to form radioactive isotopes ^{A+1}X. The γ emissions and radioactive decay paths for the corresponding isotopes are well known. It is possible to study spectra of the emissions of the radioactive sample, and to determine the original concentrations of the elements. Calculations are performed using equations:

$$\frac{dN}{dt} = N\phi\,\sigma \quad \text{or} \quad N_{{}^{A+1}_{Z}X}(t) = \frac{N_0\phi\,\sigma}{\lambda}\left(1 - e^{-\lambda t}\right) \tag{3.46}$$

with N the number or density of the nuclide ϕ the neutron flux on the sample spot, σ the neutron capture cross section, and λ the decay constant of ^{A+1}X according to:

$$^{A+1}X^{*} \rightarrow {}^{A+1}X + \gamma \qquad \{3.22\}$$

A particular advantage of this technique is that it does not destroy the sample. With the use of automated sample handling, γ-ray measurement with solid-state detectors (see Sect. 3.1.2), and computerized data processing it is generally possible to measure more than 30 elements simultaneously in most sample types without chemical processing.

INAA can be used to perform non-destructive analyses on solids, slurries, suspensions, liquids and gases with no or minimal preparation. Due to the penetrating properties of neutrons and resultant γ-rays, the technique provides a full bulk analysis. As radioisotopes have different half-lives, their counting can be delayed to allow decay of potential interferences. INAA has been the standard analytical method for multi-element analyses with detection limits in the ppb range. Accuracy of INAA is of the order of 5%, and the relative precision can often be better than 0.1%. There are two drawbacks to the use of INAA:

- The irradiated sample remains radioactive a long time after the initial analysis, requiring handling and disposal protocols for low-level to medium-level radioactive waste.
- The number of suitable activation nuclear reactors is declining; with a lack of irradiation facilities, the technique has declined in popularity and become more expensive.

INAA, being an isotopic and not an elemental method of analysis, is capable of determining a number of actinide nuclides of radioecological interest by transforming a nuclide into a radionuclide that is more quantifiable. The nuclear characteristics that favor this technique may be summarized in an advantage factor relative to radiometric analysis of the original analyte. After activation, the nuclides that are produced are detected by γ-, β- or α-spectrometry. INAA has been used widely for actinide analysis, in particular in the field of uranium prospecting. Nowadays, it is applied to determine concentrations in environmental, geological and biological samples. INAA also has the advantage of being well suited for routine analysis.

Since neutrons activate almost all elements present in the sample, it is necessary to resolve the activity generated by actinides from activities of other elements. Separations are achieved through instrumentation or chemical methods.

The application of INAA to determine ^{230}Th, ^{232}Th, ^{235}U, ^{238}U and ^{237}Np quantitatively provides an independent method that complements the conventional radiometric techniques (Byrne 1993). INAA was used to determine, in the presence of other elements, Th distribution in geological material (Castillo et al. 1996), ^{231}Pa in environmental and biological samples (Byrne and Benedik 1999), ^{235}U and ^{238}U in geological samples (Sheng et al. 1985), and ^{237}Np in sediments (Byrne 1986). Detection limit depends on many parameters, such as: activation time, irradiation time, decay time, acquisition time. In the case of neptunium, the detection limits went down to 5×10^{-13} g Np (2×10^{-15} mol) corresponding to 2.5×10^{-9} mg kg^{-1} for a 200 mL seawater sample.

INAA is currently used for solid samples. However, it has also been successfully applied to the analysis of waters. For example, the effect of humic material on the dissolution of natural trace elements, especially of actinide homologues, and on the sorption behavior of Np, Pu, Am and Cm, has been determined (Probst et al. 1995). The groundwater constituents have been analyzed by applying INAA as well as other analytical techniques. Fractions of fluid samples were obtained by ultrafiltration and by ultracentrifugation. It was demonstrated that trace actinides were associated with the colloidal phase (Kim et al. 1987).

An improved prompt gamma neutron activation analysis unit, using a focused diffracted neutron beam, has been proposed as a major analytical tool for quantifying ^{10}B in water samples. The sensitivity for ^{10}B in water is 18′750 cps per mg. (see Riley and Harling 1998).

3.2.11.2 Radiochemical Neutron Activation Analysis

If chemical separations are performed to samples after irradiation to remove interferences or to concentrate the radioisotope of interest, the technique is called **radiochemical neutron activation analysis** (RNAA). The latter technique is performed infrequently due to its high labor cost. RNAA has however been used to determine the vertical profile of Np in sea sediments. In this system, determination of ^{237}Np by radiochemical neutron activation analysis combined with extraction chromatography has been reported by Kalmykov et al. (2004).

3.2.11.3 Activation Analysis with Fast Neutrons

The techniques developed in earlier sections are performed with thermal neutrons that may limit them to shallow layers in the materials. Fast neutron-based techniques are typically used for bulk material analysis. They employ prompt gamma rays induced by a neutron probe for classification of the interrogated object using the elemental parameters extracted from the spectral data (see Barzilov and Novikov 2015). Automatic data analysis and material classification algorithms are required for applications where access to nuclear spectroscopy expertise is limited and/or the autonomous robotic operation is necessary. Data obtained with neutron-based systems differ from elemental composition evaluations based on chemical formulae due to statistical nature of nuclear reactions, the presence of shielding and cladding, and other environmental factors. Experimental data that are produced by the spectral decomposition can be expressed graphically as sets of overlapping classes in a multi-dimensional space of measured elemental intensities. Results of application of these methods to data sets obtained for a pulsed 14-MeV neutron generator-based active interrogation system were discussed in Barzilov and Novikov (2015).

3.2.12 Techniques with Neutron Interaction and Lepton Detection

With neutron interaction and neutrino detection it is possible to discriminate passive antineutrino generation from those generated by fission (Reaction 1.4) followed by short life beta decay and (Eq. 3.2) and antineutrino production. The nuclear fission process that occurs in the core of nuclear reactors results in unstable, neutron-rich fission products that subsequently beta decay and emit electron antineutrinos. However, what really distinguishes active from passive modes is the energy of the antineutrino spectra. As seen in Sect. 3.1.3, the shorter the β decay half-life the larger the β and antineutrino energies (see Fig. 3.8).

The prediction of the absolute flux and energy spectrum of the emitted reactor neutrinos hinges upon a series of seminal papers based on measurements performed in the 1970s and 1980s. Details on neutrino physics may be found in the work of Fukugita and Yanagida (2003). The recent reconsiderations of the agreement between the predicted and observed reactor neutrino flux motivates revisiting the underlying beta spectra measurements. A method of fission product beta spectra measurements for predicting reactor anti-neutrino emission has been proposed (see Asner et al. 2015). The requirements for unfolding the beta spectra measured with this method into a predicted reactor neutrino spectrum are explored.

Existing detectors

The detection methods are presented in Sect. 3.1.5. The two main detectors in the world are presented below.

KamLAND (Kamioka Liquid Scintillator Antineutrino Detector) is a 1.0 kiloton detector located at the Kamioka Observatory in Japan. Borexino is a 0.3 kiloton detector at Laboratori Nazionali del Gran Sasso near L'Aquila, Italy.

Several neutrino experiments are ongoing in various countries. Among them, Double Chooz is a reactor neutrino experiment which measures the neutrino mixing angles (see Suekane and Junqueira de Castro Bezerra 2016). Having improved the measurement of mixing angles, the Double Chooz group presented an indication of disappearance of the reactor neutrinos at a baseline of $\sim$1 km for the first time in 2011. At the Daya Bay reactor, neutrino experiments are also in progress (see Cao and Luk 2016).

Nuclear reactor results are very precious.

The KamLAND detector monitoring data from 2002 to 2012 provided by Shimitzu (2013) in Fig. 3.60 show clearly that when the Japanese reactors are in operation, e.g. through 2005–2006 the neutrino counting rate for the energy window of 2–8 MeV (see KamLAND neutrino spectrum in Fig. 3.60b) ranges between about 8× and 10 × 10^{10} cm^{-2} d^{-1}. After earthquakes, for safety reasons Japan shuts some reactors, e.g. after 16 July 2007 (detected flux down to 4 × 10^{10} cm^{-2} d^{-1}) and after the Japanese Earthquake and subsequent Fukushima accident the flux went down to 0.3 × 10^{10} cm^{-2} d^{-1} when all Japanese reactors were off Gando et al (2013) (Fig. 3.60).

More recently, the combination of antineutrino data from both KamLAND and Borexino detectors allowed tomographic reconstruction of the world image of the reactor neutrino map (see Usman et al. 2015a). Figure 3.61 shows by orthographic projection the two hemispheres of the Earth, and the various neutrino sources: the nuclear power plants (NPP) in activity revealed thanks to their neutronic activities and their specific neutrino flux. Only the NPPs are revealed, the research reactors with low power cannot be detected. Visible in the Northern Hemisphere are the European, North American reactors and Asiatic reactors in total: 430. In the Southern hemisphere, only the South African reactors (2) and South American reactors (3 Argentina, 2 Brazil) are detected. Australia is free of nuclear power plants.

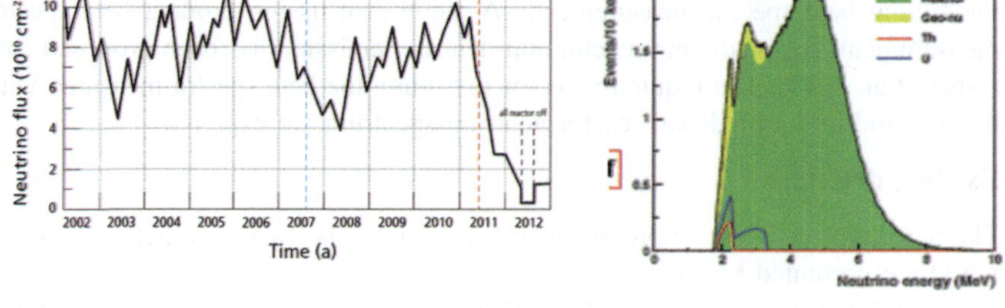

Fig. 3.60 **a** Antineutrino monitoring from KamLAND for spectra window range (2–8 MeV) and, **b** neutrino spectrum during period with reactors in operation. Adapted from Shimizu (2013) with permission © APS

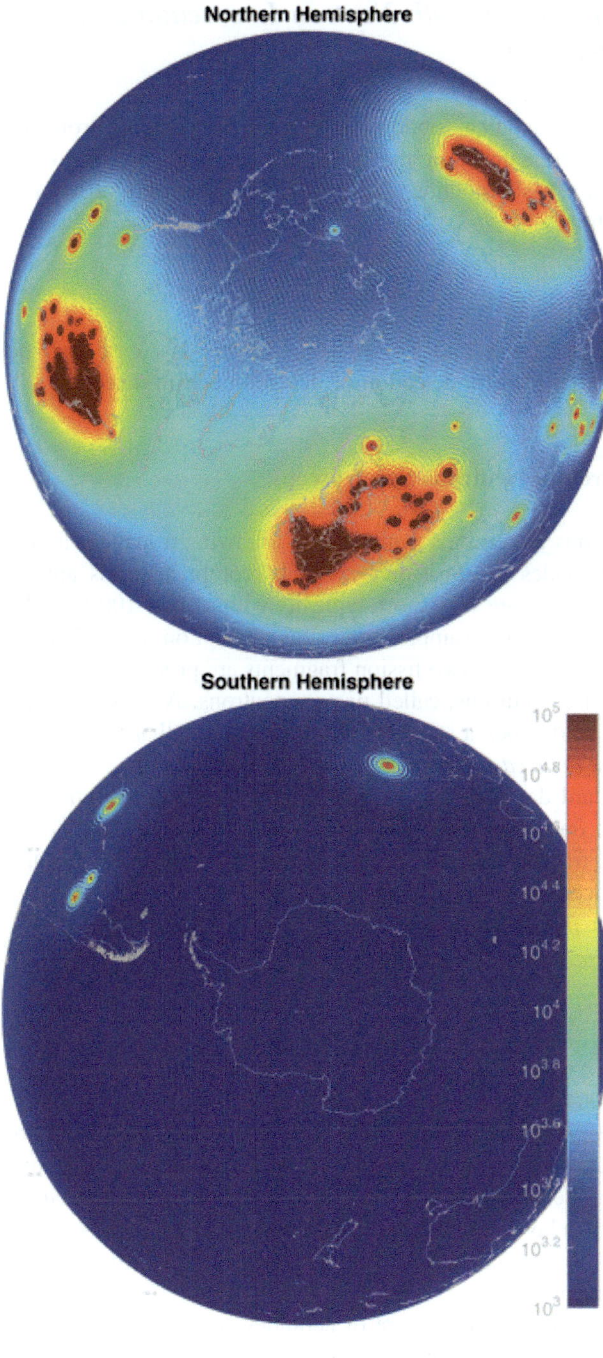

Fig. 3.61 Antineutrino tomogram gained by combining data from KamLAND and Borexino for spectra window range (3 MeV). *Note* the neutrino oscillation wave length of 100 km. Correction: global neutrino flux $cm^{-2}\ s^{-1} \Rightarrow 10^{-2}\ cm^{-2}\ s^{-1}$. Usman et al. (2015b) with permission © Scientific Report

3.2.13 Techniques with Neutron Interaction and Neutron Detection

There are two analytical options: either the incident neutron is detected after interaction in the sample or the neutron that is detected is induced by nuclear reaction from a fissile nuclide. The production reaction is then given by a fission reaction such as:

$$ {}_{Z}^{A}X + {}_{0}^{1}n \rightarrow {}_{Z_1}^{A_1}X + {}_{Z_2}^{A_2}X + n\,{}_{0}^{1}n \qquad \{3.23\} $$

with $A = A_1 + A_2 + n$ and $Z = Z_1 + Z_2$. In this last case two possibilities are reported: prompt neutron detection or delayed neutron detection.

3.2.13.1 Prompt Neutrons Spectroscopy

Prompt neutrons are emitted immediately following a nuclear fission event. There are different modes of emission by which prompt neutrons are released but the timescales involved are very short, typically within 10 or 100 fs of the fission event.

Using ^{235}U as an example, after the absorption of a thermal neutron and immediate fission event, two fission fragments are produced. These fragments emit two or three free neutrons, called *prompt* neutrons. A subsequent fission fragment occasionally undergoes a stage of radioactive decay that yields an additional neutron, called a *delayed* neutron.

The standard deviation of the final kinetic energy distribution as a function of mass of final fragments from low-energy fission of ^{234}U and ^{236}U, presents a peak around light fragment masses region and another around heavy fragment masses region produced by prompt neutron emission.

The factor β is defined as:

$$ \beta = \frac{n_{at}}{n_n + n_{at}} \qquad (3.47) $$

with n_n the number of prompt neutrons and n_{at} the number of precursor atoms. β is equal to 0.0064 for ^{235}U.

If a nuclear reactor happened to be "prompt critical" the number of neutrons and power output would increase exponentially at a high rate. The response time of control rod systems would be too slow to moderate this potential criticality surge. However, the delayed neutrons allow the reactor to keep sub-critical as far as only prompt neutrons are concerned: the delayed neutrons come later, just in time to sustain the chain reaction.

Prompt fission neutron spectra of actinides are known and reported by Capote et al. (2016), and can be used for analytical purposes.

The spectrum of neutrons emitted promptly after ^{239}Pu(n,f)—the so-called prompt fission neutron spectrum—is a quantity of significant interest, for instance, for reactor physics and global security (see Neudecker et al. 2016). However, there are only a few experimental data sets available that are suitable for evaluations. Given the possibly substantial distortion of the prompt fission neutron spectrum by multiple scattering and background effects, special care should be taken to reduce these effects in future measurements, e.g. by measuring the ^{239}Pu prompt fission neutron spectrum as a ratio to either the ^{235}U or ^{252}Cf prompt fission neutron spectrum.

3.2.13.2 Delayed Neutrons Spectroscopy

Delayed neutron decay may occur after beta decay of one of the fission products at any time from a few milliseconds to a few minutes after the fission event. Delayed neutrons are important for reactor control.

The majority of neutron emissions outside prompt neutron production are due to induced or spontaneous fission from neutron-heavy isotopes produced as the products of neutron activation in the reactor. These neutrons are sometimes emitted with a delay but the actual delay in their production is a delay waiting for the beta decay of fission products to produce the excited-state nuclear precursors that then immediately undergo prompt neutron emission. The beta decay half-lives for the precursors to delayed neutron-emitter radioisotopes are typically fractions of a second to tens of seconds.

The delayed neutron fraction (*DNF*) is defined as:

$$DNF = \frac{n_{\mathrm{dn}}}{n_{\mathrm{pn}} + n_{\mathrm{dn}}} \tag{3.48}$$

with n_{dn} and n_{pn} are the number of delayed and prompt neutrons respectively. These two factors, β and *DNF*, are not the same in case of a rapid change in the number of neutrons in the reactor. For these investigations time-of-flight scattering can be performed at either a research reactor or a spallation source (see Sect. 3.1.5).

Delayed neutron activation analysis (DNAA) applies delayed neutron detection (see Sect. 3.1) after nuclide activation by neutrons. This technique is very sensitive to the detection of fissile isotopes. It has been compared to other techniques for ^{232}Th, ^{238}U and ^{239}Pu and ^{240}Pu analysis (Jaiswal et al. 1994; Alfassi 1990). DNAA was performed for isotopic analysis of uranium in safeguard studies (Papadopulos and Tsagas 1994), and was applied to uranium analysis in sediments (Noller and Hart 1993).

3.2.13.3 Neutron Diffraction and Scattering

Neutron diffraction (ND) is the application of neutron elastic scattering to the determination of a material atomic and/or magnetic structure. The sample investigation is placed in a thermal or cold neutron beam to obtain a diffraction pattern. The data arising from this approach provide information about the material structure. ND is similar to XRD (see Sect. 3.2.3.13) but due to their different scattering properties, neutrons and X-rays provide complementary information. Detailed information on neutron scattering applications and techniques may be found in the series edited by Anderson et al. (2016).

Like other particles, neutrons can exhibit wave-like phenomena (Eq. 3.42). Diffraction is one of these phenomena; it occurs when waves encounter objects whose size is comparable with the wavelength. If the wavelength of a particle is sufficiently short, atoms or their nuclei can serve as diffraction objects. When a beam of neutrons emanating from a reactor is slowed down and selected by their speed, their wavelength lies near 100 pm, i.e. the same order of magnitude as the atomics separation in solid materials. Impinging such a beam on a crystalline sample will scatter the incident monochromatic neutrons under a limited number of well-defined angles. This technique has been used to characterize inert matrix fuel, as in Burghartz et al. (2001).

Neutron diffraction and scattering require a source of neutrons such as a nuclear reactor or a spallation source. The other components needed are a crystal monochromator as well as filters to select the desired neutron wavelength or a time-of-flight spectrometer to sort the energies of the incident neutrons. Neutron diffraction is typically performed as a powder diffraction test, which requires the sample to be in the form of a polycrystalline powder.

The main disadvantage to neutron diffraction is the requirement for a nuclear reactor. For single crystal work, the technique requires relatively large crystals, which are usually challenging to grow. The main advantages to the technique are many, such as sensitivity to light atoms, ability to distinguish isotopes, absence of radiation damage.

Neutron diffraction can be used to determine the static structure of gases, liquids or amorphous solids. Most experiments, however, are aimed at improving understanding of the structure of crystalline solids. Although neutrons are uncharged, they carry a spin, and therefore interact with magnetic moments, including those arising from the electron cloud around an atom. Neutron diffraction can therefore be used to reveal the microscopic magnetic properties of a material.

Small Angle Neutron Scattering (SANS) is the application of neutron elastic scattering to the determination of the nano structural distribution of nanopore or nano-heterogeneity. The distance between object and detector can be of the order of several meters. The detection is carried out on a 2D detector and the distance between the center of detector (intercept incident beam) and the neutronic signal on the detector (see De Geuser and Deschamps 2012).

The example on nuclear material analysis deals with small angle neutron scattering (SANS) for investigations on nano-porosity in simulated stabilized zirconia inert matrix fuel, as per Degueldre et al. (2010). The samples of $(Er_{0.07}Y_{0.10}Ce_{0.10}Zr_{0.68})O_{1.89}$ have a porosity 8% with large distribution of pores as anticipated from ceramographic observations. The investigations with 2 detector positions for 2 sample thicknesses (0.959 and 2.950 mm) to track multi-scattering

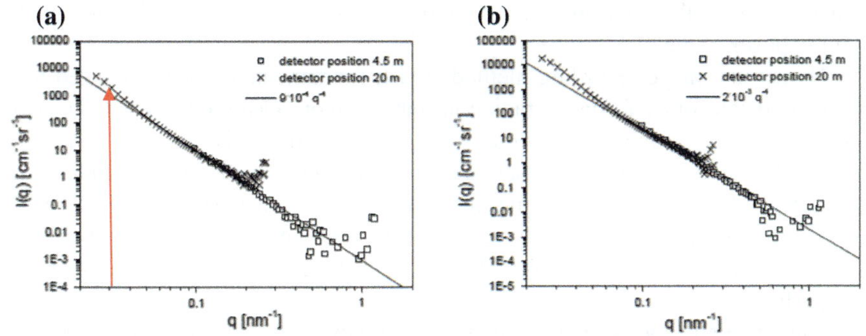

Fig. 3.62 Use of SANS for investigating nano-porosity in simulated inert matrix fuel. Samples of $(Er_{0.07}Y_{0.10}Ce_{0.10}Zr_{0.68})O_{1.89}$ with porosity 8%. Investigations with 2 detector positions for 2 sample thicknesses **a** 0.959 mm, **b** 2.950 mm. Detection of pores around 300 nm *up arrow* Degueldre et al. (2010) with permission © Elsevier

are displayed in Fig. 3.62. Detections of pores up to around 300 nm and below are possible. The investigations need to be complemented by ceramography or tomographic investigations to cover the full spectrum of pore size.

3.2.13.4 Neutron Radiography and Tomography

Neutrons may also be used for making images. The resulting image is based on the neutron absorption properties of the imaged object. The resulting images are also based on neutron attenuating properties. Some things are easily visible with neutron imaging, e.g. the presence of either absorbing or scattering elements such as boron and hydrogen respectively (Fritzsche et al. 2016). More information on neutron imaging and applications may be found in the work edited by Anderson et al. (2009).

The first demonstration of neutron radiography was made by Kallmann and Kuhn (1944); they discovered that upon bombardment with neutrons, some materials emitted radiation that could expose film. The first neutron radiographs of reasonable quality were made by Thewlis (UK) in 1955.

Around 1960, Berger (US) and Barton (UK) evaluated neutrons for investigating irradiated reactor fuel. Subsequently, research facilities were adapted. After neutrons are produced, they need to be slowed down (thermalized) to speeds (energies) suitable for imaging. This can be done while travelling, at room temperature, through water, polyethylene, or graphite to make them "thermal neutrons", i.e. in thermal equilibrium with their surroundings. To produce an image of good quality, neutrons need to be travelling parallel together. This is usually achieved with an aperture to limit the proportion of neutrons to those constrained in trajectory by the axis of the collimator. Collimator structures are made with neutron absorbing materials to eliminate neutrons that are not traveling in the desired direction.

Finally, when taking the image, a compromise exists between radiography quality and exposure time.

Numerous image recording methods exist; fundamentally, neutrons need to be converted into other forms of radiation that are more easily detected. Conversion screens may be employed, while some imaging methods use the conversion material in the image recorder itself. This may include a micro-layer of gadolinium, e.g. 25 μm Gd absorbs 90% of thermal neutrons. Other elements such as B, In, Au, or Dy may be used or materials such as LiF scintillation screens where the conversion screen absorbs neutrons and emits visible light. Image recording may then be performed.

Until recently, neutron imaging was generally recorded on X-ray film, but a variety of digital methods are now available. A neutron camera is an imaging system based on a digital camera or similar detector array. Neutrons pass through the object to be imaged by a scintillation screen that converts the neutrons to visible light and image. This light then passes through some optical processing and the image is then captured by the CCD camera. Neutron cameras can provide real-time images. This imaging system in conjunction with a rotary table can take a large number of images at different angles that can be reconstructed into a 3-D image or **neutron tomography** (NTOM).

Neutrons can also be measured with a digital detector array with small pixel sizes. The device has a 2D array of small, μm-scale channels, each with a micro-detector coated with a neutron absorber (e.g. Gd or B). The absorbing material converts neutrons in ionizing radiations that free electrons. A large voltage is applied across the detector, accelerating and amplifying the freed electrons through the small channels before their detection on a digital detector array.

The example of neutron radiography cases selected for illustration deals with fuel cladding and inert matrix fuel.

Neutron radiography and tomography have been used to investigate of the secondary hydriding of Zircaloy-4 during simulated loss of coolant nuclear accidents, as reported by Grosse et al. (2013). This study has been performed in the frame of the post-test examinations of large-scale Loss Of Coolant Accident simulation tests at the fuel rod bundle scale. The hydrogen distributions in specimens prepared from specific tests were studied by means of neutron radiography and tomography. In order to determine quantitative hydrogen concentrations, both neutron radiography and tomography were calibrated using cladding tube segments with known hydrogen concentrations. The linear dependence of the total macroscopic neutron cross-section with the H/Zr atomic ratio was determined for both methods.

The second example deals with the study of a zirconia-based inert matrix fuel under irradiation, as reported by Degueldre and Hellwig (2003). The inert matrix fuel (IMF) is an yttria-stabilized zirconia doped with erbia and plutonia. The results of experimental irradiation experiments on tests with yttria-stabilized zirconia doped with plutonia and erbia or urania pellets within research reactors are evaluated and compared. The reactor tests performed in the material tested at the Boiling Water Reactor, Halden and in the High Flux Reactor, Petten were

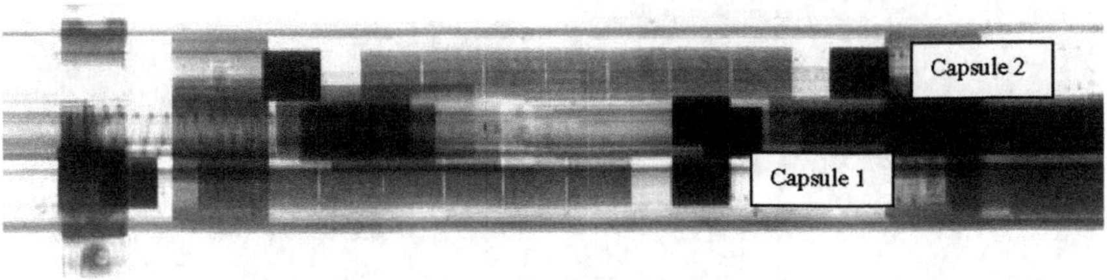

Fig. 3.63 Neutron radiography of two IMF segments during the OTTO experiment IMF composition: **(Er, Y, Am, Pu, Zr)O_{2-x} Capsule 1** and **(Y, Am, Pu, U, Zr)O_{2-x} Capsule 2**. Pellet diameter 8.00 mm, stack length 67.0 and 67.7 mm, density of plutonium fissile at beginning of life: 0.37 and 0.34 g cm^{-3}, density 5.80 and 6.02 g cm^{-3} respectively. Image obtained after one cycle. Degueldre and Hellwig (2003) with permission © Elsevier

described, with an emphasis on the IMF properties under in-pile irradiation. Neutron radiography investigations were performed on two IMF segments during the one through then out (OTTO) experiment at the Petten reactor. The IMF composition was: **(Er, Y, Am, Pu, Zr)O_{2-x}** and **(Y, Am, Pu,** U, **Zr)O_{2-x}** Capsule 1 and 2 respectively (see Fig. 3.63). Pellet diameter 8.00 mm, stack length 67.0 and 67.7 mm, density of plutonium fissile at beginning of life: 0.37 and 0.34 g cm^{-3}, density 5.80 and 6.02 g cm^{-3} respectively. This image was obtained after one cycle and revealed the excellent integrity of the fuel and the insignificant swelling of the material (Fig. 3.63).

3.2.14 Techniques with Neutron Interaction and Ion Detection

3.2.14.1 Thermal Neutron-Induced Fission Track Technique

The thermal neutron-induced fission track technique allows localization of fissile or absorbing nuclides at the micron scale in a material matrix subsurface. Based on the fission reaction {3.23}, the fission product ions are spatially detected using a sensitive film or a specific 2D detector during neutron irradiation or shortly after. Detectors for fission track analysis may be used apatite and mica (muscovite). Alpha track and fission track are somewhat similar and detectors as described in Sect. 3.1.5.2 may be used. The resolution is limited by the pathway of the fission product ions in the materials as well as in the detector. The detection limits can also be evaluated. Neutron-based analysis of fission rates and ultra-trace concentrations of ^{235}U using CR-39 as plastic track detector has been reported by Moharram et al. (2002).

Single-crystal chemical vapor-deposited diamond was investigated for its ability to act as fission-fragment detector (see Frégeau et al. 2015). In particular, timing and energy resolution were investigated for application in a simultaneous

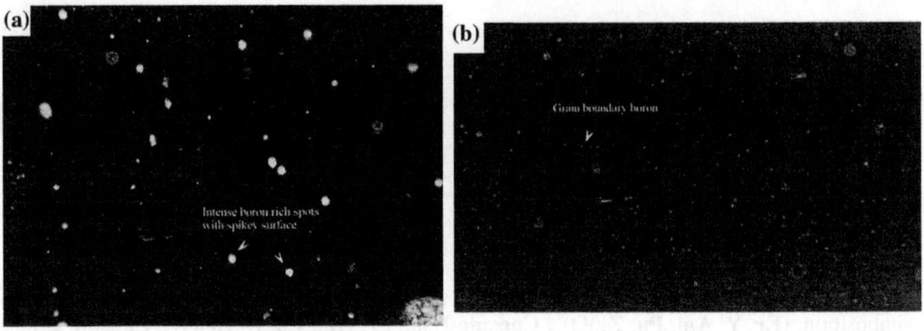

Fig. 3.64 Boron autoradiographs for: **a** VSB plate; and **b** Fortiweld plate, showing a heterogeneous distribution of B-rich particles and no evidence for grain boundary B segregation. Jones et al. (2002) with permission © Elsevier

time-of-flight and energy measurement to determine the mass of the detected fission fragment. Previous tests showed that poly-crystalline chemical vapor deposited diamonds provided sufficient timing resolution, but their poor energy resolution did not allow complete separation between very low-energy fission fragments, α-particles and noise. Present investigations proved artificial single-crystal chemical vapor-deposited diamonds can provide similar timing resolution to poly-crystalline chemical vapor deposited diamonds close to 100 ps. However, the strongly varying quality of the presently available diamond material does not allow application on a sufficiently large scale at a reasonable level of investment.

The examples on nuclear materials are numerous. Uranium trace analysis in steel and other materials has been carried out using solid state nuclear track detectors (see Chakarvarti et al. 1980). In another example, the effect of the microscale distribution of boron on the yield strength of C–Mn steels subjected to neutron irradiation has been studied by Jones et al. (2002). Figure 3.64 presents boron autoradiographs for one VSB steel plate; and a Fortiweldsteel plate. Both metals contain 1 ppm $\leq$ [B] $\leq$ 48 ppm. However, autoradiographs show a heterogeneous distribution of boron with boron-rich particles in the bulk of the steel and no evidence for grain boundary boron segregation. These particles strongly affect the yield strength of the steel in post-irradiation conditions.

3.2.14.2 Fast Neutron-Induced Fission Track

Fast neutrons may also be used to induce fission tracks and to apply techniques allowing localization of fast neutron fissile nuclides or specific nuclide recoil detection at the micron scale in sub-surfaces of nuclear materials.

Specific examples may are found in the literature. The Spatial distribution of thorium fission rate in a fast spallation and fission neutron field has been performed experimentally and compared with modeling (Monte Carlo) predictions by Borger et al. (2012).

Fast detection of neutrons in dedicated nuclear track detectors were investigated using new etching conditions (see Abdalla et al. 2015). The neutron irradiation is performed using a 5 mCi Am–Be source. Using the new etching condition, irradiated samples were etched for 4 h or 80 min. Suitable analyzing software has been used to analyze experimental data. The dependence of neutrons track density on the neutrons fluence is investigated. When etched under optimum conditions, the relationship between track density and fluence is determined, which is found to be linear. Detection efficiency has been represented for both sample types and found to be constant with fluence, which reflects the importance of using adequate detectors in the field of neutron dosimetry. Linear relationship between track density and effective dose is determined.

Neutron Elastic Recoil Detection Analysis (**NERDA**) has also been applied (theory see ERDA Sect. 3.2.18.8) to analyze light elements in specific matrices. Neutron elastic recoil detection has been applied for hydrogen isotope analysis in fusion materials (see Kubota et al. 2006, 2007). NERDA using a 14.2 MeV neutron beam has been developed to measure depth profiles of hydrogen isotopes up to several hundred micrometers under low background conditions. The hydrogen profile from the surface to 800 μm had three components, while a few counts of deuterium were detected. The detection limit of about 60 ppm was deduced for deuterium. In the experiment, the energy spectrum of recoil-charged particles from a sample irradiated by a neutron beam from the DT fusion neutron source is recorded. A depth profile is computed from the spectrum numerically processed with this response function. Hydrogen depth profiles of inner and outer baffle plates exposed to DD plasma were measured as an application of NERDA. The profiles from the surface to 800 μm were measured (Fig. 3.65). It is found that hydrogen is more concentrated in the region of more than 200 μm depth.

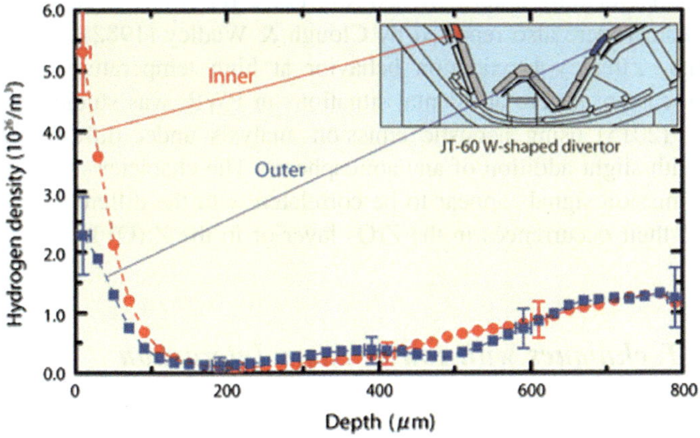

Fig. 3.65 Hydrogen depth profile of JT-60 from wall into the plasma. Kubota et al. (2006) with permission © Elsevier

3.2.14.3 Neutron Elastic Recoil Detection Analysis

Interactive acoustic emission (AE) is derived from AE (see Sect. 3.2.1) and is occurring during the processes of mechanical indentation/loading (e.g. nano- / macro-probe) or of chemical reaction on materials. The sample structure is accompanied by stress changes that generate local sources of elastic waves. This results in small surface displacements of a material produced by elastic or stress waves generated when the accumulated elastic energy in a material or on its surface is released rapidly. The waves generated by sources of AE are of practical interest in the field of quality control, system feedback and process monitoring.

Acoustic emission (AE) techniques applied to material property evaluation during indentation was tested by Faisal et al. (2013) with emphasis on application and emerging trends. There are two aspects of application of AE technique to indentation which makes it unique, i.e. (1) enhancing the understanding of the evolution of material accommodation mechanisms under loading and (2) qualitative and quantitative evaluation of mechanical properties such as fracture toughness and bond strength from the AE signal.

Acoustic emission measurements from a range of non-irradiated and advanced gas-cooled reactor (AGR) irradiated graphite fuel sleeves after discharge during three-point bend tests reported by Burchell et al. (1986). The AE information is explained by considering separately the known changes in graphite microstructure that occur in the reactor due to radiolytic oxidation and fast neutron irradiation.

Stress waves excited by impacts of rigid bodies were studied by Petersen (2009) in different applications of acoustic non-destructive testing. In the practice of AE testing, the impacts are considered as mechanical interferences and fracture process. However, since an impact itself presents a danger to the structural integrity, estimation of the impact hazard and effect is required. The latter consideration is of great interest in nuclear reactor industry.

Indentation loading studies of acoustic emission from temper and hydrogen embrittled steel were also reported by Clough & Wadley (1982).

Recently, Zircaloy-4 oxidation behavior at high temperature, which can be reached in case of severe accidental situations in PWR, was studied in situ by Al Haj et al. (2015) using acoustic emission analysis under different helium and oxygen (with slight addition of air) atmospheres. The characteristic features of the acoustic emission signals appear to be correlated with the different populations of cracks and their occurrences in the ZrO_2 layer or in the Zr(O) layer.

3.2.15 Techniques with Ion or Atom Interaction and Phonon Detection

The techniques derived from the ion reaction of the sample with particle detection make use of the potential of particle reemitted. Among the techniques listed in

Table 3.1, ion-phonon techniques concern the acoustic emission itself. The sources of ions/atoms are electrodes, ion guns, accelerators or the ions themselves in solution.

3.2.15.1 Interactive Atom-Phonon techniques

Interactive acoustic emission (AE) is derived from AE (see Sect. 3.2.1) and is occurring during the processes of mechanical loading (e.g. with atom, nano- or macro-probe) of materials. The sample structure is accompanied by stress changes that generate local sources of elastic waves. This results in small surface displacements of a material produced by elastic or stress waves generated when the accumulated elastic energy in a material or on its surface is released rapidly The waves generated by sources of AE are of practical interest in the field of quality control, system feedback, process monitoring. In applications, interactive AE is typically used to detect, locate and characterize damage, e.g. Hamstad (1986).

More recent research has focused on using AE to not only locate but also to characterize the source mechanisms, i.e. crack growth, friction, delamination, matrix cracking, etc. This would give AE the ability to tell the end user what source mechanism is present and hence allow them to determine whether or not structural repairs are necessary (see Zeng et al. 2000).

Acoustic emission measurements from a range of four unirradiated nuclear graphites during three-point bend tests are reported. The technique was applied to the testing of small beam specimens cut from irradiated Advanced Gas-cooled Reactor (AGR) graphite fuel sleeves after discharge from the reactor (see Burchell et al. 1986). The AE information is explained by considering separately the known changes in graphite microstructure that occur in the reactor due to radiolytic oxidation and fast neutron irradiation. Coarsening of the material due to radiolytic oxidation increases the total number of AE events and the proportion of events of low amplitude. Fast neutron irradiation increases the fracture stress and makes the stress-strain curve more linear. As a consequence, the number of AE events is reduced along with the proportion of events of low amplitude.

3.2.16 Techniques with Ion or Atom Interaction and Photon Detection

The analyses derived from the ion interaction with sample and particle detection benefit of the large potential of the produced particles. They could be listed as ion-photon techniques (see Table 3.1) such as: inductively coupled plasma optical emission spectroscopy (ICPOES), proton-induced X-ray emission spectroscopy (PIXE) and particle-induced gamma radiation emission (PIGE). The sources of ions are electrodes, ion guns, accelerators or the ions in solution.

3.2.16.1 Inductively Coupled Plasma Atomic Emission Spectroscopy

Optical Emission Spectroscopy (OES) is preceded by an excitation mode that can be provided by a flame, a plasma torch, an arc or a spark, from where atom emission takes place. Today because of its flexibility, the plasma torch has replaced the other sources.

Inductively Coupled Plasma Optical Emission Spectrometry (ICPOES), also referred to as Inductively Coupled Plasma Atomic Emission Spectroscopy (ICPAES), is an analytical emission spectroscopy that can be used for the detection of trace elements. It is a technique that uses ICP to excite ions and emit lines characteristic for the element analyte. It is equivalent to a flame technique with a temperature ranging from 6,000 to 10,000 K. It is also a solution technique that has also been used for single particle analysis (see in ICPMS Sect. 3.2.18.4). The intensity of this emission is indicative of the element concentration within the sample of solution. Information may be found in Thompson (1989).

The ICPOES is composed of three parts: an injector nebulizer, the plasma torch and the optical spectrometer. The ICP torch consists of 3 concentric quartz glass tubes.

A peristaltic pump delivers an aqueous or organic sample into an analytical nebulizer where it is changed into mist and introduced directly inside the plasma flame. Nebulizer converts liquids into an aerosol, and that aerosol can then be swept into the plasma to create the ions.

The output of the radio frequency generator surrounds part of the quartz torch. Argon gas is used to carry the sample and excite it in the generated plasma. The plasma used in an ICP is made by partially ionizing argon gas ($Ar \rightarrow Ar^{+} + e^{-}$). The carrier gas is sent through the central channel and into the hot plasma. The sample is then exposed to radio frequency which converts the gas into plasma. The stable high temperature (~7000 K) of the plasma causes ionization of a large fraction of the sample. This fraction of ionization approaches 100% for the elements (e.g. sodium), depending on the ionization potential. The sample immediately collides with the electrons and charged ions in the plasma and is itself broken down into charged ions. The various molecules break up into their respective atoms which then lose electrons and recombine repeatedly in the plasma, giving off radiation at the characteristic wavelengths of the elements involved.

The spectrum is recorded using an optical emission spectrometer. In the spectrum, the intensity of the emission line is theoretically given by:

$$I = h\left(\frac{c}{\lambda}\right) A N_0 \left(\frac{g_k}{g_0}\right) \exp\left(\frac{-\Delta E}{kT}\right) \tag{3.49}$$

with h Plank's constant, k Boltzmann's constant, A the Einstein probability coefficient for spontaneous emission, c the speed of light, λ the wavelength of the emitted line, g_k/g_0 the ratio of statistical weights associated to upper and ground states for the corresponding transition, ΔE their energy difference. The intensity of the line is proportional to the concentration and is a priori decreasing with temperature. Equation 3.49 can also be used to estimate the temperature of the plasma.

The examples on nuclear materials are firstly dealing with the analysis of Np and secondly with the analysis of the uranium isotopes.

A reliable and straightforward ICPOES procedure for neptunium analysis of samples potentially originating from spent nuclear fuel has been reported by Krachler et al. (2014). In contrast to most other analytical techniques employed for the determination of Np, analysis may be performed without separation of the residual elements present in the investigated solution. The Np ICPOES analytical line has been calibrated using a standardized ^{237}Np stock solution. The investigated Np emission wavelengths revealed complex ICPOES spectra with either broad peaks having widths of up to 30 pm or splitting of the ^{237}Np signal into multiplets. Np detection limits (DL) were lowered by at least one order of magnitude. The most intensive Np emission line at $\lambda = 382.92$ nm provided the lowest DL of 3.1 μg kg^{-1}; however, it may be of limited practical use because severe spectral interference from Nd, significantly present in spent fuel, at $\lambda = 382.914$ nm, overlaps the Np signal. Additional detailed interference studies helped identify potential complications during Np analysis caused by concomitant elements (Nd, Pu, Th, and U) in the analyte solution. Finally, emission signals at 410.84, 429.09 and 456.04 nm were used to analyze selected salt and deposit samples from pyrochemical separation experiments.

ICPOES analysis can also provide isotopic information (Fig. 3.66). The so-called isotopic shift is caused by atomic transitions of various isotopes of the same element emitting light at slightly different wavelengths. The pioneering work was carried out as early as 1981 (see Edelson and Fassel 1981), being later applied to the nuclear field. The isotopic shift of 25 pm between the ^{235}U and ^{238}U ICPOES emission signals at $\lambda \sim 424.4$ nm, for example, allows for the analysis of depleted, natural and enriched U samples (see Krachler et al. 2015). Benefiting from the high optical resolution of HR-ICPOES of <5 pm, baseline separated emission signals for the two U isotopes are achieved easily. As highlighted in Fig. 3.66, even the emission signals of ^{233}U, ^{235}U, and ^{238}U can be separated from each other without difficulty at $\lambda \sim 411.6$ nm (but also at $\lambda \sim 424.4$ nm) using commercial HR-ICPOES instruments.

3.2.16.2 Particle-Induced X-Ray Emission

Particle-Induced X-ray Emission (PIXE) or Proton-Induced X-ray Emission is a technique used in the determining of the elemental make-up of a material or sample. When a material is exposed to an ion beam, atomic interactions occur that give off radiation of wavelengths in the X-ray light spectrum specific to an element. PIXE is a powerful yet non-destructive elemental analysis technique now used routinely by geologists, archaeologists, art conservators and others to help answer questions of provenance, dating and authenticity (see Johansson et al. 1995).

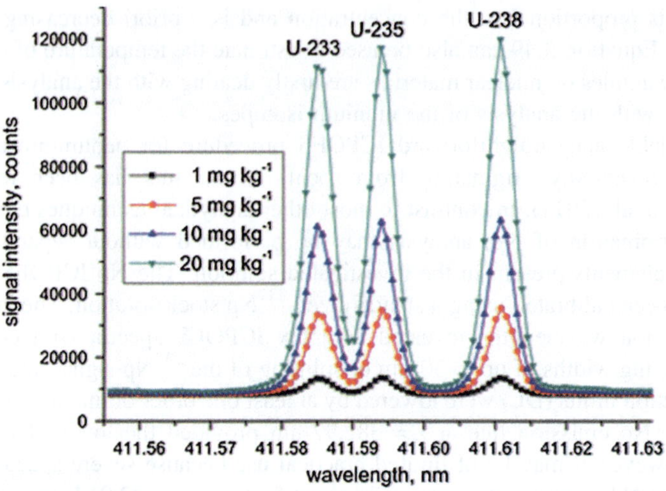

Fig. 3.66 HR-ICPOES spectra of the certified reference material EC-NRM-199 having almost identical isotope amount fractions of ^{233}U, ^{235}U, and ^{238}U. Spectra have been recorded using the optimized instrumental set-up (see text for details) at four different concentration levels. Reported concentrations refer to the total amount of U, i.e. concentrations of individual U isotopes amount to only ~1/3 of this total U concentration. Krachler et al. (2015) with permission © Springer

Extensions of PIXE using tightly focused beams (down to 1 μm) give the additional capability of microscopic analysis. This technique, called **microPIXE**, can be used to determine the distribution of trace elements in a wide range of samples.

Bombardment with ions of sufficient energy—e.g. MeV protons, produced by an ion accelerator—cause inner shell ionization of ion or atoms in a specimen. Outer shell electrons drop down to replace inner shell vacancies; however, only certain transitions are allowed. X-rays of a characteristic energy of the element are consequently emitted. An energy-dispersive detector is used to record and measure these X-rays.

Only elements heavier than fluorine can be detected. The lower detection limit for a PIXE beam is given by the ability of the X-rays to pass through the window between the chamber and the X-ray detector. The upper limit is given by the ionization cross section, the probability of the K electron shell ionization; this is maximal when the velocity of the proton matches the velocity of the electron (10% of the speed of light), and therefore 3 MeV proton beams are optimal.

The nuclear material example deals with the characterization of material for fusion device.

The PIXE technique was applied to the analysis of impurity elements deposited on the surface probe and the fixed ring limiters of a tokamak/stellarator fusion device (Amemiya et al. 1984). Various elements were identified as impurities contaminating the plasma discharge. Heavy elements were originated from the composite materials of a vacuum vessel and limiters. The Fe and Mo ion flux along the toroidal field was 5×10^{13} cm^{-2} per discharge, which was about five times larger than for an another direction in the scrape-off plasma.

3.2.16.3 Particle-Induced Gamma-Ray Emission

Proton-Induced Gamma-ray Emission (**PIGE**) analysis is based on the detection of the prompt gamma rays emitted from nuclei that are in an excited state following a charged particle-induced nuclear reaction. The energy of the gamma ray is indicative of the isotope present and the intensity of the gamma ray is a measure of the concentration of the isotope in the sample. PIGE is a rapid, non-destructive technique that, in principle, could be used in the analysis of any element (more specifically an isotope). In practice, PIGE was utilized in the analysis of light elements Li, B, and F which are often difficult to determine by other analytical techniques. Because the technique is based upon specific nuclear reactions, the sensitivity of PIGE varies greatly from isotope to isotope (Gurbic 2014). In addition, the sample matrix also influences the detection limit of an element because of variations in the background of the gamma ray spectrum. Detailed information on Proton-Induced Gamma Ray Emission can be found in the monograph proposed by Vizkelethy (2003).

The nuclear example deals with the analysis of elemental signatures in uranium dioxide samples by PIGE utilized as a tool in nuclear forensics (see Apostol et al. 2016). This non-destructive technique was applied on a Tandem accelerator with 3 or 5 MeV protons and detection of gamma-rays using high purity germanium detectors. For accurate analysis, the spectra of the analyzed samples of (HEU) UO_2 had to be compared with the non-irradiated sample target and the natural background spectrum. Determination of Li, F, from 1 to 10 ppm Na, Zn, Al, from 10 to 100 ppm and Si, Mg, Zr, Ti, Cr, Fe around 1000 ppm as well as Si, and P, Cr and Ti was carried out for 3 h irradiation and 3 h counting.

3.2.17 Techniques with Ion Interaction and Neutron Detection

The techniques derived from the ion reactions/interactions on a sample with particle detection make use of the large potential of particle type produced or reemitted. Only NRA is listed as ion-neutron technique (Table 3.1); see Sect. 3.2.18.9 for details.

Neutron-induced NRA may be used to provide information on nuclear material in the environment using the reaction between high energy ions (e.g. from cosmic ray) and "heavy metal" (e.g. Th, U) to provide information on the latter.

A pulsed proton beam from the accelerator was used to induce prompt neutron and delayed neutron in uranium sample (see Morse et al. 2007). The signals in uranium were measured with a ^{3}He detector. It was shown that the accelerator data is in good agreement with Monte Carlo radiation transport calculations and published results.

3.2.18 Techniques with Ion or Atom Interaction and Ion Detection

The techniques derived from the ion reaction/interaction into sample and with ion detection make use of the large potential of particle type produced or reemitted. They could be listed from ion–ion to atom–atom (Table 3.1). The ion–ion and/or atom-atom analytical modes are adapted in several powerful techniques such as, by increasing energy: atom force microscope (AFM), nano-indentation (NI), the inductively coupled plasma mass spectroscopy (ICPMS), the secondary ion mass spectroscopy (SIMS), the accelerator mass spectroscopy (AMS), the Rutherford back-scattering (RBS), the elastic recoil detection analysis (ERDA) and the nuclear reaction analysis (NRA). Moreover, volumetry (VoL) is added as representative for the reaction of ions with other ions in solution. The sources of ions are electrodes, ion guns, accelerators or the solated ions themselves in solution.

3.2.18.1 Volumetric Analysis

Volumetric analysis (**VoL**), also known as titrimetry, is a traditional analytical method in chemistry. It is applied to determine a concentration or a quantity of analyte. A reagent prepared as a standard solution (see Sect. 2.3) is used as titrant of well-defined concentration. A fixed volume of analyte solution reacts with a determined volume of titrant.

Titrations require fluid reagents, i.e. titrant and analyte in solution. The solids of titrant and analyte are dissolved into a protic (or aprotic) solvent such as aqueous solution, glacial acetic acid or ethanol. The volumetric tests refer to the reaction between titrant and analyte, namely: acid-base, complexation, oxido-reduction and precipitation. The volumetric analysis is followed by a colored indicator, and quantitative reaction of analyte and titrant is completed at the end point corresponding to the indicator color change. Information on advanced potentiometry, potentiometric titrations and their systematic errors can be found in the monograph by Néher-Neumann (2009).

Different methods to determine the endpoint include.

- pH or pM (-log [M]) meter: a potentiometer with an electrode whose potential depends on the amount of H^+ ion or M^{n+} presents in the solution (for detail see Sect. 3.1.5). The pH or pM of the solution is measured throughout the titration, more accurately than with an indicator; at the endpoint, a sudden change in the measured pH or pM is observed.

- Potentiometer: an instrument that measures the electrode potential, e.g. pe. These are used for redox titrations; the potential of the working electrode suddenly changes as the endpoint is reached.
- Conductivity: ion concentration can change significantly in a titration, which changes the conductivity: for example, during acid-base titration, the H^+ and OH^- ions react to form neutral H_2O.
- Amperometry: measures the current produced by the titration reaction as a result of the oxidation or reduction of the analyte. The endpoint is detected as a change in the current.
- Indicator: a compound that changes color in response to a chemical change. An acid-base indicator (e.g. phenolphthalein) changes color depending on the pH. Redox indicators are used to monitor pe. A drop of indicator solution is added to the analyte solution prior to the beginning of the titration; at the endpoint the color changes.
- Reagents color change: in some reactions, the reagents change color and indicator is not required. This is often the case in redox titrations when different oxidation states of the reagents and their products produce different colors.
- Spectroscopy: used to measure light absorption by the solution during titration if the spectra of the reactants are known. The concentrations can be determined by Beer's law (Eq. 3.8).
- Precipitation: if a reaction produces a solid that precipitates during the titration. The example is the titration between Ag^+ and Cl^- forming the insoluble AgCl. Cloudy precipitates make it difficult to determine the endpoint. To compensate, precipitation titrations often have to be done as "back" titrations.
- Thermometric or calorimetric volumetry: owing to the heat of the reaction, the endpoint is determined by the rate of temperature change or by the heat produced (or consumed).

Acid–base volumetric analysis

An **acid–base titration** allows quantitative analysis of the concentration of an unknown acid or base solution by neutralization with a base or an acid as titrant. The acid base reaction for an actinyl hydroxyde may read for example:

$$\left[AnO_2(OH)_n\right]^0 + H^+ \rightarrow \left[AnO_2(OH)_{n-1}\right]^+ + H_2O \qquad \{3.24\}$$

A simple pH meter can be used to monitor titration reactions (see Sect. 3.1).

The nuclear material example is the pH titration method that has been proposed for the simultaneous determination of uranium, nitrate and free-acid in the feed solution of the sol-gel process of nuclear fuel fabrication, by Kamat et al. (1998). The pH titration curves generated by slow addition of alkali to solutions containing varying concentrations of uranyl nitrate and nitric acid were studied using a PC-linked autotitrator. A procedure with multiple choices of equations has been developed for the estimation of free-acid, nitrate and uranium in pure uranyl nitrate solution by a single titration. The technique provides a simple single-step method with required accuracy and precision for the simultaneous estimation of the three quantities in the uranyl nitrate feed solution of the sol-gel process for making UO_3 microspheres. The relative standard deviations in the determination of uranium and nitrate were ±0.82% and ±1.52%, respectively, in 15 determinations (see Kamat et al. 1998).

Complexometric analysis

Complexometric titration is a form of volumetric analysis in which the formation of a colored complex is used to indicate the end point of a titration. Complexometric titrations are particularly useful for the determination of a mixture of different metal ions in solution. An indicator or a pM electrode is usually used to detect the end-point of the titration. The complexation reaction reads:

$$[AnO_2]^{n+} + L^- \rightarrow [AnO_2L]^{(n-1)+} \quad \{3.25\}$$

The complexometric titration may also be followed as a function of pH for a weak acid ligand Issa et al (1978).

Redox volumetric analysis

The **redox titrant** is a reducing or an oxidizing reagent when the analyte is respectively oxidizing or reducing in character. The redox reaction may read:

$$[AnO_2]^{m+} + nX^{x+} + 2yH^+ \rightarrow \left[AnO_{2-y}\right]^{(m+2y-n)+} + nX^{(x+1)+} + yH_2O \quad \{3.26\}$$

Redox titrations are usually carried out at constant pH and since some redox reactions require molecular rearrangement, energy is required for accelerating their kinetics. Therefor, some redox reactions may need heating of the analyte solution. The titration of the hot solution is carried out to enhance the reaction rate. As an example, volumetric tests with oxalate standard solutions require heating up to 60 ° C to activate the oxidation.

The nuclear examples developed below illustrate the titration of uranium and plutonium solution. The goal is to determine with accuracy and precision the concentration of U and Pu. These volumetric tests are known as Davies and Gray (1964) potentiometric titration. They are applied at the gram sample level, for all types of U compounds (few interferences) after ignition if the organics are present or for pure uranium or plutonium oxides, if impurity correction is required.

A procedure has been adapted which enables uranium to be determined without chemical separation in solutions containing iron, plutonium, nitrate and many other foreign ions, which interfere in conventional redox methods. All of the operations needed are carried out in one vessel, in the cold. An excess of iron(II) sulfate is employed to reduce uranium(VI) to uranium(IV) in a concentrated phosphoric acid solution containing sulphamic acid. The excess of iron(II) is subsequently oxidized by nitric acid in the presence of molybdenum(VI) as catalyst. After adding sulphuric acid and diluting the mixture with water, the determination is completed by titration with standard potassium dichromate solution in the usual manner, using barium diphenylamine sulphonate as indicator.

The Davies-Gray titration for the assay of uranium in nuclear materials has been studied with emphasis on its performance, (see Bickel 1997). It is based on the redox reaction:

$$nU(IV) + 2M(X) \rightarrow 2M(X - n) + n\,U(VI) \qquad \{3.27\}$$

with M(X) the oxidizing titrant and n the number of electron exchanged during the half redox reaction, e.g. Reaction {3.6}.

In their work "Determination of small amounts of uranium by redox potentiometric titration", Yakshin and Krokhin (2011) investigate the selection of oxidizing titrant. The redox titration of U in H_3PO_4 solution with an automatic titrator for determination of small amounts of U (0.1–0.2 mg) has been studied. The choice of the oxidant appreciably affects the accuracy of the U determination, which decreases in the order $KMnO_4 > K_2Cr_2O_7 > NH_4VO_3 > Ce(SO_4)_2$. Irrespective of the oxidant used, the repeatability, laboratory precision, and accuracy of the method are significantly improved when the amount of U in the sample is increased from 0.1 to 0.2 mg, and high accuracy of the U determination is attained. The relationships obtained can be used in the development of a method for quantitative determination of small amounts of U in solutions.

In their work "Potentiometric titration in a low volume of solution for rapid assay of uranium", Sahoo et al. (2012) focus on a simple, inexpensive potentiometric titration technique for the assay of uranium using small volumes of sample aliquot (25–100 μL) along with all reagents (total volume of solution being less than 2.5 mL). The technique involves modification of the well-known Davies and Gray Method recommended for assay of uranium(VI) in nuclear materials by introducing an innovative potentiometric titration device with a mini cell developed in-house. After appropriate chemical conditioning, the titration is completed within a couple of minutes with display of online titration plot showing the progress of titration. The first derivative plot generated immediately after titration provides information of end point. The main advantage of using this technique is to carry out titration with minimum volumes of sample and reagents generating minimum volume of wastes after titration. The validity of the technique was evaluated using standard certified samples. This technique was applied for assay of uranium in a typical sample collected from a fuel reprocessing laboratory. Further, the present technique was deployed in investigating the optimum conditions for efficient in situ production of U(IV). The precision in the estimation of uranium is highly satisfactory (RSD less than 1.0%) (see Sahoo et al. 2012).

In their work on the determination of uranium and plutonium by sequential potentiometric titration, Kato and Takahashi (1976) deal with a weighed sample of uranium and plutonium oxides (MOX) dissolved in a mixture of nitric and hydrofluoric acids. After the reduction of uranium and plutonium to uranium(IV) and plutonium(III) by chromium(II) sulfate, 5 ml of buffer solution (KCl-HCl, pH 1.0) is added to the solution. The solution is diluted to 30 ml with water and the pH of the solution is adjusted to 1.0–1.5 with 1 M sodium hydroxide. The solution is stirred until the oxidation of the excess of chromium(II) by air is completed. After the removal of dissolved oxygen by bubbling nitrogen through the solution for 10 min, uranium (IV) is titrated with 0.1 N cerium(IV) sulfate. Plutonium is subsequently titrated by 0.02 N cerium(IV) sulfate. When a mixture of uranium and plutonium is titrated with 0.1 N potassium dichromate potentiometrically, the potential change at the end point of plutonium is very small and the end point of uranium is also unclear when 0.1 N potassium permanganate is used as a titrant. In the present method, nitrate, fluoride and copper(II) interfere with the determination of plutonium and uranium. Iron interferes quantitatively with the determination of plutonium but not of uranium. Results obtained in applying the proposed method to 50 mg of mixtures of plutonium and uranium (7.5–50)% Pu were accurate to within 0.15 mg of each element (see Kato and Takahashi 1976).

3.2.18.2 Atom Force Microscopy

Atom Force Microscopy (AFM) is a technique widely used for probing the topography of surfaces at much higher magnifications than those achieved using a traditional optical microscope. The technique relies on the interaction of atoms on a surface with a cantilever-mounted fine sharp tip which is scanned over the sample surface to produce very small movements of the tip. Information on non-contact atomic force microscopy may be found in a work edited by Morita et al. (2015).

Atomic Force Microscopy-derived techniques are the contact mode AFM, Conductive-AFM (C-AFM), the Scanning Tunneling Microscope (STM), the Lateral Force Microscopy (LFM) and the Nano-Indentation (NI). Figure 3.67 shows various microscopy techniques derived from AFM and STM.

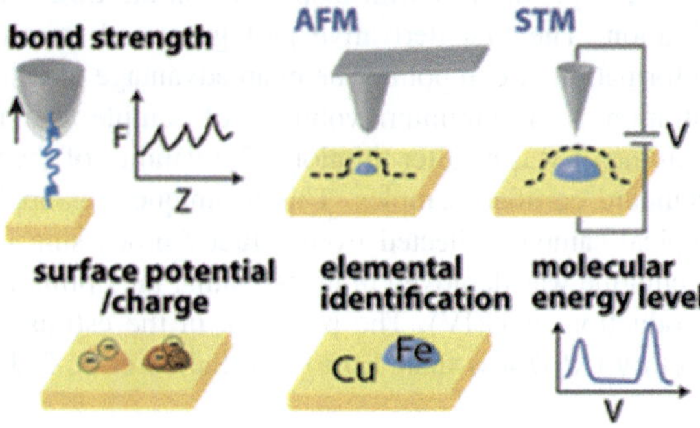

Fig. 3.67 Multi-modal scanning probe microscopy methods are able to characterize surface topography and electronic structures down to the nanometer (and in some cases atomic) scale

Several techniques are used in AFM to measure the topography and investigate the surface forces on the nanoscale: For imaging sample topography:

- Contact mode, where the tip is in contact with the substrate. Gives high resolution but can damage fragile surfaces.
- Tapping/intermittent contact mode (ICM), where the tip is oscillating and taps the surface.
- Non-contact mode (NCM), where the tip is oscillating and not touching the sample. For measuring surface properties (and imaging them).
- Lateral force microscopy (LFM), when the tip is scanned sideways it will cause cantilever twist which can be measured by the photodetector. This method is used to measure friction forces on the nanoscale.
- Force Modulation Microscopy. Rapidly moving the tip up and down while pressing it into the sample makes it possible to measure the hardness of the surface and characterize it mechanically.
- Electrical force microscopy. If there are varying amounts of charges present on the surface, the cantilever will deflect as it is attracted and repelled. Kelvin probe microscopy is normally more sensitive than measuring static deflection.

Figure 3.68 depicts the type of action and interaction between the tip and the analyzed sample.

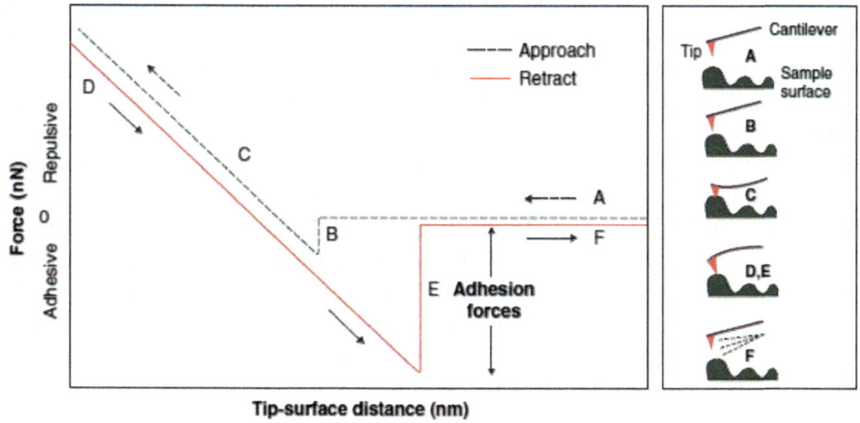

Fig. 3.68 Idealized force-distance curve describing a single approach-retract cycle of the AFM tip, which is continuously repeated during surface scanning. (*A*) The AFM tip is approaching the sample surface. (*B*) The initial contact between tip and surface is mediated by the attractive van der Waals forces. (*C*) The tip applies a constant force on the surface with indentation and cantilever deflection. (*D*) Subsequently, the tip tries to retract and to break loose from the surface. (*E*) Various adhesive forces between sample and AFM tip, which hamper tip retraction. (*F*) The tip withdraws and loses contact with the surface upon overcoming the adhesive forces

Atom force microscopy is used in the field of material physics, for example:

- an identification of atoms at a surface,
- evaluation of an interaction between a specific atom and its neighboring atoms, and
- a change in physical properties arising from a change in an atomic arrangement.

Another major application of AFM is force spectroscopy: the direct measurement of tip-sample interaction forces as a function of the gap between the tip and sample. For this method, the AFM tip is extended toward and retracted from the surface as the deflection of the cantilever is monitored as a function of piezoelectric displacement. These measurements have been used to measure nanoscale contacts, atomic bonding, Van der Waals forces, and dissolution forces in liquids and single-molecule stretching and rupture forces. Furthermore, AFM was used to measure, in an aqueous environment, the dispersion force due to polymer adsorbed on the substrate. Forces of the order of a few piconewtons (pN) can now be routinely measured with a vertical distance resolution of better than 0.1 nm. Force spectroscopy can be performed with either static or dynamic modes. In dynamic modes, information about the cantilever vibration is monitored in addition to the static deflection.

The operation modes of AFM are generally classified into image formation and the others. The image formation is a plotting method as a color mapping through changing the *x–y* position of the tip by scanning and corresponding some particular amount to each *x–y* coordinate. The color mapping is a method corresponding to a value for each coordinate. It means the colored image which especially expresses the intensity of a value. The correspondence between the intensity of a value and a tinge is translated by a color scale. The other group includes various matters such as force spectroscopy and potential mapping.

The detector (see Fig. 3.68) of AFM measures the deflection of cantilever and converts it into an electrical signal. So, during the oscillating motion, the output of the detector of AFM is a Time-Intensity curve. The Intensity will be proportional to the displacement of cantilever. The motion of cantilever is regarded mostly as a sinusoidal vibration in the steady state. Consequently, the output is seen as a sine wave. The wave is characterized by three factors: frequency, amplitude, and phase. The AFM controller extracts at least one value of frequency, amplitude and phase as output signal of the AFM detector.

Several theoretical approach are reported (see Capella and Dietler 1999; Müller et al. 2011). When the cantilever is bent by a point force acting in the z-direction, F_N the normal force at the tip will deflect distance $z(x)$ from the unloaded position along the *x*-axis as

$$z(x) = \frac{1}{2}\frac{F_N L}{EI}\left(x^2 - \frac{1}{3L}x^3\right) \tag{3.50}$$

with cantilever length L, Young's modulus E, and moment of inertia I.

AFM operation is usually described as one of three modes, according to the nature of the tip motion:

- contact mode, also called static mode (as opposed to the two dynamic modes)
- tapping mode, also called intermittent contact, AC mode, or vibrating mode
- non-contact mode, or, again after the detection mechanism, frequency modulation AFM.

The nuclear material examples deal with AFM characterization of Generation V reactor surfaces exposed to high flux plasma.

The first example concerns deuterium plasma interaction with graphite. Almqvist et al. (1995) reported the results of scanning tunneling (STM) and atomic force microscopy (AFM) studies of D^+ irradiated graphite and graphite-silicon composites. The microscopes were used for studying surface topography and for measuring the surface roughness. The substrates were exposed at various temperatures (60 °C and 700 °C) to different doses of deuterium ions in simulators of plasma–surface interactions and in a tokamak. The initial stages of radiation damage, nanometer-sized bubbles/blisters, were found in plasma-eroded surfaces. These structures only appeared in the graphite phase on the multi-component material. The microroughness of the surfaces was measured. AFM was also used for probing the thickness of the plasma-modified layers. Moreover, the AFM reveals the co-deposited layers formed on surfaces facing the tokamak plasma. The appearance of these layers is clearly correlated to the amount of co-deposited atoms.

The second example is reported by Hong-Yan et al. (2015), and deals with the effect of helium implantation on silicon carbide and graphite. They revealed that silicon carbide has the possibility of replacing graphite as plasma facing materials. Pressureless sintered silicon carbide and graphite were implanted with He^+ ions of 20 and 100 keV at different temperatures and different fluences. The He^+ irradiation-induced microstructure changes were studied by electron microscopy and atomic force microscopy (AFM). Figure 3.69 gives the original AFM images of SiC before the He^+ implantation, as well as the morphologies of SiC after 20 keV He^+ implantation with the fluence of 2×10^{17} cm^{-2} at RT and 600 °C. The results reveal that the arithmetic mean roughness rises from 42.3 nm to 46.6 nm at RT and to 54.2 nm at 600 °C after the He^+ implantation. AFM data of SiC further demonstrate the slight surface profile changes.

Using an atomic-force microscope (AFM), it is possible to measure a roughness of a sample surface at a high resolution, to distinguish a sample based on its mechanical properties (for example, hardness and roughness) and, in addition, to perform a microfabrication of a sample (for example, an atomic manipulation).

3.2.18.3 Micro- Nano-Indentation (NI)

Micro- and **nano- Indentation** (**nI**) are indentation tests applied to small volume of investigated material. Indentation is a commonly applied mechanical test; its nano-version was developed in the mid-1970s to measure the hardness of material

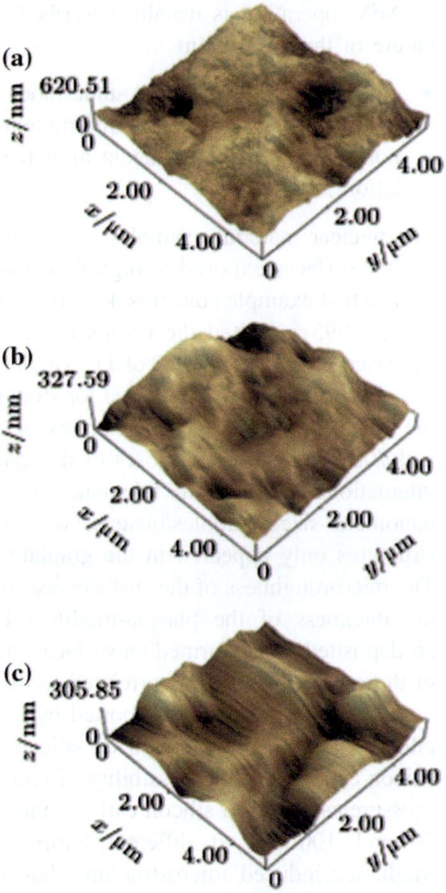

Fig. 3.69 AFM images of bulk SiC before and after He^+ implantation: **a** before He^+ implantation, **b** He^+ implantation at RT, 20 keV, fluence 2×10^{17} cm^{-2}, **c** He^+ implantation at 600 °C. The results reveal that the mean roughness (Ra) rises from 42.3 nm. **a** to 46.6 nm, **b** at RT and to 54.2 nm, **c** at 600 °C after the He^+ implantation (see Hong-Yan et al. 2015) with permission © Chinese Physics. IOP

in sub-grain phases. The indentation head is posed on the surface and pressed onto the surface then released, displaced to the vicinity for a new mechanical cycle and repeated along the sample surface. The indentations are recorded for data analysis. This method is based on the acquisition of force curves recorded at each pixel of the topographic image. The force curves are analyzed instantaneously. Then, provided that parameters such as cantilever spring constant and tip geometry are calibrated, the quantification of the nano-mechanical properties is possible. In the present configuration this analysis is done through the Derjaguin et al. (1975) fit model of the retracing curve for a spherical indenter, which accounts for the adhesion effects between the sample surface and the tip, according to the equation:

$$F = \frac{4}{3} E^{*} \sqrt{Rd^{3}} + F_{\text{adh}} \tag{3.51}$$

Thus, F is the applied force, R is the tip radius, d is the deformation value at a given force, F_{adh} the maximum adhesion force, and E^* is known as the effective elastic modulus. Details on micro- and nano-mechanical testing of materials and devices may be found in Yang and Li (2008).

The nuclear material examples are given below.

Small-scale characterization of irradiated nuclear materials was demonstrated by advanced micro- nano-mechanical testing. Nano-indentation and micro-cantilever testing of ion irradiated nuclear material is reported by Armstrong et al. (2015) for iron–12 wt% chromium implanted with iron ions and tungsten implanted with both tungsten ions and tungsten and helium ions.

A high temperature composite material made of TiAl with 2 at.% W has been investigated as a potential structural material for high-temperature gas-cooled nuclear reactors. The tests are performed with thin samples irradiated by 24 MeV $^4He^{2+}$ ions; and investigated toward its microstructure and its hardness (see Pouchon et al. 2009). The hardness was investigated using a micro-hardness tester and a nano-indenter. From room to moderate irradiation temperatures, the radiation-induced hardening of the material slowly vanishes until the material completely recovers at about 943 K. Beyond this temperature, He-bubble formation seems to harden the material again, until beyond 1200 K a steep increase in hardening is detected. The results are consistent with microscopic investigation and supporting the microstructural development as a function of temperature.

ODS steels are also being considered as possible material for use in future fusion applications. Investigations on stability of yttria particle dispersions in pure Fe matrix during irradiation by single Fe and dual Fe and He ion beams, up to 100 dpa per He at 600 °C are reported by Robertson et al. (2012). Irradiation-induced evolutions were studied by nano-indentation measurements. A significant evolution of the nano-indentation response is observed with a micro-mechanical evolution ascribed to enhanced strain localization, associated with irradiation-induced particle size changes. Specific effect of implanted He is the augmentation of the micro-mechanical yields stress of the ODS material.

Zirconium hydride phases such as encounter in rods during reactor operation are difficult to measure for their mechanical properties because of the difficulties in preparing macro-size samples. Consequently, nano- and micro-indentations may be used to study mechanical properties of these zirconium hydrides ($ZrH_{1.83}$) phases (see Xu and Shi 2004). Hardness, apparent Young's modulus, yield stress and fracture toughness of zirconium hydrides were measured, using e.g. nano-indentation (see Fig. 3.70), and compared to the values for a zirconium sample.

3.2.18.4 Inductively Coupled Plasma Mass Spectrometry

Inductively Coupled Plasma Mass Spectrometry (**ICPMS**) is a variety of mass spectrometry which detects ions from elements by ionizing the sample with ICP as described in Sect. 3.2.16.1 and using subsequently MS to separate and quantify the

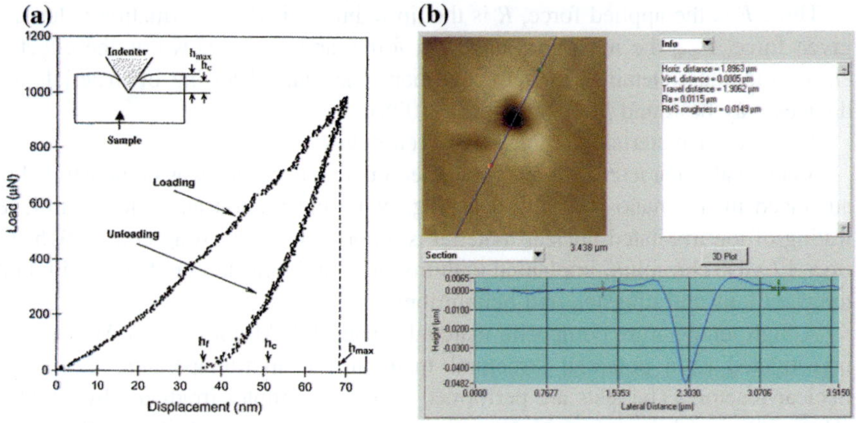

Fig. 3.70 Investigation of mechanical properties of ε-zirconium hydride using micro- and nano-indentation techniques. **a** Typical load-displacement curve in nano-indentation: h_r: residual depth, h_c: contact depth, h_{max}: maximum depth; **b** nano-indentation observations and results. Xu and Shi (2004) with permission © Elsevier

generated ions. Information on inductively coupled plasma mass spectrometry may be found in Jarvis (1992).

The ICPMS allows determination of elements with atomic mass ranging from 6 to 250 e.g. Li to Cf, and sometimes of the element/isotope cluster ions. Some masses are prohibited such as 40 due to the abundance of argon ($^{40}Ar^{+}$) in the sample. Other blocked regions may include mass 80 (due to the $^{40}Ar_2^{+}$), and mass 56 (due to $[^{40}Ar^{16}O]^{+}$), the latter of which greatly hinders ^{56}Fe analysis unless the instrumentation is fitted with a reaction chamber. Such interferences can be reduced by using a High Resolution ICPMS (HR-ICPMS) which uses two or more slits to constrict the beam and distinguish between nearby peaks. This comes at the cost of transmission, for example to distinguish iron from argon by upgrading the resolution power up to 10,000.

A single collector ICPMS may use a multiplier in pulse counting mode to amplify very low signals, an attenuation grid or a multiplier in analogue mode to detect medium signals, and a Faraday cup to detect larger signals. A multi-collector ICPMS may have more than one detector, normally Faraday cup which are much less expensive. With this combination, a dynamic range of 9 orders of magnitude, from 1 ppt to 100 ppm is possible (see Linge and Jarvis 2009).

The first step in analysis is the introduction of the sample. This has been achieved in ICPMS through a variety of means. The most common method is the use of analytical nebulizers. Nebulizer converts liquids into an aerosol and this aerosol is then injected into the plasma to create the ions. Details on the nebulizer are given in Sect. 3.2.14.1. A desolvating nebulizer may also be used; this uses a long heated capillary, coated with a fluoropolymer membrane, to remove most of the solvent and reduce the load on the plasma. Matrix removal introduction systems are sometimes used for samples, such as seawater, where the species of interest are at trace levels, and are surrounded by much more abundant contaminants (see Hein et al. 2017).

The plasma used in an ICPMS is made by partially ionizing argon gas ($Ar \rightarrow Ar^+ + e^-$). The energy required for this reaction is obtained by pulsing an alternating electric current in wires that surround the argon gas. After the sample is injected, the plasma's extreme temperature causes the sample to separate into individual atoms (atomization). Next, the plasma ionizes these atoms ($M \rightarrow M^+ + e^-$) so that they can be detected by the mass spectrometer. The carrier gas is sent through the central channel and into the very hot plasma. The sample is then exposed to radio frequency which converts the gas into plasma. The high temperature of the plasma is sufficient to cause a very large portion of the sample to form ions. This fraction of ionization can approach 100% for some elements (e.g. sodium), but this is dependent on the ionization potential. A fraction of the formed ions passes through a ~1 mm hole (sampler cone) and then a ~0.4 mm hole (skimmer cone), the purpose of which is to allow a vacuum that is required by the mass spectrometer.

The vacuum is created and maintained by a series of pumps. The first stage is usually based on a roughing pump, most commonly a standard rotary vane pump. This removes most of the gas and typically reaches a pressure of around 133 Pa. Later stages have their vacuum generated by more powerful vacuum systems, most often turbomolecular pumps. Older instruments may have used oil diffusion pumps for high-vacuum regions.

A sector ICPMS commonly has four sections: an extraction acceleration region, steering lenses, an electrostatic sector and a magnetic sector. The first region takes ions from the plasma and accelerates them using a high voltage. The second may use a combination of parallel plates, rings, quadropoles, hexapoles and octopoles to steer, shape and focus the beam so that the resulting peaks are symmetrical, flat topped and have high transmission. The electrostatic sector may be before or after the magnetic sector depending on the particular instrument, and reduces the spread in kinetic energy caused by the plasma. The geometry of the instrument is chosen so that the combined focal point of the electrostatic and magnetic sectors is at the collector, known as Double Focusing.

The basic equation for mass spectroscopy is

$$\frac{m}{z} = \frac{B^2 R^2}{2V} \tag{3.52}$$

where m is the mass of the ion, z its charge, B the magnetic field, V the accelerating potential and R the radius of the ion trajectory in the mass spectrometer (see Aston 1919). For B, V and z constant the radius of trajectory R is proportional to the square root of m. For B in gauss, V in volt and R in cm the coefficient becomes 4.825×10^{-5} instead of ½ for mks units. This may allow estimating the resolution in term of mass.

The detection limit in terms of concentration is currently low (ppt level when no interference is co-recorded) and as low as one part in 10^{15} (part per quadrillion, ppq) on non-interfered low-background isotopes and for specific measurements. The later DL may be obtained by Quadrupole High resolution ICPMS for example.

Nuclear material examples deal with ICPMS of waters into and in the surrounding of a generation 0 reactor (georeactor, see Sect. 1.2.1). The concentration and the role of colloids in the transport of elements in the environment of a fossil reactor at Bangombé, Gabon, were assessed. Filtered fluids and colloid samples collected on membranes and resuspended in solution were analyzed by ICPMS and HR-ICPMS in order to examine the element association in the colloid phase within the size range 3000 to 400 to 50 nm. The colloid concentrations for the size 400 to 50 nm range from 80 to 300 ng ml^{-1}. They consist of silica particles associated with ferrihydrite coated with organics. Inside the reactor (BAX03: depth 11.9–12.5 m) the uranium 235 isotopic abundance is somewhat depleted confirming the former activities of the geo-reactor activity (see Degueldre and Laaksoharju 2014), while outside the reactor zone (BAX04: depth 8.9–10.2 m) the isotopic signature is normal. In both, water uranium is partially associated to the groundwater colloids (see Fig. 3.71). Trace element results show that metals including Pb, Sc, Y, La, Ce, Pr, Nd, Bi, Th and U are associated to various degrees with the colloid phase (Fig. 3.71). The distribution ratios of these trace elements between the water and the colloid phase (*Kp*) were experimentally determined. The high Pb distribution ratios of 10^{+7} ml g^{-1} are specifically discussed. Values range from 10^{+6} to 10^{+5} ml $g^{-1}$for the trivalent elements (Sc, Y, La, Ce, …, Bi). For uranium, a *Kp* of the order of 10^{+5} ml g^{-1} may be calculated and compared with data gained using the surface complexation model. These *Kp* values suggest that the uranium is partially sorbed or associated with groundwater colloids. Measurements from the reactor

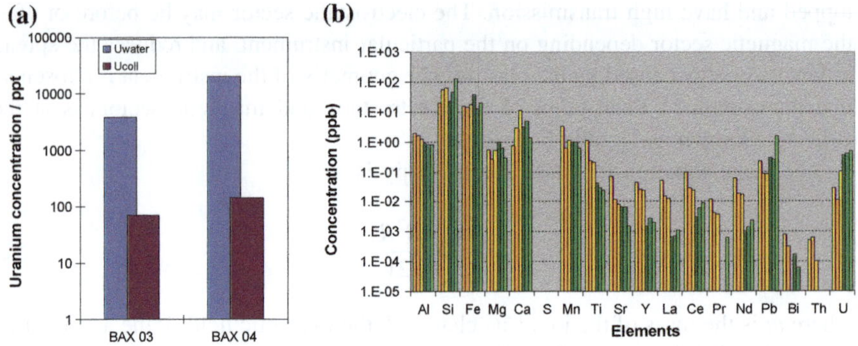

Fig. 3.71 Element association with groundwater colloids from the Bagombe fossil reactor and from its environment. **a** HR-ICPMS of uranium, inside reactor (BAX03; $^{235}U/^{238}U$ ratio: 0.7091 ± 0.0005) outside reactor (BAX04, $^{235}U/^{238}U$ ratio: 0.7270 ± 0.0005). **b** ICPMS for trace elements. Conditions: inside reactor zone cut off >3000 nm (*dark orange) filled square*, >400 nm (*orange) filled square*, >50 nm (*yellow) filled square*, outside reactor zone cut off >3000 nm (*dark green) filled square*, >400 nm (*green) filled square*, >50 nm (*pale green) filled square*. Adapted from Degueldre and Laaksoharju (2014) with permission © Elsevier

zone show that about 2–4% of the uranium is associated with the colloid phase, which contributes partially to the uranium transport.

Single Particle-Inductively Coupled Plasma Mass Spectrometry (SP-ICPMS)

The protocol of utilization of the ICPMS unit in a single particle analysis mode is adapted to determine the amount of one element isotope in single particle or single colloid to derive their size distribution (isotope per isotope), the number of particles per volume unit as well as the size of the single particle knowing their composition. The flash of ions M^+ due to the ionization of a specific element/isotope colloidal particle in the plasma torch can be detected and measured in a time scan for M^+ or $[MO]^+$ according to the sensitivity required by the mass spectrometer. The peaks of the recorded intensity of the MS signal (at the fixed mass of M^+) can be analyzed as a function of time the particle size (with M major component) or fraction of the studied element in the colloid phase. The Equation describing the phenomena is reported in Degueldre and Favarger (2003). The number of ion N_M for a single $M_\xi M'_{\xi'} O_{\zeta/2}$ particle of size d is given by:

$$N_M = \xi \frac{\pi d^3 \rho N_{Av}}{6 M_{MO_{\zeta/2}}} = \frac{1}{\eta_A \eta_c} \int_t^{t+\Delta t} s_A(t)\, dt \qquad (3.53)$$

with ρ the colloid density N_{Av} the Avogadro constant and $\Sigma\xi = 1$ (metal stoichiometry); and with the signal $s_A(t)$ detected for the isotope $^{A}M^+$ (with abundance η_A and counting efficiency η_c) as function of time t with the peak lasting Δt. The frequency ($f(s_A)$) of the flashes is directly proportional to the concentration of particles N_{col} in the colloidal suspension and is given by

$$f(s_A) = N_{col}\, q_{col} \eta_{ne} \qquad (3.54)$$

with q_{col} the colloid sample injection rate, and η_{ne} the efficiency of nebulization. This technique is very promising for the speciation of elements/isotopes in colloidal particle in fluids. Since its development at the beginning of this century, SP-ICPMS has retained interest as recently reported in Laborda et al. (2016).

The **nuclear material example** is the thorium colloid analysis by single particle ICPMS (see Degueldre and Favarger 2004). The flash of ions due to the ionization of a thorium colloidal particle in the plasma torch can be detected and measured in a time scan for $^{232}Th^+$ or $^{248}[ThO]^+$ (see Fig. 3.72) according to the sensitivity required by the mass spectrometer. Tests performed on thorium dioxide colloidal particles show that the intensity of the ion flashes generated in the plasma torch signal can be exploited for deriving size distribution of these refractory particles. This feasibility study also describes the experimental conditions and the limitation of the plasma design to detect thorium colloids in a single particle analysis mode down to about 10 fg corresponding to 80 nm colloid of ThO_2.

Colloid analysis by single particle inductively coupled plasma-mass spectrometry was also performed on UO_2 and U_3O_8 particles, (e.g. Degueldre et al. 2006b).

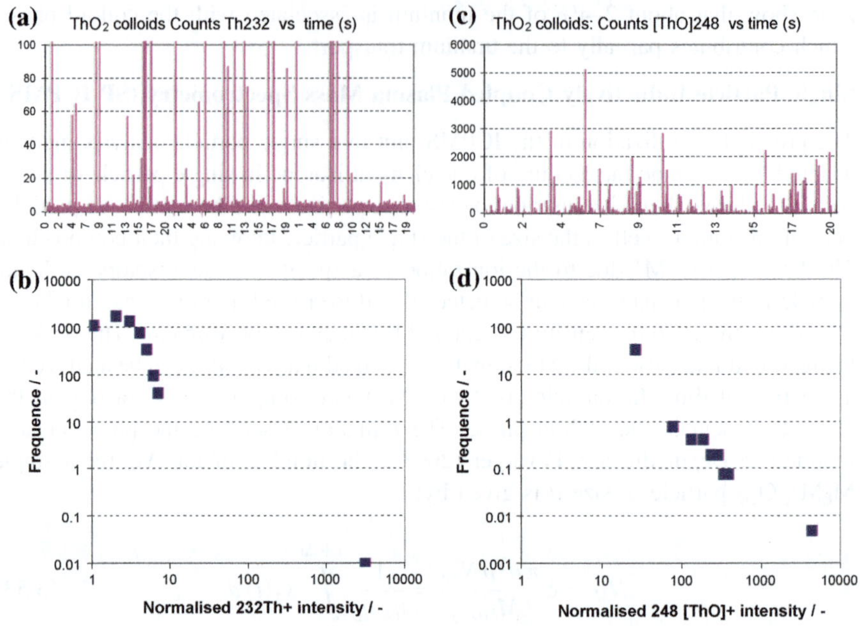

Fig. 3.72 Thoria colloid analysis by Single Particle-ICPMS mode, size detection limit ~80 nm. **a** and **c** signal for $^{232}Th^{+}$ and $^{248}[ThO]^{+}$ recorded in time scan, particle sample record during 20 s. **b** and **d** ICP signal distribution for $^{232}Th^{+}$ and $^{248}[ThO]^{+}$; conditions: 10 ms detection time for $^{232}Th^{+}$ and $^{248}[ThO]^{+}$ ion detection. Data from Degueldre and Favarger (2004) with permission © Elsevier

3.2.18.5 Secondary Ion Mass Spectrometry

Secondary Ion Mass Spectrometry (**SIMS**) is generally used to analyze the isotope composition of solid surfaces by sputtering the sample surface with a focused primary ion beam and analyzing the ejected secondary ions. The mass/charge ratios of the secondary ions are measured with a mass spectrometer to determine the isotopic and subsequently the elemental or molecular composition of the sample to a depth of 1 to 2 nm. Due to the large variation in ionization yields among different materials, SIMS is usually considered to be qualitative. Detailed information on secondary ion mass spectrometry may be found in Benninghoven et al. (1986). Details on the fundamentals of mass spectrometry may be found in Hiraoka (2013).

Typically, a secondary ion mass spectrometer consists of:

- a primary ion source generating the primary ion beam
- a primary ion sub-unit, accelerating and focusing the beam onto the sample
- a high-vacuum sample chamber holding the sample
- a secondary ion extraction lens
- a mass analyzer separating the secondary ions for their mass to charge ratio
- an ion detection unit.

Three types of primary ion **sources** are typically utilized.

- First, gaseous ions are produced with plasmatrons or by ionization. Noble gases ions ($^{40}Ar^+$, Xe^+) and ions of atomic or molecular oxygen ($^{16}O^-$, $^{16}O_2^+$, $^{16}O_2^-$), or even ionized molecules such as SF_5^+ (produced from SF_6) or C_{60}^+ (fullerene ion). This type of ion gun is easy to operate and generates roughly focused but high-current ion beams.
- A second source type, the surface ionization source, generates $^{133}Cs^+$ primary ions. Cesium atoms vaporize through a porous tungsten plug and are ionized during evaporation. Depending on the gun design, fine focus or high current can be obtained.
- A third source type, the liquid metal ion gun, operates with metals or metallic alloys, which are liquid at room temperature or slightly above. The liquid metal covers a tungsten tip and emits ions under influence of an intense electric field.

A gallium source is able to operate with elemental gallium, recently developed sources for gold, indium and bismuth use alloys which lower their melting points. The liquid metal ion gun provides a tightly focused ion beam (<50 nm) with moderate intensity and is additionally able to generate short pulsed ion beams. It is therefore commonly used in static SIMS devices.

Detection depends on the SIMS unit. There are three basic analyzers available: sector, quadrupole, and time-of-flight.

- A sector field mass spectrometer uses a combination of an electrostatic analyzer and a magnetic analyzer to separate the secondary ions by their mass to charge ratio.
- A quadrupole mass analyzer separates the masses by resonant electric fields, which allow only the selected masses to pass through.
- A time-of-flight mass analyzer separates the ions in a field-free drift path according to their velocities. It requires pulsed secondary ion generation using either a pulsed primary ion gun or a pulsed secondary ion extraction.

A Faraday cup measures the ion current hitting a metal cup, and is sometimes used for high-current secondary ion signals. With an electron multiplier an impact of a single ion starts off an electron cascade, resulting in a pulse of 10^8 electrons which is recorded directly.

A microchannel plate detector is similar to an electron multiplier, with lower amplification factor but with the advantage of laterally resolved detection. Usually it is combined with a fluorescent screen, and signals are recorded, e.g. with a CCD-camera. While only charged secondary ions emitted from the material surface are used to analyze the material, these represent a small fraction of the particles emitted from the sample. The secondary ion intensity is given by:

$$I_s = I_p Y(E)_m \alpha^+ \theta_m \eta \tag{3.55}$$

with I_p the primary ion intensity, Y the sputtering yield, α the ionization probability, θ_m a matrix constant and η the transmitivity in system. SIMS is one of the most sensitive surface analysis techniques. The isotope detection limits range from parts per million (ppm) to parts per billion (ppb). The detection limits in term of atom density for trace isotopes are between 10^{12} and 10^{16} cm^{-3} depending on the matrix material, the instrumentation, the primary ion beam used and other factors.

In surface analysis, it is usual to distinguish between static SIMS and dynamic SIMS.

- **Static SIMS** is the mode involved in surface atomic or molecular monolayer analysis, for example with a pulsed primary ion beam and a time-of-flight MS.
- **Dynamic SIMS** is the mode involved in bulk analysis, associated to the sputtering process.

The nuclear material examples deal with SIMS analysis of an UO_2 fuel irradiated at "low temperature" to 65 MW d kg^{-1} as reported by Walker et al. (2009). During burn-up, volatile fission products such as Xe and Kr may be found in nano- to micro-phases. The so called "gas bubbles" in the high burn-up structure (periphery of the fuel) are revealed by ion maps for ^{133}Cs and ^{85}Rb, the decay products of ^{133}Xe and ^{85}Kr (see Fig. 3.73). Xe and Kr are "naturally" together in the "bubbles". Other volatile ion maps showing the distributions of ^{129}I, ^{130}Te and ^{137}Cs (^{137}Ba) in high burn-up structure at fuel rim.

Secondary Ion Mass Spectrometry routinely provides radial isotopic distribution of actinides in irradiated nuclear fuel samples but the measurements are mainly qualitative because of many isobaric interferences. A simple method to solve the

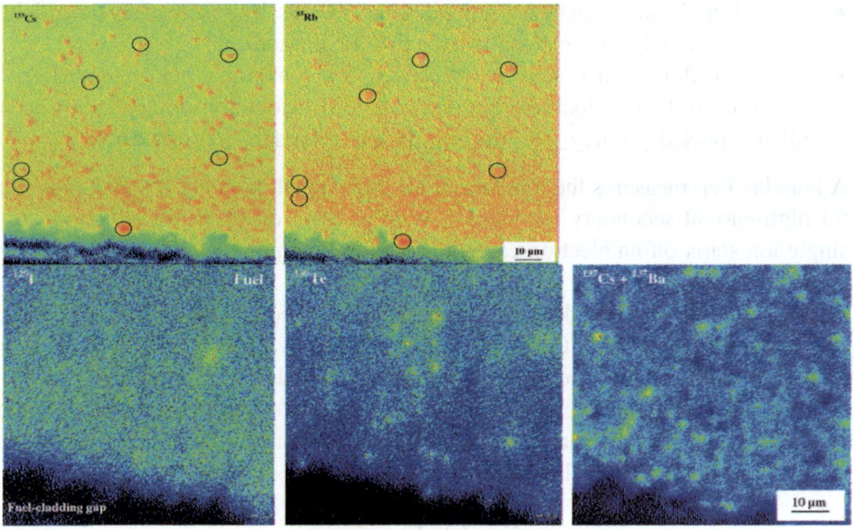

Fig. 3.73 Gas bubbles in the high burn-up structure as revealed by ion maps for ^{133}Cs and ^{85}Rb, the decay products of ^{133}Xe and ^{85}Kr. Xe and Kr are together in the bubbles. Conditions: Local ion intensity: *Black* is low; *red* is high. Ion maps showing the distributions of ^{129}I, ^{130}Te and ^{137}Cs (^{137}Ba) in high burn-up structure at fuel rim. Walker et al. (2009) with permission © Elsevier

isobaric interferences between Am and Pu for mass 241 and 242 is proposed by Portier et al. (2012). The method utilizes isotope ratio measurements recorded for the $^{A}An^{+}$ ions as well as for the $^{(A+16)}[AnO]^{+}$ cluster ions. The results are obtained by solving a system of 4 equations with 4 unknowns. The method is applied on thorium-based fuels, which represent a sustainable alternative to the classical uranium-based fuels commonly used in most power plants around the world. This method could be used for solving the $^{A}An^{+}$ isobaric interference in UOX or MOX by solving 9 equations and 9 unknowns including the fractions of $^{A}An^{+}$ $^{(A+16)}[AnO]^{+}$ and $^{(A+32)}[AnO_2]^{+}$.

Scanning ion microscope

High Voltage **Scanning Ion Microscope** (HV**SIM**) proceeds with a focused ion nano-beam used to sputter solid targets and to cause the ejection of secondary ions (see Ireland 2014). These are analyzed through a MS where individual mass abundances may be determined allowing both chemical and isotopic information to be obtained. This is the basis for secondary ion mass spectrometry and ion microprobe or ion microscope analysis. The instruments require high mass resolution mass spectrometers in order to discriminate both atomic and molecular isobaric interferences. The ion microprobe has been described as the geochemist's ultimate tool with the key strength of ion microprobe analysis being the high sensitivity with good ion yields for many elements and low inherent backgrounds. However, the extremely small amount of material often sampled can limit analytical precision.

In an HVSIM design, the ion optical system is based on a high-brightness ion source (see Magilin et al. 2015). Specifically, the ion optical system is divided into two components: an ion injector and a probe-forming system that consists of an accelerating tube and a multiplet of quadrupole lenses. The cross-over is formed and controlled by the injector, which acts as an object collimator, and is focused on the image plane by the probe-forming system. The ion microprobe beam has a size of 100 nm and 2 MeV energy. When the influence of the chromatic and third-order aberrations is theoretically taken into account, the HVSIM forms an ion microprobe.

3.2.18.6 Accelerator Mass Spectrometry

Accelerator Mass Spectrometry (**AMS**) differs from other MS in that it accelerates ions to very high kinetic energy before mass analysis. The special strength of AMS among the mass spectrometric system is its potential to separate a rare isotope from an abundant neighboring mass (e.g. ^{14}C from ^{12}C) because of the accelerated ion high voltage, *V*, and the radius, *R* (see Eq. 3.52). The method also suppresses molecular isobars completely because negative ions of metals are rather unstable, the striper that makes use of foil or gas jet eliminates neutral (in tandem accelerator) … In many cases AMS can separate atomic isobars (e.g. ^{14}N from ^{14}C) or also cluster ions such as $[^{235}UH]^{3+}$ and $^{236}U^{3+}$ (Lachner et al. 2013). This makes possible the detection of naturally occurring, long-lived radio-isotopes such as ^{10}Be, ^{36}Cl, ^{26}Al and ^{14}C. Their typical isotopic abundance ranges from 10^{-12} to 10^{-18}. AMS can compete with techniques of decay counting for isotopes for which the half-life is long enough (>1000 a). Details on the AMS may be found in Litherland (1980), Lunney et al. (2001).

AMS can be used to detect very small amounts of man-made isotopes such as ^{129}I (fission product) and ^{236}U (as anthropogenic actinide), as well as Pu, and for isotope tracing for nuclear safeguards and nuclear forensics (Hotchkis et al. 2002). The long-lived radioisotope ^{236}U ($\tau = 23.4$ Ma) is produced predominantly by neutron capture on ^{235}U, building up to high levels in nuclear fuel. It is a potentially useful tracer of irradiated uranium for a variety of applications, due to its virtual absence in natural samples (^{236}U:^{238}U ratio $<10^{-10}$ in uranium ore). ^{236}U has been distributed in the environment as a result of nuclear activities including nuclear explosions, accidents at nuclear plants, dumping of nuclear waste and releases from nuclear facilities. It is also present in depleted uranium, due to the recycling of uranium fuel in enrichment plants, and has been dispersed in areas where depleted uranium munitions have been used. ^{236}U has been measured in a variety of environmental media by AMS. Such measurements are possible using only a few grams of environmental materials, yielding a few micrograms of uranium for analysis by AMS. The resulting 236/238 isotopic ratios are typically in the range 10^{-6}–10^{-13}. More recently, the measured $^{129}I/^{236}U$ ratios of samples collected in the North Sea in 2009 are in reasonable agreement with the expectations from documented or estimated releases from the two major nuclear reprocessing plants, suggesting that the $^{129}I/^{236}U$ ratio can be used as a water mass tag in the North Atlantic region (see Christl et al. 2015). However, the results indicate that, in contrast to ^{129}I, additional contributions of bomb produced ^{236}U cannot be neglected in the North Sea region (see Table 3.15).

Table 3.15 Comparison of isotopic ratio gained by AMS in anthropogenic and natural media

Ratio	Natural	Atlantic	North Sea	Model	Discharge
^{129}I:^{236}U	10–20#	n r	400	350	800–2000°
^{236}U:^{238}U	10^{-13}	10^{-9}	10^{-8}	2×10^{-8}	2*–4°× 10^{-3}

Data from Christl et al. (2015), *Springfield, UK; °LaHague, F (indicative values). From Casacuberta et al. (2016) #Pacific Ocean, n r: not reported

3.2.18.7 Rutherford Back-Scattering Spectrometry

Rutherford Back-scattering Spectrometry (**RBS**) is utilized in materials science to determine the composition and structure of materials by measuring the back-scattering of high energy ions, usually $^1H^+$ or $^4He^{2+}$ impinging on a sample. Detection is carried out using ion detectors (see Sect. 3.1.6.2).

The energy loss of an ion that undergoes back-scattering depends on two processes: the recoil energy lost to sample nuclei in scattering events, and the energy continuously lost to small-angle scattering from the electrons while traversing the sample material. The first is fixed by the mass ratio of the projectile to sample nucleus and the initial energy of the projectile ion. For a given angle between the detector and the incident beam, sample nuclei of different masses scatter the ions under unambiguous energies which are characteristic for their mass. For a thin layer of material, they produce separate peaks on an intensity vs. energy plot. Concentrations can be calculated from the peak integrals. For a thick sample, the energy of projectiles back-scattered from greater depth is shifted to lower energies due to the continuous energy loss in the material on the way in and out of the sample. Therefore, the RBS energy spectrum can be interpreted as a concentration depth profile. The depth range of the profile is on the order of a few micrometers. However, depth profiles of all elements present in the sample are overlaid in the same spectrum, which can make data interpretation cumbersome. The detection limit largely depends of the matrix and isotope analyzed and is generally between at.% (light elements in heavy matrix) and 10^{-5} (heavy elements in light matrix).

RBS can be used to measure surface layers or the bulk composition (within the penetrated depth range) of a sample material. For mono-crystals, it is possible to gain information about the structure of the sample surface. When the incident beam or the scattered projectiles are aligned to symmetry planes or axes, the crystal quality (defect density as a function of depth) and the position of certain elements in the lattice can be determined.

Conventional RBS is performed using helium ions, while heavy ion RBS has the advantage of a better mass resolution for heavy elements, being able to separate isotopes in some cases because the relative mass separation increases for higher masses. Depth resolutions down to 1 nm can be obtained with heavy ion detectors in optimized set-ups, leading to quantitative studies of surface interface mixing and inter-diffusion.

As nuclear material example, the study of the oxide layer formed on stainless steel exposed to Boiling Water Reactor (BWR) conditions by ion beam techniques may be presented (see Degueldre et al. 1998b). Until that time, no attempt to monitor the build-up of stainless steel oxidation layers in-line, i.e. under BWR operating conditions, had been made (see DRS monitoring Sect. 3.2.3.3 and Fig. 3.23). The in situ analysis requires however calibration. This has been performed using RBS or its associated technique resonant scattering spectrometry where the beam energy is raised close to the Coulomb barrier of the investigated nucleus and specific elastic scattering resonances are excited. This spectroscopy is an absolute technique that may be used off-line for corrosion layer calibration.

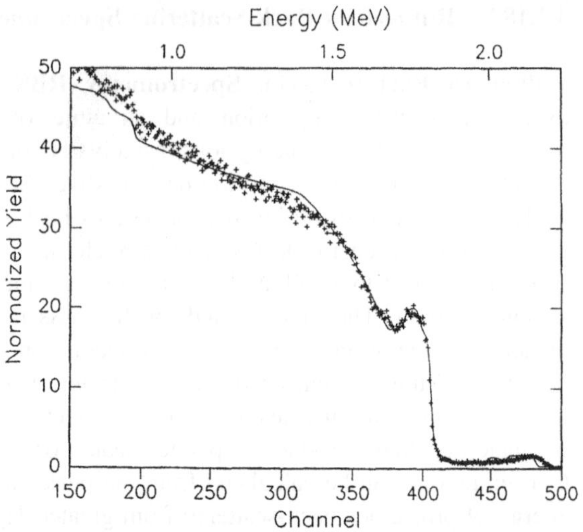

Fig. 3.74 Typical Macro-RBS spectrum recorded for austenitic stainless steel samples after exposure to the loop water. Conditions: water temperature, 563 K; pressure, 9 MPa (90 bar); water velocity, 0.038 m s^{-1}; 400 ppb O_2; exposure time, 31 days; +experimental counting rate; − modeled spectrum. Note the depression in the metal RBS-spectrum due to the oxide layer. Degueldre et al. (1998b) with permission © Elsevier

A typical Macro-RBS spectrum recorded for austenitic stainless steel samples after exposure to the loop water simulating BWR water is plotted in Fig. 3.74. The stainless steel sample was exposed for 31 days to hot water (563 K) at a pressure of 9 MPa (90 bar) with 400 ppb O_2 and a velocity of 0.038 m s^{-1}. The depression in the metal RBS-spectrum was found to be due to the oxide layer and the hump around channel 180 due to O resonance which allows quantifying this light element even in a heavy matrix. Results obtained off-line allowed absolute calibration of the oxide layer.

3.2.18.8 Elastic Recoil Detection Analysis

Elastic Recoil Detection Analysis (ERDA) relates to forward "recoil" scattering of sample nuclei. Similar to RBS, ERDA is an ion beam analysis technique in materials science to obtain elemental concentration depth profiles in thin layers of samples. In this nuclear technique an energetic heavy ion beam is directed at a sample to be characterized, and elastic Coulomb interactions between the ions of beam and the atoms of the target are induced. In contrast to backscattering, the recoiling sample atoms are detected in this case. Again, the energy of the recoil is well determined by the mass ratio of the projectile and the sample nucleus. As in RBS, continuous energy loss in the materials causes an energy shift depending on the depth of the interaction. Since all elements in the sample can be scattered forward the detected ions have to be identified in a spectrometer by discriminating mass or nuclear charge. As a result of this identification, separate depth profiles are obtained for each element and ambiguities in data interpretation are removed. For very light sample elements the discrimination between recoiling nuclei can be accomplished by simple range foils in front of a charged particle detector which completely stop scattered beam particles and heavier sample atoms. The foil technique is mainly used for the analysis of hydrogen. ERDA helps to quantify the elemental composition of a shallow layer below the sample surface (a few hundred nm). Information about ERDA may be found in the work of Steinhausen et al. (1994) and the limitations to depth resolution in high-energy, heavy-ion elastic recoil detection analysis are provided by Elliman et al. (1998).

As mentioned above, the principles of ERDA are similar to RBS. The projectile energy range for light incident ions is 0.5–3.0 MeV per nucleon. For medium heavy ion beams, e.g. ^{36}Cl, a common total energy of approximately 30 MeV is adequate; for heavier projectile ions such as ^{127}I the energy range is generally between 60 and 120 MeV. The energy E_2 transferred by projectile ions of mass m_1 and energy E_1 to sample atoms of mass m_2 recoiling at an angle ϕ, with respect to the incidence direction is given by the following equation.

$$E_2 = \frac{(4m_1m_2)}{(m_1+m_2)^2}\left(E_1\cos^2\phi\right) \tag{3.56}$$

For sample nuclei much heavier than the projectile ($m_2/m_1 < <1$), all recoiling ions have similar velocities. The maximum scattering angle of the projectile ions, $\theta_{\max}$, can be deduced from simple kinematics as:

$$\theta_{\max} = \arcsin\frac{m_2}{m_1} \tag{3.57}$$

Thus the maximum scattering angle for beam particles is limited by the heaviest mass m_1 present in the sample. If placed at an angle greater than this limiting angle $\theta_{\max}$, scattered projectiles cannot reach the detector, which greatly reduces background count rates and prevents degradation of the detector.

The ERDA instrumentation was originally developed for detection of hydrogen or other light element (He, Li, C, O, Mg, K) profiling with an absorber foil in front of the energy detector for beam suppression. For hydrogen analysis, silicon diode detectors are commonly used. For mass or nuclear charge discrimination in heavy ion ERDA, time-of-flight (ToF) detectors, magnetic spectrometers or multi-anode gas ionization detectors are necessary. These detectors are described in Sects. 3.1.2 and 3.2.16.

Depth resolution of concentration profiles is a key parameter that characterizes recoil spectrometry. Calculation requires an incoming and an outgoing ion beam, depth of collision, two trajectories with incoming and outgoing paths and the given scattering and recoil angles.

ERDA in transmission geometry, where only the energy of the recoiling sample atoms is measured, was extensively used for contamination analysis of thin target foils for nuclear physics experiments. This technique is excellent for discerning different contaminants of foils used in sensitive experiments. Using ^{127}I ion beam, a profile of various elements can be obtained and the amount of contamination can be determined.

In the nuclear industry, quantification of hydride distribution in irradiated Zircaloy-4 cladding is reported by ERDA using hydrogen content mapping. The detrimental effect of hydride blisters, localized high hydrogen concentration zones, on the cladding resistance was observed in integral tests in dedicated research reactors. A cold spot technique was developed to generate hydride blisters on Zircaloy-4 cold work stress relieved non-irradiated pre-hydrided claddings tubes (see de Menibus et al. 2012). The hydrogen concentration field in and around the blister was measured by Micro Elastic Recoil Detection Analysis (μ-ERDA) and with micro- and nano-hardness techniques. Hydrogen concentration was found to decrease from the center of the blisters toward their borders.

The last example deals with the quality control of boron carbide used in neutron detectors and H in B_4C. The isotope ^{10}B has a high absorption cross-section for thermal neutrons, and reacts in either of the reactions:

$$^{10}\mathrm{B} + \mathrm{n} \rightarrow {}^{7}\,\mathrm{Li}\,(0.84\ \mathrm{MeV}) + {}^{4}\mathrm{He}\,(1.47\ \mathrm{MeV}) + \gamma\,(0.48\ \mathrm{MeV}) \qquad \{3.28\}$$

or

$$^{10}\mathrm{B} + \mathrm{n} \rightarrow {}^{7}\,\mathrm{Li}\,(1.02\ \mathrm{MeV}) + {}^{4}\mathrm{He}\,(1.78\ \mathrm{MeV}) \qquad \{3.29\}$$

when a neutron is captured. A ^{10}B-based neutron detector uses these reactions to convert neutrons into α and γ particles. Triethylboron is the single-source precursor for chemical vapor deposition of B_xC thin films. A series of species of triethylboron-based products was examined and found to be dominated by *β*-hydride eliminations of C_2H_4 to yield BH_3. Films grown at temperatures below 600 °C contain high amounts of H, while temperatures above 1000 °C result in C-rich films. ERDA has been used (Pedersen et al. 2012) to analyze the produced light element matrices and light analyte, e.g. H in B_4C. Figure 3.75 presents the ERDA analysis depth profile recorded after thermal treatment of the produced sample.

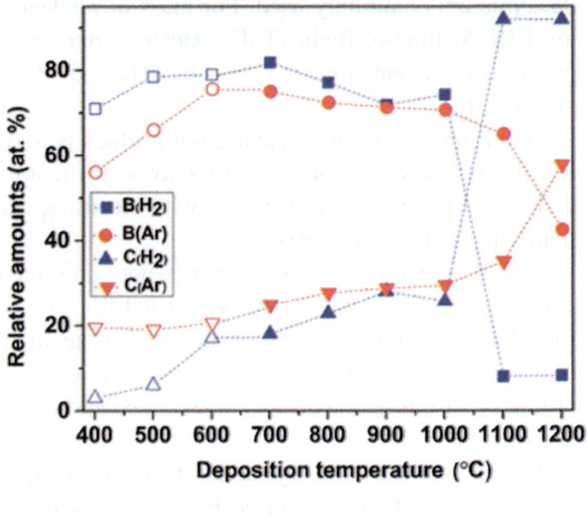

Fig. 3.75 ToF-ERDA data showing the relative amounts of B and C of films deposited in H_2 and Ar atmospheres, respectively, as a function of deposition temperature (see Imam et al. 2015). Open symbols correspond to data points from Pedersen et al. (2012) with permission © RSC

3.2.18.9 Nuclear Reaction Analysis

Nuclear Reaction Analysis (NRA) is based on the interaction of energetic (from a few 100 keV to several MeV) ions with light nuclei. Every nuclear analytical technique that uses nuclear reactions has specific isotope sensitivity. NRA is insensitive to matrix effects (maybe due to the large energy of the incident particles), and there is less interference than for methods where signals from different isotopes overlap. In NRA the nuclear reaction can produce charged particles, γ-rays and neutrons. Any of the reaction products can be detected to identify the target nucleus. If γ-ray emission is observed the technique is often referred to as PIGE (Particle-induced Gamma Emission, see PIGE Sect. 3.2.16.3). NRA measures the concentration and depth distribution of isotopes in the surface layer (~1–10 μm) of the sample but is unable to provide information about the chemical states and bonds of the elements or about the microscopic structure in the sample. Detailed information on NRA may be found in Cherniak and Lanford (1992) in the *Encyclopedia of Materials Characterization* or, and specifically for hydrogen, Fukutani (2002).

If irradiated with select projectile nuclei at kinetic energies E_{kin}, specific target isotopes can undergo a nuclear reaction under resonance conditions for a sharply defined resonance energy. To obtain depth information, the incident kinetic energy of the projectile nucleus–which must exceed the resonance energy and its energy loss per distance traveled in the sample—has to be known. To contribute to the nuclear reaction the projectile nuclei have to slow down in the sample to reach the resonance energy. Thus, each initial kinetic energy corresponds to a depth in the sample where the reaction occurs (the higher the energy, the deeper the reaction). If the used nuclear resonance is broad the reaction takes place over a large depth range. In this case the energy loss of emitted charged particles (e.g. alpha lines from the decay of target nuclei) can be used to conveniently extract the depth information.

The technique is limited by the available nuclear reactions, e.g. Table 3.16. Generally, NRA can be used for only light elements, up to calcium. This is the basis for an important property of these methods: the nuclear reaction technique is one of the few analytical techniques that can quantitatively measure hydrogen profiles in

Table 3.16 Some typical nuclear reactions used in NRA

Reaction	Incident particle energy/cross section/
$^{7}Li + {}^{1}H \rightarrow {}^{4}He + {}^{4}He$	$E \gg 3$ MeV, $\sigma \gg 3$ mb sr^{-1}, $Q = 17.3$ MeV
$^{12}C + {}^{2}H \rightarrow {}^{1}H + {}^{13}C$	$E \gg 1 - 3$ MeV, $Q = 2.3$ MeV
$^{9}Be + {}^{3}He \rightarrow {}^{1}H + {}^{11}B$	$E \gg 1 - 5$ MeV, $Q = 10.3$ MeV
$^{31}P + {}^{4}He \rightarrow {}^{1}H + {}^{34}S$	$E \gg 3$ MeV, $Q = 0.63$ MeV
$^{7}Li + {}^{1}H \rightarrow {}^{1}n + {}^{7}Be$	$E \gg 2$ MeV
$^{1}H + {}^{15}N \rightarrow {}^{12}C + {}^{4}He + \gamma$	$E \gg 6.4$ MeV, resonance
$^{19}F + {}^{1}H \rightarrow {}^{4}He + {}^{16}O + \gamma$	$E \gg 340$, 484, 872 keV, resonance
$^{27}Al + {}^{1}H \rightarrow {}^{28}Si + \gamma$	$E \gg 991$ keV, narrow resonance

solids. Combined with channeling, NRA can provide information about the location of the measured element in a crystalline lattice. NRA sensitivity and depth resolution depend on the specific nuclear reaction. Although NRA is quantitative, in most cases standards have to be used. This technique requires a particle accelerator capable of accelerating ions up to several MeV, limiting its availability.

The **nuclear material example** treated here concerns hydrogen isotopes in reactor graphite (see Le Guillou et al. 2015). Graphite has been used worldwide as moderator in the first generation of gas-cooled reactors, e.g. CO_2-cooled reactors. The first graphite-moderated nuclear facilities are now being decommissioned and several thousand tons of irradiated graphite waste are waiting for long-term disposal. Nuclear grade graphite is porous and made of synthetic polycrystalline graphite consisting of micrometer-sized well-graphitized petroleum coke grains blended by less graphitized coal tar pitch. The coke grains are made of several more or less oriented crystallites. Irradiated nuclear graphite contains activation products such as ^{14}C, ^{36}Cl and ^{3}H, whose behavior during dismantling and waste management has to be considered for safety reasons because of their activity and/or their half-life.

Tritium, with its short half-life ($\sim$12 a), contributes with ^{14}C to most of the initial radioactivity of the irradiated graphite waste. It was produced mainly by neutron activation of ^{6}Li impurities according to $^{6}Li + {}^{1}n \rightarrow {}^{4}He + {}^{3}H$. Tritium may potentially be released by isotopic exchange with water or water vapor. To anticipate its behavior, its inventory, localization and speciation in the irradiated graphite need to be analyzed.

Graphite samples from a reactor block were implanted by $^{2}H^{+}$ simulating ^{3}H. The effect of thermal treatments on the mobility of the implanted ^{2}H was then investigated at temperatures ranging from 200 °C to 1200 °C, in inert atmosphere or in a gas simulating the reactor coolant. The $^{2}H^{+}$ profiles were analyzed using the nuclear reaction:

$$^{2}H + {}^{3}He \rightarrow {}^{1}H + {}^{4}He \qquad \{3.30\}$$

with an incident ^{3}He beam a 4 MV. $^{3}He^{+}$ ions react with the implanted ^{2}H, producing 13 **MeV** protons and 2 **MeV** $^{4}He^{+}$ particles. Incident $^{3}He^{+}$ particles of 900 keV were used to adjust the maximum of the nuclear reaction cross section ($\sigma \sim$ 60 mb sr^{-1} for $E(^{3}He) \sim$ 620 keV) with the deuterium penetration range of $\sim$750 nm.

Results are displayed in Fig. 3.76, showing the evolution of the of proton spectra issued from the nuclear reaction on deuterium measured on samples after 4 h annealing at temperatures ranging from 200 to 1200 °C. The annealing process results in a decrease of the peak areas with increasing temperature. The main process driving the thermal behavior of deuterium is its release through the graphite porous structure. The release that begins roughly at 600 °C and is completed at 1200 °C can be quantified. Above 700 °C, an enlargement of the peak is observed which could correspond to a slight diffusion of deuterium toward the surface of the graphite.

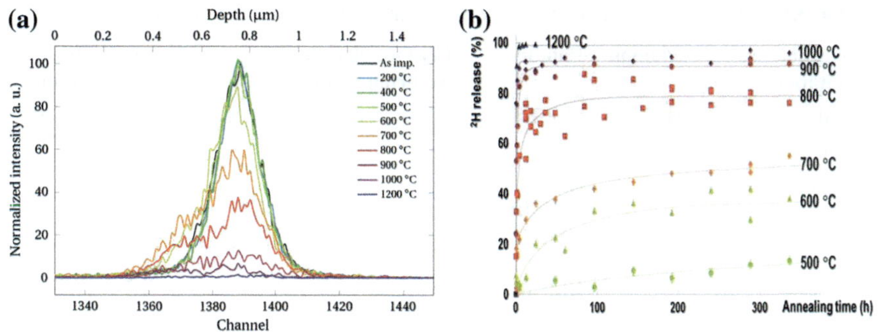

Fig. 3.76 NRA data recorded on carbon sample for various thermal treatments. **a** NRA spectra and ^{2}H concentration depth profile as recorded from an implanted sample for various temperature. **b** Deuterium releases as a function of annealing temperature. Courtesy of Dr Toulhoat

Figure 3.76 presents also the deuterium releases as a function of annealing from 2 min to more than 300 h. Release regimes were identified such as: detrapping of ^{2}H located near the free surfaces; transient regime corresponding to the thermal detrapping and diffusion; and a saturation regime. The main process driving the thermal migration of deuterium in nuclear graphite is its release through the porous structure. In the perspective of purification prior to disposal, temperatures higher than 1200 °C should allow the tritium decontamination.

3.3 Material Characterization by Techniques Under Specific Field

The analytical techniques presented in Chap. 2 are applied field free. The ones described in this chapter are passive and interactive techniques applied under fixed field that can be magnetic, electric or gravitic, constant or pulsed. Based on the classification proposed in Table 3.1, the field-specific analytical techniques adapted here are listed in Table 3.17.

3.3.1 Techniques with Field or Phonon Detection

3.3.1.1 Barkhausen Noise Analysis Under Magnetic Field

The spectrum of **Barkhausen Noise (BN)** is a noise in the magnetic output when the magnetizing field applied to a sample is changed or suppressed. Discovered by Barkhausen (1919), it is caused by rapid changes of size of magnetic domains, locally magnetically oriented atoms in ferromagnetic materials. Barkhausen's work

Table 3.17 Detection—excitation: passive versus interactive under specific field

Detection excitation	Phonon	Photon	Lepton: electron Neutrino Muon	Neutron	Ion atom
Field	BN	MoBES	–	–	Grav
Phonon	–	–	–	–	–
Photon	–	NMR SSNMR NMRTom, EPS, XMCD, PeEM, MoBAS,	–	–	–
Lepton: electron muon	–	μSR	μSR EMCD,	–	CE, Coul CV, DPV, DCP, DPP, APT
Neutron	–	–	–	–	–
Ion	–	–	–	–	–

Analytical techniques including excitation (if any) and detection for isotope, element, or species characterization (see list of abbreviations in the glossary) (*Note* excitations or detections are performed with phonons, photons, electrons, neutrons, ions (or atoms or molecules) considering the particles (plain, solvated or cluster) or their associated waves)

in magnetism and acoustics led to the discovery of BN, which evidenced that magnetization affects domains of a ferromagnetic material, rather than individual atoms alone. The Barkhausen effect is a series of sudden changes in the size and orientation of ferromagnetic domains that occurs during a continuous process of magnetization or demagnetization.

The nuclear material example deals with ferromagnetic RPV steels. Barkhausen Noise (BN) signals obtained from a group of samples formed by Fe—1 wt% Cu were investigated by Ortega Paredes et al. (2013). These authors simulate possible processes that change the microstructure of the low ferritic steel alloy doped with Cu impurities, used as pressures vessels material of nuclear reactors in service. Twelve samples were made and modified by different thermal and mechanical treatments. The BN signals obtained from those samples were processed with emphasis on treatment optimization for defect/damage reparation.

Information on the technique may be found in "Nondestructive Evaluation Techniques on the Radiation Damage of Reactor Pressure Vessel Steel Due to Neutron Irradiation" by Kim et al. (1997). The non-destructive evaluation of RPV steel with irradiation damages including micro-structure interstitial, void and dislocation may be performed by studying the changes in the hysteresis loop, Barkhausen noise amplitude and the harmonics frequency due to a neutron irradiation were measured and evaluated. The Mn–Mo–Ni low alloy steel of RPV was irradiated to a neutron fluence of 2.3×10^{19} cm^{-2} ($E \geq 1$ MeV) at 288 °C. The saturation magnetization of neutron irradiated metal did not change. The neutron irradiation caused the coercivity to increase, whereas susceptibility decreased. The amplitude of Barkhausen noise parameters associated with the domain wall motion decreased by a neutron irradiation. The spectrum of Barkhausen noise is analyzed with an applied frequency of 4 and 8 Hz, sampling time of 50 and 20 μs. The harmonic frequency shows 4, 8, 12 and 16 Hz reflected from a non-irradiated specimen. On the contrary, the harmonic frequency disappeared on the irradiated specimen.

The results suggest that the magnetic measurements may be used as a promising non-destructive evaluation method in monitoring degradation of the reactor material such as a SA-508 of PWR-RPV steel due to a neutron irradiation. Information on the technique may be found in harmonic frequency analysis of acoustic BN on neutron irradiated material by Sim et al. (2000).

3.3.2 Techniques with Photon Detection

3.3.2.1 Mössbauer Emission Spectroscopy Under Magnetic Field

MossBauer Emission Spectroscopy under magnetic field (**MoBES**) is similar to the technique described in Sect. 3.1.2.4. However, it is now performed under specific magnetic field to discriminate the electronic energy level of the atoms and molecules for fine structure analysis.

Detailed information may be found in the Tutorial "Mössbauer Spectroscopy" edited by Yoshida and Langouche (2013). Since the application of a constant magnetic field on the oscillating source may be an issue (see Fig. 3.2) the technique is normally applied in absorption mode as described in Sect. 3.3.4.6.

3.3.3 Techniques with Atom/Ion Detection

3.3.3.1 Gravimetry

Gravimetric (GRAV) analysis describes a method applied in analytical chemistry, **under gravitic field**, for the quantitative determination of stoichiometric analytes based on the mass of a reference solid. A simple example is the measurement of the weight of a solid powder suspended in a water sample: after homogenization, a known volume of water is processed for separation by filtration or centrifugation, and the collected solid is weighed after drying.

The analyte must be converted to a well-defined and crystallized solid by precipitation with a specific reagent. The precipitate collected is washed to eliminate residual amounts of reagents, dried to remove moisture and weighed until constant weight. The analyte concentration in the original sample is calculated from the volume of solution or suspension, the mass of precipitate and its chemical composition.

In other cases, it may be easier to remove the analyte by evaporation. The analyte might be collected—e.g. in a cryogenic trap or on some absorbent material such as activated carbon—and measured directly. Or, the sample can be weighed before and after it is dried; the difference between the two masses gives the mass of analyte lost. This is especially useful in determining the water content of complex materials such as foodstuffs.

Gravimetric analysis may be a precise and accurate method. Gravimetric analysis has been used to determine the atomic masses of elements within accuracy of six figures or more. Gravimetry provides insignificant instrumental error and does not require a series of standards for running tests. The method does not require expensive equipment (balance). Gravimetric analysis can also be used to calibrate other instruments in lieu of reference standards.

Actinide gravimetric tests quoted for Th and U in Table 3.18 may be used for ore samples analysis, after dissolution and separation. For example, at pH 5–9, uranium gives a red precipitate with oxine, $UO_2(C_9H_6NO)_2 \cdot (C_9H_7NO)$ (Hecht and Reich-Rohrwing 1929) that can be used for its gravimetry. The oxinate should be washed with oxine solution (Claassen and Vissen 1946) with ethanol, then dried. This technique is relatively easy to apply but it suffers from a lack of specificity. The gravimetric determination of uranium and thorium alone, as well as in the presence of about ten times the amount of cerium (III) and lanthanum, can be readily carried out by using 1-hydroxy-xanthone as complexing agent (see Dev and Jain 1961). Thorium may be determined using Alizarin-s (see Sarma and Raghava Rao 1955) or 3-Oximinomethylsalicylic acid, (see Ray 1963).

Table 3.18 Gravimetric reagent and pH required for the analysis of Th and U

GRAV analyte	Reagent name	Reagent formula	pH	References
Th(IV)	Alizarin-s	$C_{14}H_7NaO_7S$	2.7–3.5	Sarma and Raghava Rao (1955)
Th(IV)	3-Oximinomethylsalicylic acid	$2\text{-HO-3-(CH:NOH)}C_6H_3COOH$	4–5	Ray (1963)
Th(IV)	1-hydroxy-xanthone	$(C_{13}H_8O_3)$	2.6–3.5	Dev and Jain (1961)
U(VI)	1-hydroxy-xanthone	$(C_{13}H_8O_3)$	5–7	Dev and Jain (1961)
U(VI)	Oxine	$(C_9H_6NO)\ (C_9H_6NO)_2 \cdot (C_9H_7NO)$	5–9	Hecht and Reich (1929)

3.3.4 *Techniques with Photon Interaction and Photon Detection*

3.3.4.1 Nuclear Magnetic Resonance

Nuclear Magnetic Resonance (**NMR**) may be used for fluid samples as well as for solid samples. This technique can be used as a spectroscopic method on the bulk of the sample or as an imaging method to observe in a 3D mode the distribution of an isotope and its chemical species in the sample. General information may be found in the monograph "Understanding NMR spectroscopy" by Keeler (2010).

NMR is a spectroscopy which deals with absorption and reemission of electromagnetic radiation (microwaves) by material nuclei in a magnetic field. These phenomena take place at a specific resonance frequency which depends on the magnetic field strength and the magnetic properties of the isotope nuclei. Under external magnetic field, the nucleus, initially is in its nuclear ground state, splits the nuclear energy levels. The energy difference (ΔE) of a particular level becomes:

$$\Delta E = -\mu \cdot B_0^{\infty} = h\nu \tag{3.58}$$

where B_0 is the external magnetic field and μ the magnetic moment, and where h is the Planck's constant and ν the corresponding photon frequency. Typical frequencies are 60–1′000 MHz (radio frequencies). NMR allows also the observation of specific magnetic properties of the nucleus. The molecular physics of crystals and non-crystalline materials may be understood using NMR data. NMR is also routinely used in advanced medical imaging techniques, using magnetic resonance imaging (MRI).

All isotope nuclei that contain an odd number of protons and/or of neutrons have an intrinsic spin, magnetic moment and angular momentum. All nuclides with even numbers of both have a total spin of zero. The most commonly studied nuclei are ^{1}H and ^{13}C, although nuclei from isotopes of many other elements (see Table 3.19) have been and are studied by high-field NMR spectroscopy as well.

Nuclear magnetic resonance (NMR) can be used to explore electronic states via the interaction between nuclear and electron spins. NMR has been applied to a wide range of fields, such as molecular structure analysis in chemistry and for materials science.

In media with molecular mobility (fluid liquid solutions, soft matter), only intramolecular spin interactions between nuclei can be considered. In a classical liquid-state NMR experiment, Brownian motion leads to an averaging of anisotropic interactions. In such cases, these interactions can be neglected on the time-scale of the NMR experiment and only intramolecular interactions are observed. The work edited by Tycko (1994) provides information on nuclear magnetic resonance probes of molecular dynamics.

Since the resonance frequency of a nuclide in a sample is proportional to the applied magnetic field, the signal can also be exploited in **imaging** NMR, see

Table 3.19 NMR 'active' isotope spin and natural abundance

Nuclide	Spin	Abundance (%)	Nuclide	Spin	Abundance (%)	Nuclide	Spin	Abundance (%)
^{1}H	1/2+	99.985	^{17}O	5/2+	0.038	^{125}I	5/2+	100
^{2}H	1+	0.015	^{19}F	1/2+	100	^{129}Xe	1/2+	26.40
^{6}Li	1+	7.59	^{23}Na	3/2+	100	^{131}Xe	3/2+	21.23
^{10}B	3+	19.8	^{29}Si	1/2+	4.683	^{133}Cs	7/2+	100
^{11}B	3/2−	80.2	^{31}P	1/2+	100	^{195}Pt	1/2−	33.83
^{13}C	1/2−	1.11	^{35}Cl	3/2+	75.77	^{235}U	7/2−	0.72
^{14}N	1+	99.634	^{83}Kr	9/2+	11.49	^{239}Pu	1/2+	*
^{15}N	1/2−	0.366	^{113}Cd	1/2+	12.22	^{243}Am	5/2−	*

*Radioisotope

Sect. 3.3.4.3; if an object is placed in a non-uniform magnetic field the resonance frequencies of the specimen nuclei depend on where these nuclei are located in the field. Since the imaging resolution depends on the magnitude of magnetic field gradient, the effectiveness of NMR can also be improved using 2-D and 3-D multi-frequency techniques. Detailed information on modern NMR methodology may be found in the work of Heise and Matthews (2013).

The nuclear material example deals with speciation of complex extractants using NMR. In actinide science, in general, NMR studies have been forced to limit their scope to nuclei, e.g. ^{1}H, ^{13}C, ^{31}P, associated with ligands.

Amide grafted polymer has been developed using a chloromethylated resin, as inert solid support, anchored with 4-ethoxy-4-ethyl-*N*,*N*-bis-2-ethylhexylbutanamide (EEBEHBA), for the selective extraction of actinides from lanthanides in acidic matrices. The polymer has been characterized using among others ^{13}C-NMR spectroscopy (see Maheswari and Subramanian 2005). The fabricated polymeric sorbent shows greater selectivity in extracting U(VI) over Th(IV) and other lanthanides, even under high acidities (2–4 M HNO_3). The grafted polymer offers good distribution ratio values for actinides of interest over a wide range of acidity.

As an example from the organic extraction phases complexes, the work of Hirose et al. (1993) may also be cited. The coordination and complexation behavior of Zr-dibutylphosphate was investigated by ^{31}P-NMR measurement, in order to clarify the mechanism of the previously reported phenomena that the precipitates once formed at low HDBP/Zr mole ratio of 2 in an organic solution disappears with increasing HDBP/Zr. It was found that DBP^{-} (or HDBP) ligands in these species bridge two Zr ions to form 3D network structure and with interaction of HDBP in the solvent. Thanks to the NMR analysis, the mechanism was elucidated for the re-dissolution of Zr-DBP precipitates with sufficient free HDBP in highly viscous oily liquid.

3.3.4.2 Solid-State Nuclear Magnetic Resonance

Solid-State Nuclear Magnetic Resonance (**SSNMR**) spectroscopy is a kind of nuclear magnetic resonance (NMR) spectroscopy, characterized by the presence of anisotropic (directionally dependent) interactions. Basic information on SSNMR is provided in the work of Apperley et al. (2012) and detailed information on Magnetic Resonance in Solid-State Sciences may be found in the monograph of Slichter (1996).

A nuclear spin interacts with an external but also with local intra- or extra-molecular magnetic field. Interactions between nuclei are functions of their interatomic distance. Actually, these interactions depend on their orientations. In media without or with restricted mobility such as fine crystal or glass powders, or molecular aggregates, anisotropic interactions have a potential effect on the behavior of the nuclear spin system. Anisotropic interactions modify the nuclear spin energy levels and consequently the resonance frequency of all sites in a molecule, and often contribute to a line-broadening effect in NMR spectra.

High-resolution conditions in solids can be established using macroscopic sample orientation, enhancement of mobility by highly viscous sample conditions, and a variety of radio frequency irradiation patterns. While the latter allows decoupling of interactions in spin space, the others facilitate averaging of interactions in real space. Appropriate methods of sample preparation are required.

The NMR nuclear material examples deal with actinide compounds and adsorbing materials such as nano porous zeolite.

SSNMR analysis of the actinide nuclei has not been possible, except until recently. Among the actinide elements, a ^{235}U NMR study on UO_2 has been reported including hyperfine field of the 5-f electrons. Pu has attracted much attention owing to its rich chemistry. Pu is important for nuclear fuel but the electronic state of Pu ions has not been well understood. For such reasons, scientists around the world have been searching for the ^{239}Pu NMR signal. Americium oxide (^{243}Am) has also been investigated, however, the damage yielded by a recoil altered the signal. The review of Walstedt et al. (2014) on NMR studies of actinide oxides may be used for an overview.

Investigation on **^{239}Pu NMR** can be used to characterize the structure of plutonium materials. The observation of ^{239}Pu NMR has been reported by Yasuoka et al. (2012). NMR measurements of $^{239}PuO_2$ were performed, for a wide range of external magnetic field values at a temperature of $T = 4$ K. By mapping the external field dependence of the measured resonance frequency, the nuclear gyromagnetic ratio was determined to be 2.856 ± 0.001 MHz T^{-1} for $^{239}PuO_2$. Figure 3.77 shows the ^{239}Pu NMR spectrum measured for high purity PuO_2. This spectrum can be compared with the ^{239}Pu NMR spectrum measured in PuO_{2-x}, which has oxygen defects. It can be seen that the resonance frequencies are different, yielding the observed shift. This result implies that the ^{239}Pu NMR analysis has high resolution

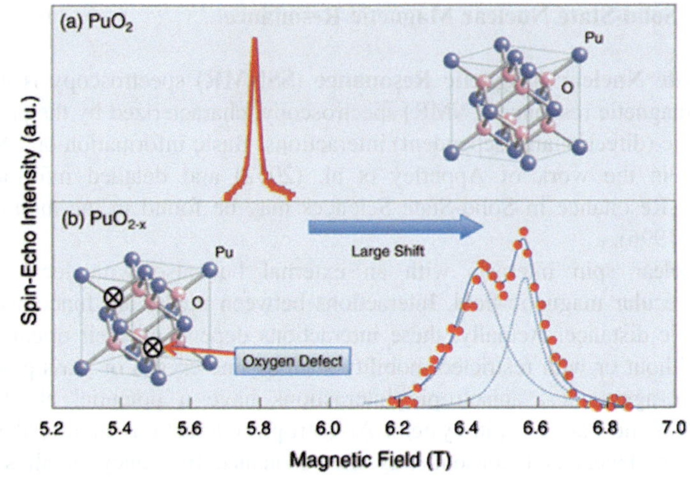

Fig. 3.77 ^{239}Pu NMR signals observed for PuO_2 and PuO_{2-x}, (Yasuoka et al. 2012, Permission © Science.)

and is very sensitive to the differences in the coordination state of oxygen and the structure difference in the solid state material.

The pore of adsorbing materials such as nano porous zeolite may be probed using ^{129}Xe NMR. **Molecular sieves** are micro-pore networks (pore size <2 nm) embedded in aluminosilicates (zeolites) or aluminophosphates. They are used to decontaminate fluids. Zeolites are built from corner-sharing $[SiO_4]^{4-}$, $[PO_4]^{3-}$ and $[AlO_4]^{5-}$ tetrahedra linked by oxygen atoms. These structures form a framework of high internal surface with cavities, cages, and channels able to host specific atoms or molecules with proper size and shape, such as noble gases. Sorbed Xe can carry information about the microporous structure by changes in the NMR spectrum versus the free bulk gas. The NMR of the isotope ^{129}Xe adsorbed gas is a convenient way to probe the nanoporous structure (see Jameson 2004). The NMR chemical shift of adsorbed ^{129}Xe is known to depend on the size and shape of the pore, the composition of the porous material, and the nature and concentration of other adsorbed atoms or molecules. Due to fast exchanges between the gas phase and adsorption sites, the ^{129}Xe NMR signals are averaged (Fig. 3.78).

^{129}Xe NMR is a powerful probe of void space in materials since the Xe chemical shift is sensitive to the pore local environments. Xe spectra contain information about the channels and pores, as reviewed by Weiland et al. (2016). Static ^{129}Xe NMR spectra of adsorbed xenon in most of the materials usually reveal a temperature and loading-dependent anisotropic and asymmetric nuclear-shielding tensor. Besides ^{129}Xe, other noble gases can be used such as ^{21}Ne, ^{83}Kr and ^{131}Xe. NMR characterizations have also been carried out on gases adsorbed in various molecular sieves.

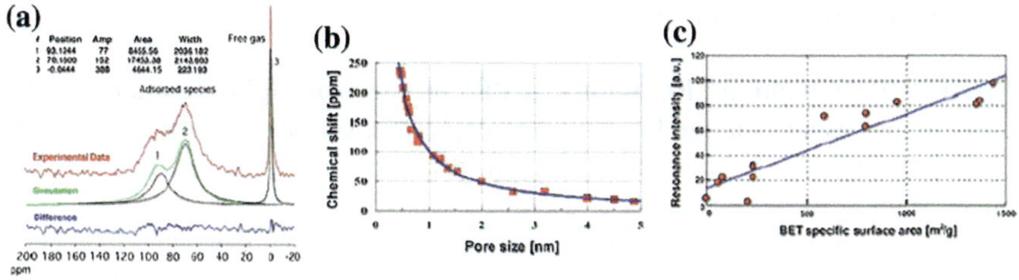

Fig. 3.78 ^{129}Xe NMR spectroscopy provides detailed information about porous materials and their void spaces. **a** Spectral analysis. **b** Adsorption pore size and chemical shift. **c** Specific surface area and NMR intensity. Adapted from Toyoko Kagaku Co. Ltd, All Rights Reserved 1997–2015

3.3.4.3 NMR 3D-Tomography

Nuclear Magnetic Resonance Imaging (**NMRI**), or **Magnetic Resonance Tomography** (**MRT**) is used to image the object using strong magnetic field gradients. A strong magnetic field, B_0, first aligns the spins of each atom of the object. Hydrogen is used in classical NMRI. As the magnetic field is directed down the center of the MRI instrument, the hydrogen protons align either towards the object head or feet, with approximately 50% going either way, effectively cancelling each other out. Next, a radio frequency pulse, which is specific to hydrogen, is applied by the MRI instrument toward the part of the body being examined. Algorithms are used to reconstruct the 3D image of the object according to the density of the absorbing ^{1}H nuclides.

The example reported here deals with the environment of nuclear material, e.g. the geosphere sample in the vicinity of a repository. NMR has been applied to image 2D and recently 3D subsurface structures. Although limited resolution is reported for imaging deep 3D structures, high resolution is obtained for shallow 2D structures, (see Jiang et al. 2015). A concept of separated transmitter and receiver loops has been adapted to obtain increasing resolutions for imaging 3D shallow structures. To evaluate 3D approach, a field campaign including surface NMR with various layouts was conducted on top of a frozen artificial hydrosystem in local environment with well known geometry.

Porous media experimental characterizations, network constructions and applications to reactive transport have been recently reviewed by Xiong et al. (2016) focusing on pore network modeling. These authors emphasize among others the use of NMR to image the interior of rock to obtain the pore structure spatial distribution across a larger scale than the microscopic techniques e.g. Blümich et al. (2009). Since the water molecules in the porous system move randomly, they probe well the pore structure. Therefore, in addition to the information produced by its imaging, the NMR also provides information on the pore size distribution by measuring the transverse relaxation time.

3.3.4.4 Electron Paramagnetic Resonance

Electron Paramagnetic Resonance (EPR) or **Electron Spin Resonance (ESR)** spectroscopy is a technique for analyzing unpaired electrons in materials. The EPR basic concepts are similar to those of NMR, but electron spins are excited instead of nuclei spins. EPR spectroscopy is useful for studying specific metal complexes or radicals. Each electron has a spin number $s = 1/2$, with magnetic momentum $m_s = 1/2$. In a magnetic field of strength B_0, the electron's magnetic moment aligns either parallel ($m_s = -1/2$) or antiparallel ($m_s = +1/2$) to the field, these alignments having a specific energy dictated by the Zeeman effect:

$$E = m_s \, g_e \, \mu_B \, B_0, \tag{3.59}$$

where: $g_e = 2.0023$ is the free electron's scaling factor, μ_B is the Bohr magneton Zeeman (1897). The difference between the lower and the upper energy state is $\Delta E = g_e \, \mu_B \, B_0$ for unpaired free electrons, implying that the splitting of the energy levels is proportional to the magnetic field e.g. Assenheim (2014).

An unpaired electron can move between the two energy levels by either absorbing or emitting a photon of energy hv such that the resonance condition is obeyed:

$$hv = \Delta E = g_e \, \mu_B \, B_0. \tag{3.60}$$

Equation (3.60) combines frequency and magnetic field values. EPR measurements are however traditionally carried out with microwaves in the 9–10 GHz region and with fields corresponding to about 3,500 G (0.35 T). EPR spectra can be generated by either varying the frequency of the incident photon on a sample or fixing the magnetic field constant. In practice, it is usually the frequency that is kept constant. Paramagnetic compounds, such as free radicals, exposed to microwaves at a fixed frequency exhibit EPR responses by increasing an external magnetic field corresponding to unpaired electrons level change between their two spin states. Since there are typically more electrons in the lower state, there is a net absorption of energy, and it is this absorption that is monitored and converted into a spectrum. Detailed information on EPR spectroscopy applications may be found in the work edited by Drescher and Jeschke (2012).

The application to solid-state physics and chemistry are the determination of the:

- valence state and coordination environment of induced paramagnetic ions;
- nature and structure of paramagnetic ion-radicals defects in solids;
- nuclear spin for some nuclei;
- mechanism of chemical reactions including reactions under irradiation.

The applications in nuclear materials are determination of valence state and coordination environment of actinides, fission and corrosion products in nuclear fuels, nuclear waste forms, and transmutation targets. EPR allows also the study of radiation damage and radiation effects in solids including fuels, waste forms (nuclear waste glasses and ceramics), transmutation targets, repository host rocks, etc.

Detailed information on the applications of EPR in radiation research may be found in the work edited by Lund and Shiotani (2014).

The nuclear material examples presented in this section concern actinides and fission products. They include:

- Studies of Radiation-Induced Paramagnetic Centers (RIPC) have been performed in borosilicate and aluminophosphate nuclear waste glasses and connections between the nature of RIPC and the structure of glasses (see Bogomolova et al. 1993).
- Determination of the nature of RIPC in e^- and γ-irradiated complex borosilicate, borophosphate, aluminophosphate, and aluminoborophosphate glasses and correlation between the type of RIPC and the structure of glasses, e.g. Fig. 3.79.
- Determination of valence state and coordination environment of paramagnetic ions of transition elements (Ti^{3+}, V^{4+}, Cr^{3+}, Mn^{2+}, Fe^{3+}, Cu^{2+}, Mo^{5+}), and U^{5+} in glasses, e.g. Bogomolova et al. (2001).
- Investigations of the structure of complex aluminoborosilicate, aluminophosphate, aluminoborophosphate, and aluminosilicophosphate glasses [e.g. Bogomolova et al. (1997)] using V_2O_5 dopant as a spin probe.
- Identification of SO_4^- ion-radical in sodium alumino-sulfate-phosphate glasses. Sulfate-ions are mainly located in the voids of aluminophosphate glass network (e.g. Bogomolova et al. 1997).
- Identification of impurity paramagnetic ions in zirconolite-, sphene-, and kosnarite-based (NZP) ceramics and determination of their positions in the crystal lattice, see Knyazev and Stefanovsky (1997).
- Study of radiation defects in ion-implanted silicate, phosphate, and fluoride glasses e.g. Stefanovsky and Stefanovskaya (2014).

The conventional materials currently enlisted for the deposition of nuclear waste are borosilicate glasses, glass ceramics, or ceramic composites. Whilst the latter has

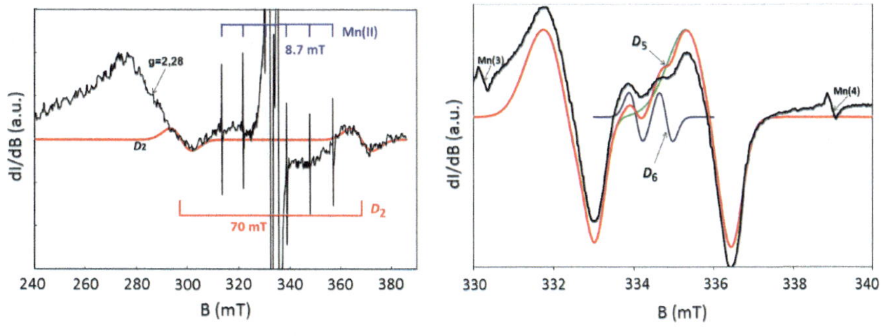

Fig. 3.79 ESR spectrum (*left*) and its fragment (*right*) of glass (mol%) 40 Na_2O, 15 Al_2O_3, 5 Fe_2O_3, 40 P_2O_5 electron irradiated to a dose of 1.0 MGy (*left*) and computer simulation of the D_2 (*left*) and D_5 and D_6 doublets (*right*), courtesy of Stefanovsky et al. (2016)

many benefits, glasses and glass ceramics remain the most widely used as they can be easily modified to incorporate a wide array of waste streams on a large scale. Within these structures, several structural modifications will result from internal radiation caused by radioactive decay of fission products. Inelastic events resulting from internal radiation can be replicated on an accelerated timescale using MeV-range β-irradiation. This approach has been used to observe the following structural changes such as:

- an increase in the polymerization and plasticity of the glassy matrix (see Yang et al. 2013);
- the formation of paramagnetic defects, namely silicon peroxy radicals ≡ Si–O–O· (see Dutt et al. 1991), E' centers ≡ Si^- (see Weeks 1994), HC_1 centers ≡ Si–O · Na^+ (see Griscom 1984; Kordas et al. 1982) and boron oxygen hole centers (BOHC) ≡ B–O · (see Griscom 1974);
- the clustering of alkalis and other charge compensators, Boizot et al. (1998);
- the radiolysis of bonds to create molecular oxygen, Neuville et al. (2003);
- the reduction of rare earths and transitional metals, see Taurines and Boizot (2011), Malchukova and Boizot (2010).

3.3.4.5 X-Ray Magnetic Circular Dichroism

X-ray Magnetic Circular Dichroism (XMCD) is a difference spectrum of two X-ray absorption spectra (XAS) taken in a magnetic field, one taken with left circularly polarized light, and one with right circularly polarized light. XMCD measures the difference in the absorption coefficient ($\mu_c = \mu^+ - \mu^-$) near an X-ray absorption edge, with the helicity of the incident circulary polarized X-ray parallel (μ^+) and antiparallel (μ^-) to the sample magnetization direction. XMCD spectra yield two basic types of information on the absorbing atom spin and orbital magnetic moment. Detailed information may be found in the work "Magnetism: a synchrotron radiation approach" edited by Beaurepaire et al. (2008).

One XMCD example on nuclear material is summarized below. Since BWR deposit materials (CRUD e.g. Fig. 3.33) are magnetic they can be studied by XMCD to characterize the spinel's part of these products. XMCD investigations on zinc ferrite nanoparticles of size ~12 nm have been carried out to study the irradiation effect of 100 MeV O^{7+} ions to an ion fluence of 5×10^{13} cm^{-2} (see Singh et al. 2015). Although as-deposited sample exhibits well-ordered cubic spinel phase, a slight degradation in crystalline phase is observed after irradiation. This behavior is analogous to that reported for zinc ferrite with similar size and can be explained on the basis of distortion in the atomic position in effect of deposited energy inside the material. XAS at O K-edge, Fe L-edge, and Zn L-edge of these samples were measured. Major changes occur in the pre-peak region for irradiated samples. The spectral analysis of the Fe L-edge reveals that the Fe 2p core-level gives rise to state splitting with $2p_{3/2}$ and $2p_{1/2}$ multiplets around ~708 and 721 eV in effect of spin-orbit coupling. The octahedral crystal field lifts the degeneracy of

$2p_{3/2}$ and $2p_{1/2}$ levels. Presence of these spectral features indicates that Fe remains in the ferromagnetic high spin 3d configuration. Intensities of various spectral features for XAS spectra at different edges were reduced after irradiation. This has been associated with the increase of broken bonds in the irradiated samples. Bond breaking in irradiated materials is the consequence of energy deposition in the form of thermal spikes. Further, to investigate magnetic properties, the Fe L-edge transitions have been recorded under magnetic field induced by right and left circularly polarized light. XMCD to XAS signal intensity ratio is estimated as 1.96 to 2.98 employing increase of 3d magnetic moment after irradiation.

3.3.4.6 PhotoElectron Emission Microscopy Spectroscopy

PhotoElectron Emission Microscopy spectroscopy (PEEM), also called photoelectron microscopy, is a used type of emission microscopy. PEEM uses local modulations in electron emission to create image contrast. The excitation is provided by UV light, synchrotron radiation or X-ray interaction of the investigated specimen. PEEM measures the electrons indirectly by collecting the emitted secondary electrons generated in the electron cascade that follows the creation of the primary core hole in the absorption process. PEEM is a surface sensitive technique because the emitted electrons originate from a very shallow layer (the nm); see Mundschau (1991).

The example of nuclear material deals with structural material of a Generation IV reactor investigated by XMCD. Iron chromium alloys are widely investigated as nuclear material for the vessel of these advanced reactor systems such as HTR or fusion reactors. Understanding of the role of nanomagnetism in the Fe–Cr alloy system using synergistic of PEEM experiments and first principle ab initio calculations are precious (see Froideval et al. 2007). For the case of FeCr alloys, the magnetic moments may be imaged locally with resolution of the order of the μm (see Fig. 3.80).

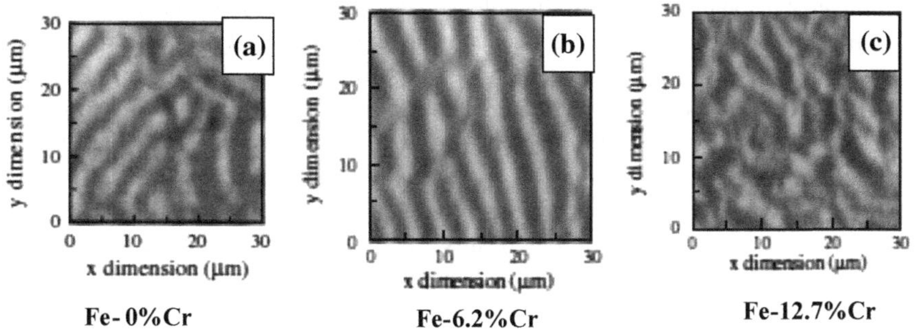

Fig. 3.80 PEEM images revealing the individual local atomic structure and magnetic moments in Fe–Cr model systems (Froideval et al. 2007) with permission © APS

3.3.4.7 Mössbauer Absorption Spectroscopy

MossBauer Absorption Spectroscopy (MoBAS) investigates nuclei in states with an angular momentum quantum number $I > 1/2$ that have a non-spherical charge distribution. This produces a nuclear quadrupole moment. In the presence of an **asymmetrical electric field** produced by an asymmetric electronic charge, distribution or ligand arrangement splitting of the nuclear energy levels is observed. The charge distribution is characterized by a single quantity called the Electric Field Gradient (EFG).

In the case of an isotope with $I = 3/2$ excited state, such as ^{57}Fe or ^{119}Sn, the excited state is split into two sub-states $m_I = \pm 1/2$ and $m_I = \pm 3/2$; this gives a two line spectrum or 'doublet'.

The magnitude of splitting, Δ, is related to the nuclear quadrupole moment, Q, and the principle component of the EFG, V_{zz}, by the relation: $\Delta = eQV_{zz}/2$.

In the presence of a **magnetic field** the nuclear spin moment experiences a dipolar interaction with the magnetic field called Zeeman splitting (see Sect. 3.3.4.3). There are many sources of magnetic fields that can be experienced by the nucleus. The total effective magnetic field at the nucleus, B_{eff} is given by:

$$B_{\mathrm{eff}} = \left(B_{\mathrm{contact}} + B_{\mathrm{orbital}} + B_{\mathrm{dipolar}}\right) + B_{\mathrm{applied}} \tag{3.61}$$

the first three terms being due to the atom's own partially filled electron shells. $B_{contact}$ is due to the spin on those electrons polarizing the spin density at the nucleus, $B_{orbital}$ is due to the orbital moment on those electrons, and $B_{dipolar}$ is the dipolar field due to the spin of those electrons.

The primary characteristics of Mössbauer spectra are usually the isomer shift, the quadrupole splitting and the magnetic splitting, alone or in combination.

The nuclear example is provided by the study of Bogé et al. (1986) on Mössbauer spectroscopy of ^{237}Np doped borosilicate glasses. Mössbauer and magnetization measurements have been carried out on borosilicate glasses containing neptunium. Different "melting" and annealing times, in various crucibles have been investigated. Using the 59.5 keV provided by a ^{241}Am source as Mössbauer resonance in ^{237}Np the valence states of neptunium ions in glasses and the hyperfine interactions have been characterized. Roughly, absorption spectra show two sites easily differentiated above 50 K by two lines. Isomer shifts (relative to $NpAl_2$) are respectively: -10.0 ± 1.0 and $+32.5 \pm 0.5$ mm s^{-1}. These resonance lines can be assigned, the first at Np^{4+}, the second at Np^{3+} species. The intensity ratio Np^{4+}/Np^{3+} varies with elaboration conditions. At 4.2 K, Np^{4+} ions exhibit a hyperfine splitting arising from relaxation phenomenon Fig. 3.81.

Hen et al. (2015) present the results obtained by superconducting quantum interference device magnetometry, specific heat, and Mössbauer spectroscopy measurements carried out on Np_2Ni_{17} polycrystalline samples. Long-range magnetic order, with a moment μ(2b) of 2.25 μB, are shown to occur below the Neel temperature $T_N = 17.5$ K on the Np (2b) sites. A non-trivial situation is observed in

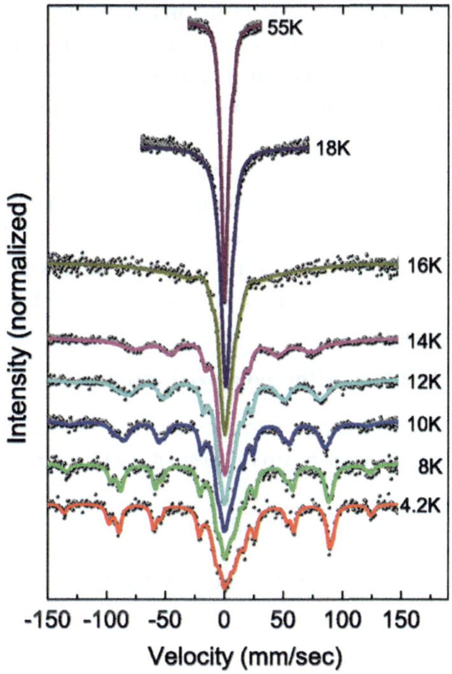

Fig. 3.81 Mössbauer spectrum of $^{237}Np^{4+}$ showing hyperfine splitting at low temperature, Bogé et al. (1986), with permission © Springer

that the other Np sites (2d) do not take part to the order transition and carry only an induced moment of about 0.2 μB below T_N.

3.3.5 *Techniques with Photon Interaction and Lepton Detection*

This class of technique under specific field is unoccupied today. An analytical potential may be identified for the development of such technique sooner or later.

3.3.6 *Techniques with Lepton Interaction and Photon Detection*

3.3.6.1 Muon Spin Spectroscopy

Muon spin spectroscopy (**muSR**) is an experimental technique based on the implantation of spin-polarized muons in matter and on the detection of the influence of the atomic or molecular surroundings on their spin motion. The motion of the muon spin is due to the magnetic field experienced by the particle. It may provide information on its local environment in a similar way to other magnetic resonance techniques. Detailed information on muSR may be found in the review of Cox (1987) or in the work edited by Carretta and Lascialfari (2007).

muSR is an atomic and molecular experimental technique dedicated to condensed matter. Here the particle is used as a probe. Its two main features are the local nature of the muon probe, due to the short range of its interactions with matter, and the time-window (10^{-13}–10^{-5} s) of the dynamical processes in the condensed media that can be investigated by this technique.

The production and detection of the muon proceed through several nuclear reactions.

Producing the polarized muons

The collision of an incident proton beam, e.g. 600 MeV, with a nucleus of a target produces a positive pion (π^+) via the possible reactions:

$$p + p \rightarrow p + n + \pi^+ \quad \text{and/or} \quad p + n \rightarrow n + n + \pi^+ \qquad \{3.31\}$$

From the subsequent weak decay of the pions (mean lifetime $\tau_\pi + = 26.03$ ns) positive muons (μ^+) are formed via the two body decay:

$$\pi^+ \rightarrow \mu^+ + \nu_\mu \qquad \{3.32\}$$

Parity violation in the weak interactions implies that only left-handed neutrinos exist, with their spin antiparallel to their linear momentum. Since the pion is spinless both the neutrino and the μ^+ are ejected with spin antiparallel to their momentum. This is the key to provide spin-polarized muon beams. According to the value of the pion momentum different types of μ^+-beams are available for μSR measurements.

Implanting the muons in the sample

The muons are implanted into the sample where they lose energy very quickly (faster than 100 ps), which is much shorter than a μSR time window (up to 10–20 μs), and all the processes involved during the deceleration do not affect the muon spin, so that the muon is thermalized without any significant loss of polarization. The μ^+ particles usually nest in interstitial sites of the crystal lattice. In most metallic samples the muon is in a so-called diamagnetic state and behaves like a free muon. In insulators or semiconductors a collective screening cannot take place and the muon usually picks-up one electron and forms an atomic structure called muonium (Mu = $\mu^+ + e^-$), which has similar size, reduced-mass and ionization energy to the hydrogen atom.

Detecting the muons

The decay of the positive muon into a positron and two neutrinos occurs after a mean lifetime of $\tau_\mu = 2.197034 \pm 0.000021$ μs (see Table 1.1):

$$\mu^+ \rightarrow e^+ + \nu_e + \bar{\nu}_\mu \qquad \{3.33\}$$

This three body decay leads to an anisotropic distribution of the positron emission with respect to the μ^+ spin direction. The positron finally annihilates on an electron (reaction {3.4}) generating 2 gamma-photons that are detected.

The muon spin motion is measured over a time range of the muon decay. The asymmetry in the muon decay correlates the positron emission and the muon spin directions. The simplest example is when the spin direction of all muons remains constant in time after implantation. In this case the asymmetry shows up as an imbalance between the positron counts in two equivalent detectors placed in front and behind the sample, along the beam axis. Each of them records an exponentially decaying rate as a function of the time t elapsed from implantation, according to:

$$N_\alpha(t) = N_0 \exp\left(-t/\tau_\mu\right)(1 + \alpha A) \qquad (3.62)$$

with $\alpha = \pm 1$ for the detector looking toward and away from the spin arrow, respectively. Actually, the huge muon spin polarization is completely outside thermal equilibrium. A dynamical relaxation toward the equilibrium unpolarized state typically is noted as an additional decay factor in front of the experimental asymmetry parameter, A. A magnetic field parallel to the initial muon spin direction probes the dynamical relaxation rate. This experimental arrangement is called Longitudinal Field (LF) μSR.

Another simple example is when implanted all muon spins precess coherently around the same magnetic field B, perpendicular to the beam axis, causing the count unbalance to oscillate at the corresponding Larmor frequency ω between the same two detectors, according to

$$N_\alpha(t) = N_0 \exp\left(-t/\tau_\mu\right)(1 + \alpha A \cos \omega t) \qquad (3.63)$$

The frequency spectrum obtained by means of this experimental arrangement provides a direct measure of the internal magnetic field intensity distribution.

Applications

Muon spin rotation and relaxation are mostly performed with positive muons. They are well suited to the study of magnetic fields at the atomic scale inside matter, such as those produced by various kinds of magnetism. Other important fields of MuSR application exploit the fact that positive muons capture electrons to form muonium atoms which behave as a light isotope of the hydrogen atom. This allows investigation of kinetic isotope effect in some chemical reactions.

The nuclear material example treated for MuSR is the investigation on Zircaloy samples.

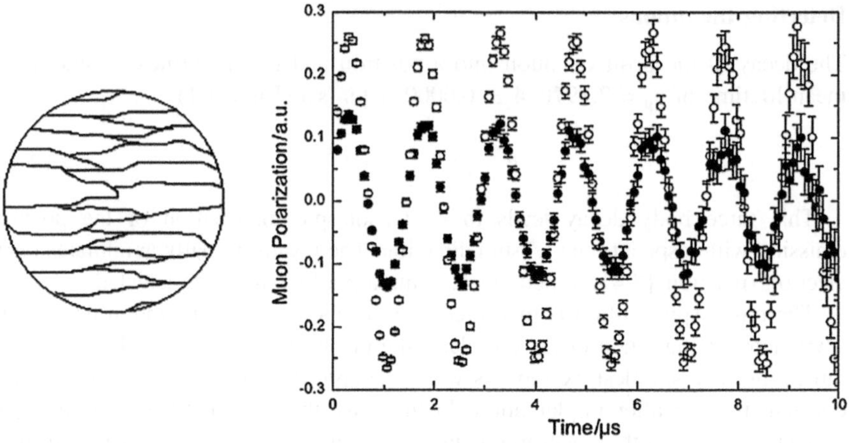

Fig. 3.82 MuSR analysis of as received (o: ε 0%) and cold rolled (•: ε 100%) Zry samples. Conditions: $T = 5$ K, traverse-field 50 Gauss. Note: reduction of signal in the deformed sample, prompt muonium (μ^+e^-) formation could not be evidenced. Degueldre et al. (2006a), with permission © Elsevier

The defects generated by plastic deformation of Zircaloy by cold rolling (as analogue of irradiation) are investigated using results provided by MuSR analysis. The study of muon properties in the material offers the possibility of examining different states of the defect centers over the sub-nanometric scale (see Degueldre et al. 2006a). Muon spin relaxation analysis (Fig. 3.82) provides information about the local order around the muon and about possible muon diffusion in the material. Combining these results allows completion of the complex picture yields by the defect creation yielded by plastic deformation due to cold rolling in Zircaloy.

3.3.7 Techniques with Lepton Interaction and Lepton Detection

3.3.7.1 Energy-Loss Magnetic Chiral Dichroism

Energy-loss Magnetic Chiral Dichroism (**EMCD**) **is** the equivalent of X-ray magnetic circular dichroism (XMCD) (see Sect. 3.3.4.4) in the electron microscope.

EMCD is the difference of two EELS spectra taken in magnetic fields with opposite polarities (see Hébert et al. 2008). By closely analyzing the difference in the EMCD spectrum, information can be obtained on the magnetic properties of the atom, such as its spin and orbital magnetic moment.

The signal-to-noise ratio in EMCD must be optimized with respect to the detector shape, size and position. An important increase in signal-to-noise ratio over previous experiments can be obtained when taking much larger detector sizes. The ideal shape of the detector was determined. Round apertures were shown to be a good compromise if placed in their optimal position (see Verbeeck et al. 2008). The theory for a simple analytical description of the EMCD experiment was developed and then applied to dynamical multibeam wave calculations and to an experimental data set. In all cases it is shown that a significant and welcome improvement of the signal-to-noise ratio is possible.

The EMCD example on nuclear material deals again with Crud from BWR. In normal conditions without specific injection, the PVR stainless steel corrodes slowly increasing both iron (Fe^{3+}) and nickel (Ni^{2+}) concentration in water. Consequently, a small amount of $NiFe_2O_4$ precipitates as secondary phase mainly on the fuel rod claddings (the larger surface in the reactor core). Due to its impurities (e.g. Mn^{2+}, Co^{2+}) this ferrite spinel is magnetic. A challenging issue when analyzing magnetic materials in TEM is to obtain quantitative magnetic parameters at the nanometer scale (see Song et al. 2015). The dynamical diffraction effects of electrons in the periodic crystal structures make however quantitative EMCD technique complicated through deviations closely related to diffraction geometry. Here, the asymmetry existing in three-beam geometry of EMCD experiments was quantitatively demonstrated and discussed for its impact on experimental measurements via the calculations of dynamical diffraction theory. Lastly, a comparison of the quantitative magnetic parameters was made with and without considering this effect based on the experimental data of spinel structure $NiFe_2O_4$.

3.3.8 *Techniques with Lepton Interaction and Ion Detection*

These techniques include mainly electron base analytical methods: the so-called **electroanalytical techniques**. In contrast with the passive electroanalytical techniques (see Sects. 3.1.3.1 and 3.1.5.1) most of the interactive electroanalytical techniques make use of electric field during operation. The techniques are potentiostatic or amperostatic that means operating at constant potential or at constant current. They make use of a potential control or a control of the current. The following techniques belong to this important class of analytical techniques:

- electrical conductivity (EC)
- coulometry (Coul)
- voltammetry and the derivative techniques such as cyclic voltammetry (CV) differential pulse voltammetry (DPV)
- direct current polarography (DCP), and the derivative techniques such as normal pulse polarography (NPP) and differential pulse polarography (DPP).

The list is exhaustive and some of them are applied on nuclear materials as shown for specific examples through this section. Detailed information may be found in the work "Electroanalytical Methods" edited by Scholz (2010).

The electroanalytical techniques make use of at least two electrodes: a working electrode and a counter electrode that can be a reference electrode. To optimize the system, three electrodes are used in today's practice: a working electrode, a counter electrode and a reference electrode. For conductivity two electrodes, i.e. two working/counter electrodes, are used with AC potential/current signal. The other techniques utilize a potentiostat: an electronic instrument that controls the potential difference between the working electrode and a reference electrode.

3.3.8.1 Electrical Conductivity

Electrical Conductivity (**EC & CE**) of an electrolyte solution is a measure of its ability to conduct electricity. Conductivity measurements or monitoring are used routinely in industrial and environmental applications as a fast, inexpensive and reliable way of measuring the ionic content in a solution.

Conductivity is often linked to the Total Dissolved Solids (TDS). The SI unit of conductivity is Siemens per meter (S m^{-1}). High quality deionized water has a conductivity as low as 5 μS m^{-1}, drinkable tap water typically has a conductivity in the range of 5–50 mS m^{-1}, and sea water about 5 S m^{-1}. The interpretation of ionic conductivity in dilute electrolyte solutions, concentrated solutions, low-temperature or glass-forming molten salts, ionic melts, molecular fluids, and fluids of geological and industrial interest is discussed in a single monograph "The Interpretation of Ionic Conductivity in Liquids" by Smedley (1980).

Conductivity is traditionally determined by measuring the AC resistance of the solution between two electrodes. The unit of conductivity is S m^{-1} refers to 25 °C when not indicated to another temperature. Sometimes encountered is mho (reciprocal of ohm): 1 mho m^{-1} = 1 S m^{-1}. Resistance, R, the reciprocal of the conductance, is proportional to the distance, l, between the electrodes and is inversely proportional to the cross-section area of the sample, A. Writing ρ for the specific resistance (or resistivity),

$$R = \frac{l}{A}\rho \tag{3.64}$$

In practice the conductivity cell is calibrated by using solutions of known specific resistance, so the quantities l and A need not be known precisely. If the resistance of the calibration solution is, a cell-constant, C, is derived.

The specific conductance, κ is the reciprocal of the specific resistance.

$$\kappa = \frac{1}{\rho} = \frac{C}{R} \tag{3.65}$$

Conductivity is also temperature-dependent. Dilute solutions follow Kohlrausch's Laws (1874) of concentration dependence and additivity of ionic contributions. Onsager (1926, 1927) gave a theoretical explanation of Kohlrausch's law by extending Debye–Hückel theory (1924).

This type of measurement is not ion-specific; it can sometimes be used to determine the TDS if the composition of the solution and its conductivity behavior are known. It should be noted that conductivity measurements made to determine water purity will not respond to non-conductive contaminants, therefore additional purity tests are required.

For example, conductivity measurements are used to monitor quality in BWR water. Conductivity measurements in BWR reactor water are part of the extensive efforts made in the field of water chemistry to achieve the targets of radiation field control, to keep the integrities of piping and fuel elements and to reduce the radioactive waste generation (see Ishigure 1996). BWRs have experienced extensive intergranular stress corrosion cracking in their austenitic stainless steel reactor coolant system piping, resulting in serious adverse impacts on plant capacity factors. These phenomena depend on reactor water chemistry, particularly on the concentrations of ionic impurities and oxidizing radiolysis products as well as on material conditions. Intensive monitoring of the reactor water chemistry is carried out (see Fig. 3.83); in particular, the water conductivity is measured on-line in situ at elevated temperature as well as at room temperature at the sampling point (see Uchida et al. 2008).

Finally, cement is foreseen as waste form matrix and its low permeability is desired. The electrically based control of the moisture flow in cement-based materials relies on an 1D measurement. Imaging of moisture distribution by

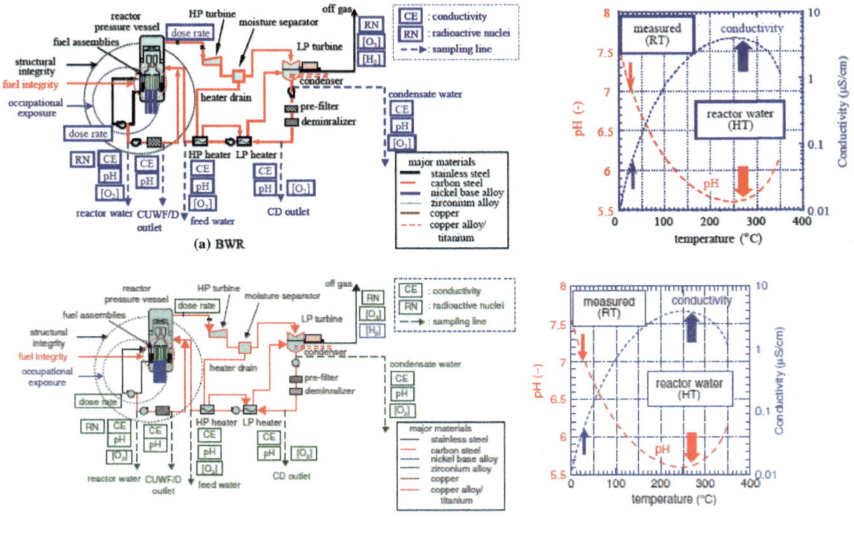

Fig. 3.83 Key analysis feature and their location in BWR system, and conductivity and pH at room and elevated temperature for pure water, adapted from Uchida et al. (2008) with permission

Electrical Resistance Tomography (**ERT**) has been reported by Hallaji et al. (2015). The aim was to study whether ERT could give information on unsaturated moisture flows in cases where the flow is non-uniform. In the experiment, the specimens were successfully monitored by ERT during the water ingress.

3.3.8.2 Coulometry

Coulometry (**COUL**) is the term for electroanalytical techniques that quantify the amount of matter reacting during electrolysis by measuring the amount of electricity (in coulombs) consumed or produced. There are two categories of coulometric techniques. The **potentiostatic coulometry** that requires a constant potential (E) during the electrolysis ($I(t)$) uses a potentiostat. The other, called coulometric titration or **amperostatic coulometry**, keeps the current constant (I) requiring an amperostat and a measure of the potential ($E(t)$).

For a half reaction as given by reaction {3.6}, the amount of redox sensitive reagent (the oxidizing or the reducing reagent) undergoes by redox its reduction or oxidation with the exchange of n electrons. During the electrolysis (coulometry), the charge Q exchanged corresponding to the number of mole N of oxidizing or reducing compound produced by quantitative electroreaction, is given by:

$$Q = n\,F\,N \tag{3.66}$$

In potentiostatic conditions:

$$Q = \int I\,dt \tag{3.67}$$

In amperostatic conditions:

$$Q = I\,t \tag{3.68}$$

Figure 3.84 summarizes both coulometric conditions in a comparative way.

The nuclear material example deals with the coulometry of uranium, first in potentiostatic, secondly in amperostatic conditions. Precise and accurate determination of **uranium** (U) and plutonium (Pu) in nuclear fuels is an essential requirement in nuclear fuel cycle for chemical quality.

The potentiostatic coulometric determination of uranium is suggested with a platinum working electrode by Davies et al. (1970). The experimental conditions require first the reduction of uranium(VI) to uranium(IV) at a platinum working electrode, by controlled-potential. The procedure has been used successfully as a subsidiary method in the routine determination of uranium in pure uranyl nitrate solutions. The platinum electrode has several important practical advantages. The consecutive determination of iron(III) and uranium(VI), or plutonium(IV) and uranium(VI) can be carried out with the same working electrode in the same

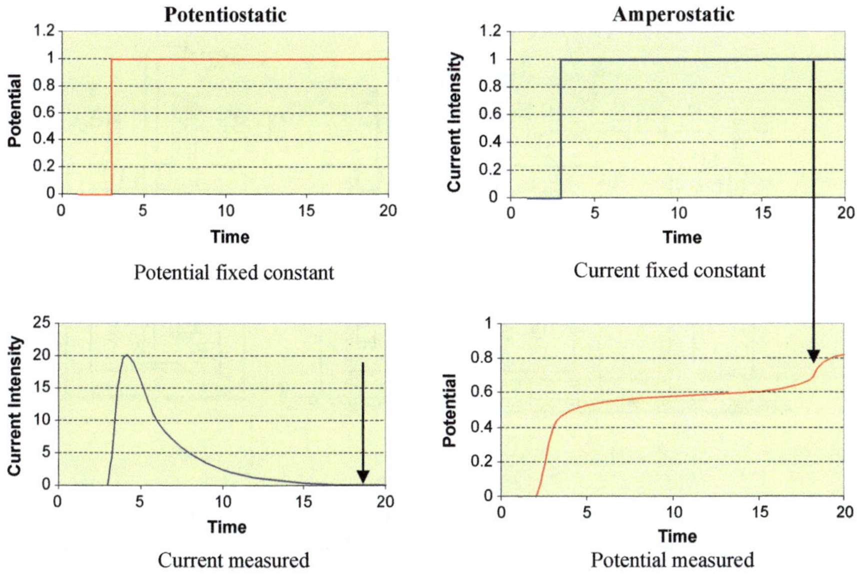

Fig. 3.84 Comparison of the $E - I$ conditions required for both coulometric investigations

solution and the coulometric oxidation of uranium(IV) to uranium(VI) is practicable. The rate of stirring of the cell liquor is much less critical in the case of the platinum electrode. Two main problems had to be overcome before a practical procedure could be achieved; hydrogen evolution during the uranium(VI)-(IV) reduction had to be eliminated so that 100% current efficiency could be obtained for the desired reaction, and electrode-surface poisoning phenomena had to be controlled so that reaction times could be kept reasonably short.

The constant-current coulometric determination of uranium in the pure metal has been reported by Merciny et al. (1981), revisiting the Degueldre (1978) thesis on this topic. The method is proposed for the highly precise and accurate constant-current coulometry of uranium in high-purity uranium metal. Precisely weighed amounts of uranium and pure iron are dissolved in 7 M sulfuric acid containing some hydrogen peroxide (40% vv). The solution is quantitatively transferred to the coulometric cell by rinsing with 1 M H_2SO_4, saturated with cerium(III) sulfate. The first step is the quantitative electro-chemical reduction to U (IV), Fe(II) and Ce(III) on a gold gauze electrode at constant current (100 mA) until evolution of hydrogen is observed. The hydrogen is then removed by flushing the solution with very pure nitrogen until the potential of a platinum gauze electrode reaches a constant value. Oxidation on the gold gauze electrode is carried out under precisely controlled constant current; after the quantitative oxidation of U(IV) to U (VI) and Fe(II) to Fe(III), and crossing the end-point. This end-point is determined very precisely potentiometrically through back-titration by successive current injections of 10 mA during 1 s. The method was tested on a National Bureau of

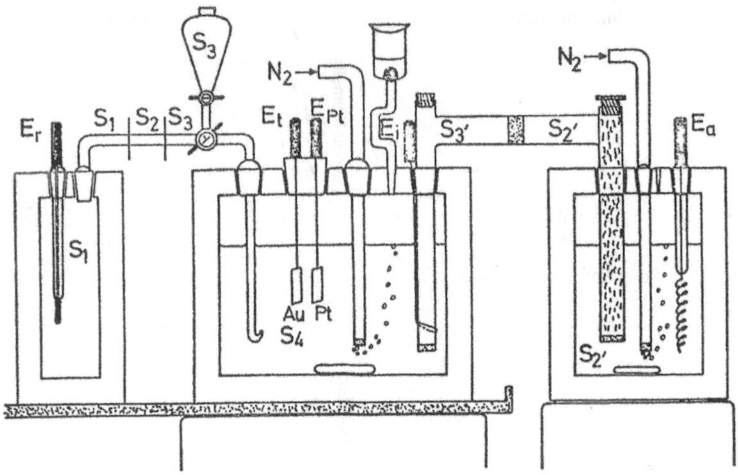

Fig. 3.85 Coulometric cell utilized by Degueldre (1978). S for solution. E_r: Reference electrode S_1 KCl 1 m, E_t: Working electrode S_2 K_2SO_4 saturated, E_{pt}: Pretreatment electrode S_3 S_3' H_2SO_4 1M, E_i: Indicative electrode S_4 H_2SO_4 1M, $Ce(SO_4)_2$–U–Fe, E_a: Auxiliary electrode S_2' H_2SO_4 0.5M, K_2SO_4

Standard reference material, uranium (NBS 960). Figure 3.85 shows the coulometric cell utilized for these quantitative tests.

3.3.8.3 Voltammetry

Cyclic Voltammeter (**CV**) is a type of potentiodynamic electrochemical measurement. In a cyclic voltammetry test, the working electrode potential is scanned linearly versus time. In a CV potential program, after a first linear scan has reached the end potential, the working electrode's potential is scanned in the opposite polarity to return to the initial potential. Cycles of potential scan may be repeated as several times. The current at the working electrode is recorded versus its applied potential to yield the cyclic voltammogram. Cyclic voltammetry is used to study the electrochemical properties of an electrochemical analyte in solution. Detailed information on cyclic voltammetry may be found in the work of Kissinger and Heineman (1983).

In cyclic voltammetry the electrode potential (E) ramps linearly versus time in cyclic phases (see Fig. 3.86). The rate of voltage change over time is called scan rate (V s^{-1}). Typical scan rates are in the range 20–2000 mV s^{-1}. The electroanalytical data are plotted as current (i) vs. potential (E). During the initial forward scan an increasingly reducing potential is applied; thus the cathodic current will initially increase over this time period due to a reducible analyte. At some point (cathodic peak at E_{pc}, i_{pc}) after the reduction potential of the analyte is reached, the cathodic current will decrease as the concentration of reducible analyte is depleted.

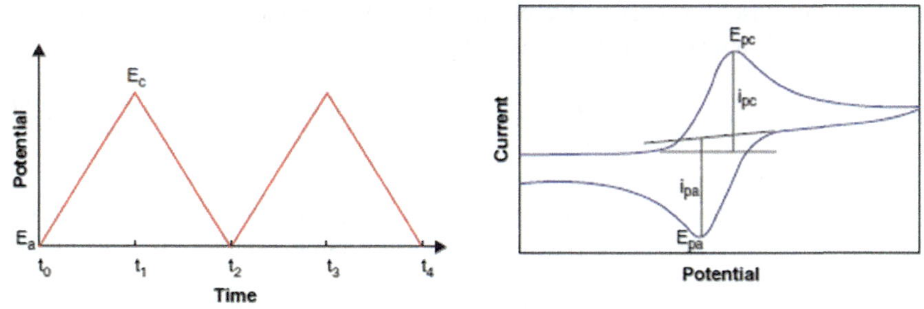

Fig. 3.86 Cyclic voltammetry: potential scan and intensity signal recorded for a redox reagent

If the redox couple is fast (reversible), the reduced analyte will start to be re-oxidized during the reverse scan, giving rise to an anodic current (with an anodic peak E_{pa}, i_{pa}); see Fig. 3.86. The more reversible the redox couple is, the more similar the oxidation peak will be in shape to the reduction peak. Consequently, the cyclic voltammogram provides information about redox potentials and electrochemical reaction rates. Detailed information on the theory of the stationary electrode is to be found in the key work of Nicholson and Shain (1964).

The utility of cyclic voltammetry is dependent on the electrochemical properties of the analyte. The analyte has to be electroactive within the scanned potential window. It is also desirable for the analyte to display a reversible CV wave in these conditions quantification of the analyte is easier.

Even a reversible couple may contain polarization (diffusion) and overpotential (kinetic) factors. An hysteresis between the reduction (E_{pc}) and oxidation peak (E_{pa}) potentials is observed. This overpotential emerges from a combination of analyte diffusion rates and intrinsic activation barrier of transferring electrons from the electrode to the electroactive species. A theoretical description of polarization is described by the Cottrell (1903) equation. In an ideal system, for a potential step, the current is given by:

$$i = \frac{nF\,A\,c_j^0\,\sqrt{D_j}}{\sqrt{\pi\,t}} \tag{3.69}$$

with n the number of electrons (to reduce/oxidize one molecule of analyte j), F the Faraday constant, 96,485 K mol^{-1}, A the area of the (planar) electrode in (cm^2), c_j^0 the concentration of the electroactive analyte j (mol cm^{-3}), D_j the diffusion coefficient for species j (cm^2 s^{-1}) and t the time (s).

Deviations from linearity in the plot of i versus $t^{-1/2}$ sometimes indicate that the redox event is associated with other processes, such as complication or a change in molecular geometry. Usually the Cottrell equation is simplified to i = k $t^{-1/2}$, where k is a constant. In addition, the scan rate square root $(\nu)^{1/2}$ may be used in place of

$t^{-1/2}$. In this case, for a reversible electroactive analyte, the difference between the reduction and oxidation peaks is given by:

$$E_{pa} - E_{pc} = 56.5/n \text{ mV} \quad (3.70)$$

for a process involving n electrons.

Reversible couples have a ratio of oxidation (i_{pa}) and reduction (i_{pc}) peak currents near to unity ($i_{pa}/i_{pc} = 1$). Even with a reversible couple this ratio may be away from unity in the presence of subsequent chemical reactions.

In a stirred solution, a Nernst diffusion layer $\sim 10^{-2}$ cm thick, lies adjacent to the electrode surface. Beyond this region is a laminar flow region, followed by a turbulent flow region which contains the bulk solution. Because diffusion is limited to the narrow Nernst diffusion region, Nernstian equilibrium is maintained and diffusion-controlled currents can be obtained. In this case, Fick's Law for mass transfer diffusion can be simplified (Randles 1948; Ševčík 1948) to give at 25 °C (298.15 K) the peak current:

$$i_p = (2.69 \cdot 10^5)\, n^{3/2} A\, D_j^{1/2}\, v^{1/2}\, C_j \quad (3.71)$$

Here, (n) is equal to the number of electrons exchanged in the electroreaction, (A) is the surface area of the working electrode in cm^2, (D_j) is the diffusion coefficient ($cm^2\ s^{-1}$), (v) is the sweep rate ($V\ s^{-1}$), and (C_j) is the molar concentration of j in the bulk solution ($mol\ cm^{-3}$).

Cyclic voltammetry has become an important and widely used electroanalytical technique in many areas of chemistry. It is often used to study electroanalytical processes, to determine the stability of reaction products and the formal reduction potential of an analyte, which can be used as an identification tool. In addition, because concentration is proportional to current in a reversible reaction, the concentration of an analyte can be determined; a calibration curve is however needed.

A number of **pulse voltammetric** techniques are available, which differ in their potential **pulse** wave forms, the number of sampling points, and whether a solid electrode (**voltammetry**) or a mercury drop electrode (polarography see Sect. 3.3.8.4) is used. Figure 3.86 presents the most used pulsed voltammetry potential forms namely: the **Normal Step Voltammetry** (**NSV**), the **Normal Pulse Voltammetry** (**NPV**), the **Differential Pulse Voltammetry** (**DPV**) and the **Square Wave Voltammetry** (**SWV**). The current registered by applying pulsed voltammetry passes from a tailed to a Gaussian like peak. In addition, the superposition of the pulse allows reading of the signal at an optimal time when the capacitive current becomes negligible, reducing the detection limit by about 2 orders of magnitude. Detailed information on the theory and applications of pulse voltammetry may be found in the work of Molina and González (2007).

Alternatively, **Alternating Current Voltammetry** (**ACV**) may be used instead of normal pulsed voltammetry to enhance the sensitivity of voltammetry, e.g. Bond and O'Halloran (1976).

The applications of voltammetry for actinide determination are numerous. Potentiostatic voltammetric method has been used successfully by Kuperman et al. (1989) for the analysis of sub-microgram quantities of neptunium and plutonium in solution. This analysis is rather easy and fast and can be performed with a precision of 5% with standard addition.

Finally, **Electrochemical Impedance Spectroscopy** (**EIS**), a non-destructive technique, measures the impedance (reciprocal of capacitance) of a system over a range of AC potential frequencies. The frequency of the current is displayed, e.g. on Nyquist plot (Huang et al. 2016). Analysis can be performed on-line providing time-dependent information about the properties but also about ongoing processes such as corrosion or any other electrochemical process.

The nuclear material example treated here deals with the actinide electrochemistry in molten salt. This is a very important domain since pyro-reprocessing in molten salt is foreseen as a key step in the recycling of fissile material in the Gen IV + reactors (molten salt reactors).

Electrochemical oxidation of U(IV) has been studied in molten $LiF–BeF_2–ZrF_4$ (65.6–29.4–5.0 mol%) at 500 °C by means of voltammetry (scan rate, v, 0.01–100 V/s), chronopotentiometry, and chronoamperometry at unsheathed pyrolytic graphite and Pt − 10% Rh electrodes. The disproportionation of the electrochemically produced U(V) has been observed when testing the uranium solutions in molten salt using the utilized electrochemical techniques (see Manning and Mamantov 1968).

Electrochemical studies on plutonium in molten salts have been reported by Bourgès et al. (2007). Plutonium was studied in ternary eutectic mixture $NaCl–KCl–BaCl_2$, equimolar mixture NaCl–KCl and pure $CaCl_2$ as well as in liquid gallium at 1073 K. The formal, or apparent, standard potential of Pu(III)/Pu redox couple in eutectic mixture of $NaCl–KCl–BaCl_2$ at 1073 K determined by potentiometry was found to be −2.56 V versus Cl_2, 1 Atm/Cl^- reference electrode. In NaCl–KCl eutectic mixture and in pure $CaCl_2$ the formal standard potentials deduced from cyclic voltammetry are −2.54 and −2.51 V respectively. These potentials led to the calculation of the activity coefficients of Pu(III) in the molten salts.

The applicability of square wave voltammetry (SWV) and normal pulse voltammetry (NPV) to on-line monitoring of actinide concentrations in molten chlorides was investigated for use in pyrometallurgical processes (see Iizuka et al. 2001). For SWV curves for plutonium, the concentration dependence of the peak current was not linear at high concentration. However, the limiting currents for uranium and plutonium determined by NPV were proportional to their concentrations up to about 1.7 wt%. The fission product elements are not likely to affect the accuracy in determination of actinide concentrations because the limiting currents for uranium and plutonium were not influenced by the addition of gadolinium. It was found to be difficult to determine neptunium concentration in the presence of plutonium due to the overlapping of the current waves. From those results, NPV was considered more promising, although the optimization of applied potential waveform is necessary for use at higher concentrations.

Cyclic voltammetry has been tested for the quantitative analysis of actinides in molten salts (see Tylka et al. 2015a, b). Actinide voltammograms recorded in chloride salt show well-defined peaks for U(III) and Pu(III), see Fig. 3.89, that can be exploited for analytical purposes. Equation 3.73 describes the peak current as a function of concentration for reversible reactions involving the formation of an insoluble product.

$$i_p = 0.61\,AC\sqrt{\frac{n^3F^3D_i v}{RT}} \tag{3.72}$$

Here A is the area of the working electrode (cm^2), v is the scan rate ($V\ s^{-1}$), T is the temperature of the melt (K), D_i is the diffusion coefficient of the electroactive species ($cm^2\ s^{-1}$), n is the number of electrons involved in an electrode reaction, and R is the universal gas constant ($J\ mol^{-1}\ K^{-1}$). Initial electrochemical measurements using CV on various salt concentrations were taken under the same experimental conditions as those used previously (see Tylka et al. 2015a, b), for salt concentrations, a 0.075-cm-diameter working electrode and 50 $mV\ s^{-1}$ scan rate.

In a previous work "Method Development for Quantitative Analysis of Actinides in Molten Salts", Tylka et al. (2015a, b) demonstrated that by following an optimized procedure, cyclic voltammetry can be used for very precise, real-time quantitative concentration measurements of actinides in molten salts. The suitability of the procedures was tested over a wide range of concentrations comparable to those expected in the normal operation of an electrorefiner used in the pyrochemical processing of irradiated nuclear fuel. It was found that electrochemical methods can be used for very precise concentration measurements up to approximately 2 wt%. For higher actinide concentrations, the diffusion coefficient was found to decrease. The application of voltammetry was also investigated for analyzing systems containing U^{3+} and Pu^{3+}.

3.3.8.4 Polarography

Direct Current Polarography (DCP) is a type of voltammetry. A dropping mercury electrode (DME) is the working electrode. DME is adopted for its renewable surface electrode and for its wide cathodic ranges owing its large hydrogen surtension. DCP was invented by Heyrovský (see Heyrovský 1922) and presented in his work "Electrolysis with a dropping mercury cathode, Part I: deposition of alkali and alkaline earth metals" (Heyrovský 1922), for which he received the Nobel Prize in 1959.

DCP is a controlled electrolysis whose response is dictated by the diffusion mass transport of an electroactive species. DCP is originally based on the electrolysis of electroactive species in solution with two electrodes, one polarizable (DME) and one unpolarizable (reference with large surface). The DME consists of mercury dropping regularly from a capillary in the studied solution. Polarography potential follows a linear-sweep from the initial potential to the final potential. The linear potential sweep controls the electroactivity itself limited by the diffusion of the electroactive species. Consequently, the plot of the polarographic current vs. potential record of an experiment has the typical sigmoidal shape. A plot of the current vs. potential in a polarography experiment shows also current oscillations corresponding to the drop surface increase followed by their falling from the capillary. A sigmoidal curve results from the connected maxima current from each drop. The limiting current on the sigmoid plateau is called the diffusion current because diffusion is the principal contribution to the flux of electroactive material for these potentials (see Henze 2001).

There are various limitations in particular for the classical polarography experiment for quantitative analytical measurements. If the current is continuously measured during the growth of the Hg drop, a contribution from capacitive current is also measured. As Hg flows from the capillary, there is an initial large increase in the surface area ($\delta A/\delta t$) and consequently an initial current of capacitive origin. Toward the end of the drop life, the change in the surface area diminishes and the contribution of capacitance reduces. Simultaneously, any redox process which occurs will result in faradaic current that decays approximately as the square root of time. Since the exponential decay of the capacitive current is faster than the faradaic current decay; the faradaic current is larger at the end of the drop life. Because the potential is changing during the drop lifetime, a small continuous contribution capacitive current ($\delta E/\delta t$) is recorded, even at the end of the drop when $\delta A/\delta t$ is small. With the usual signal to noise ratio DCP allows detection limits of only approximately 10^{-5} or 10^{-6} M.

Quantitative information

The Ilkovic equation is a relation used in polarography relating the diffusion current (I_d) and the concentration of the depolarizer (c), which is the substance reduced or oxidized at the dropping mercury electrode. The Ilkovič (1934) equation has the form:

$$I_d = k\,n\,F\,D^{1/2} m^{2/3} t^{1/6} C \tag{3.73}$$

where k is a constant which includes π and the density of mercury, and with the Faraday constant (F) has been evaluated at 708 for max current and 607 for average current, D is the diffusion coefficient of the depolarizer in the medium ($cm^2\ s^{-1}$), n is the number of electrons exchanged in the electrode reaction, m is the mass flow rate of Hg through the capillary ($mg\ s^{-1}$), and t is the drop lifetime in seconds, and C is depolarizer concentration in mol cm^3. The dependence of limiting currents on the diffusion constant, on the rate of dropping and on the size of drops was originally published by Ilkovič (1934). The potential corresponding to the half of the sigmoid wave is called $E_{1/2}$. It corresponds to the redox potential of the redox couple when the redox reaction is fast (reversible).

Differential Pulse Polarography (DPP) is a technique which superposes discrete potential pulses (e.g. 5 or 10 or 20 mV) over a linear potential ramp to obtain the experimental polarogram (see Fig. 3.87). The experimental parameters for DPP

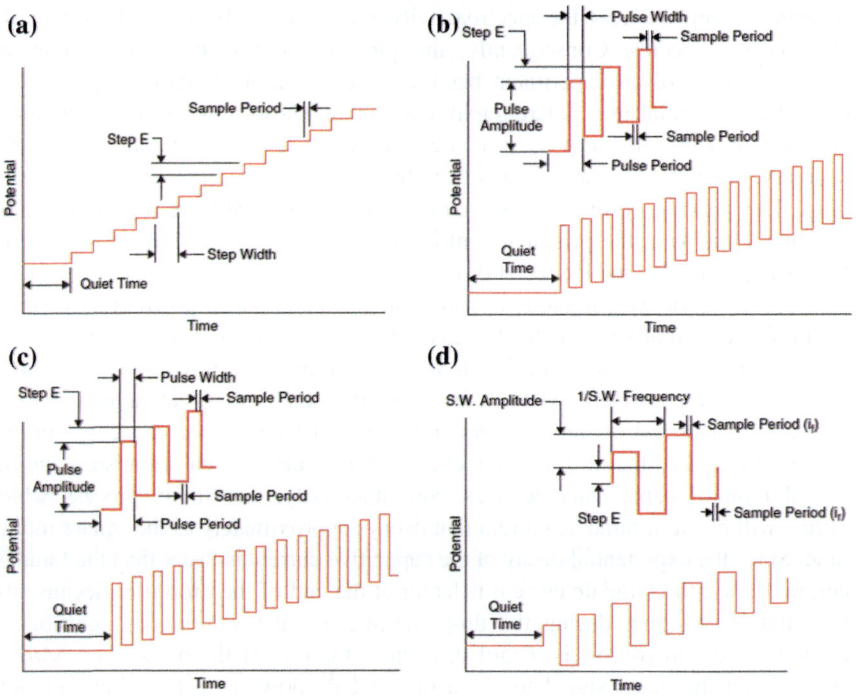

Fig. 3.87 Potential wave form for sampled current voltammetry or polarography, **a** for normal step voltammetry, **b**. for normal pulse voltammetry, **c** for differential pulse voltammetry and, **d** for square wave voltammetry. Adapted from © 2000–2009 Bioanalytical Systems (BASi Epsilon)

are accurately timed drop lifetimes and potential step duration of 50–100 ms at the end of the drop lifetime. With DPP, each potential step has the same amplitude, and the return potential after each pulse is slightly negative (for cathodic) of the potential prior to the step (Fig. 3.88).

The differential pulse polarogram is obtained by measuring the current immediately before the potential step, and again just before the end of the drop lifetime. The analytical current in this case is the difference between the current at the end of the step and the current before the step. This differential current is then plotted versus the average potential before the step and the step potential to obtain the differential pulse polarogram. Because this is a differential current, the polarogram in many respects is the differential of the sigmoidal polarogram. Consequently, the DPP record is peak-shaped.

The DPP theory has been developed by Barker and Gardner (1958) and revisited by Barker (1958) as well as by Parry and Osteryoung (1965). The equations describing the peak properties (E_p and I_p) are:

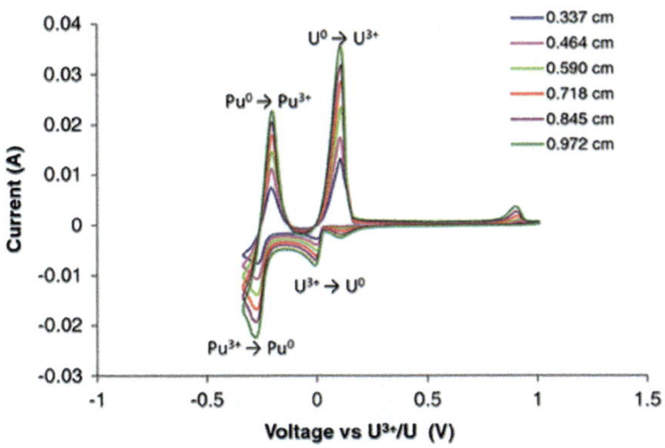

Fig. 3.88 Cyclic voltammograms obtained at different immersion depths of working electrode. Conditions: salt composition 0.45 ± 0.05 wt% U and 1.42 ± 0.14 wt% Pu, v = 50 mV s^{-1}. From Tylka et al. (2015a), with permission from Argone National Laboratory and US DOE from work performed under N° DE-AC02-06CH11357

$$E_p = E_{1/2} \pm \Delta E/2 \tag{3.74}$$

$$\Delta i_{\mathrm{Max}} = I_p = n^2 \mathrm{F}^2/(4\ \mathrm{R}T) A\, C\, \Delta E (D/\pi\, \tau)^{1/2} \tag{3.75}$$

A detection limit may be evaluated for each given experimental polarogram using Eq. 1.3:

$$C_{\mathrm{DL}} = 3\, \sigma_b / I \quad \text{with} \quad I = i\, C^{-1} \tag{3.76}$$

Differential pulse polarography has even better ability to discriminate against capacitive current because it measures a difference current. Limits of detection with DPP are 10^{-8} – 10^{-9} M.

The characterization of actinides by differential pulse polarography can be performed when the potential is compatible with the redox window of the solvent. In water, uranium has been analyzed by DPP at the ultra-trace level using the nitrate reduction catalytic wave following extraction separation Keil (1978).

Along the nuclear fuel cycle several extraction processes are found at various steps such as: for the uranium extraction from the phosphates using uranium extraction by TOPO—kerosene (or alkanes), and for the spent fuel reprocessing using uranium extraction by TBP—kerosene (or alkanes). To study the potential of electroanalytical techniques, DCP and DPP were tested on extraction phases (aprotic solvent) mixed in protic solvent i.e. alcohol, e.g. Degueldre and Taibi 1996) or aprotic solvent (dimethylformamide: DMF) able to dissolve support electrolyte. In these conditions the electroreduction of U(VI) in U(IV) may be used to analyze uranium directly in the extraction phases.

As an example, DC polarogram of U(IV) in the organic extraction phase 50% DMF—30% TBP—20% n-hexane were successfully recorded in the 30% TBP—

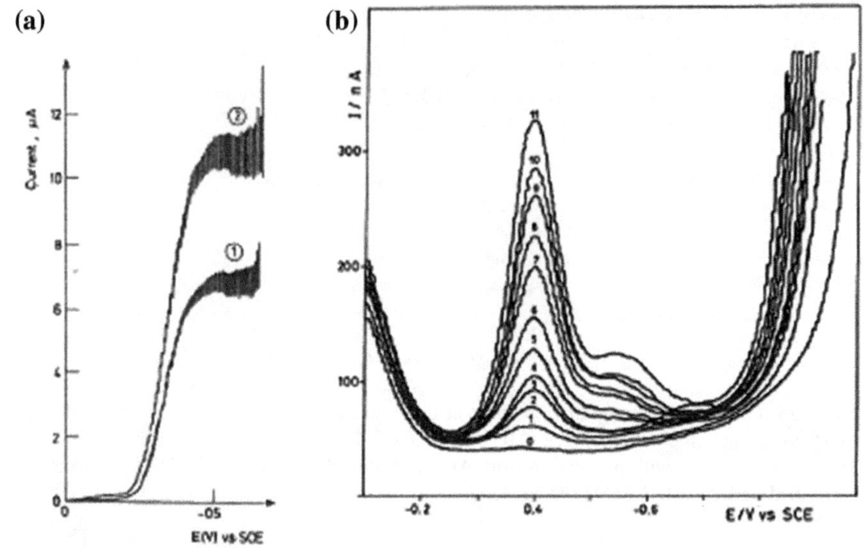

Fig. 3.89 **a** Polarogram of U(IV) in the organic extraction phase 50% DMF—30% TBP—20% n-hexane. Solution 1–57 ml of DMF—TBP solution 20 ml (30% TBP—70% n-hexane from 50 ml extraction phase in equilibrium with 50 ml HNO_3 solution plus 100.77 mg uranyl nitrate. 2-idem plus standard addition (23.23 mg) of solid uranium nitrate); record: E versus SCE, H 65 cm, drop life time: 2 s, calculated extraction yield for U(VI): 95 ± 2%. Degueldre and Meklati (1983), with permission © Elsevier. **b** U(VI) polarogram, conditions: 50% DMA 15% TBP 35% n-hexane (v/v %) 0.15 M HNO_3, $T = 25$ °C, DPP 0–11 with respectively: 0.00, 0.50, 0.99, 1.48, 1.96*, 2.44, 3.38, 4.76, 5.66, 6.98, 7.83 and 9.09 × 10^{-5} M with * test with diss. O_2 peak 4. Record: $\Delta E = 25$ mV, 40 ms, $E_s = -0.100$ V Vs SCE, $m = 1.562$ mg $s^{-1,}$ $t_{dp} = 1$ s, sweep = 5 mV s^{-1}. Degueldre and Meklati (1984), with permission © Oldenburg

70% n-hexane from 50 ml extraction phase in equilibrium with 50 ml HNO_3 solution plus 100.77 mg uranyl nitrate dissolved in DMF (see Fig. 3.89). Standard addition was carried out to estimate the extraction yield for U(VI): 95 ± 2% (see Degueldre and Meklati 1983).

Polarographic determination of uranium (VI) has also been studied in the organic extraction phase TBP—diluent along with a selected aprotic solvent (i.e. dimethyl sulfoxide or N,N-dimethylformamide). DMF was found more suitable because it dissolves the organic extraction phase better than DMSO. U (VI) extracted in TBP-kerosene, n-hexane, cyclohexane, n-dodecane, benzene, from nitric acid medium can be determined in an organic solution (e.g. 50% DMF-30% TBP-20% kerosene) where it gives a well-defined wave. In the organic solution, nitric acid added and/or extracted from the aqueous phase was found as an excellent supporting electrolyte for uranium determination (see Degueldre and Meklati 1984a). Similar results gained by DPP are presented in Fig. 3.89 for the DMA—extractant solution.

The determination of U(VI) in organic extraction phases: hydrocarbon-diethyl-2-hexyl phosphoric acid (DHPA)—trioctylphosphine oxide

(TOPO) and hydrocarbon-tri-*n*-octylamine, diluted by an alcohol has been studied by direct current polarography, as well as by differential pulse polarography (DPP). In the solutions of sulphuric acid as a supporting electrolyte in the alcohol—extractant—hydrocarbon phases, U(VI) gives rise to well defined polarograms. Its electroreduction involves a single electron step in all cases. The DPP peak height is proportional to U(VI) concentration between 2×10^{-6} and 2×10^{-3} M, and the 3σ detection limit is $\sim 10^{-6}$ M.

Similarly, hexavalent uranium and molybdenum were also investigated from their trioctylphosphine oxide extract by differential pulse polarographic by Degueldre and Meklati (1984b). The extraction of molybdenum(VI) and uranium (VI) was carried out in 4 M HCl media using 0.1 M TOPO-cyclohexane with for DPP investigation cyclohexanone, ethanol and HNO_3.

3.3.8.5 Atom Probe Tomography

Atom Probe Tomography (**APT**) is a kind of sputtering technique applied to a nanotip-electrode under defined electric field. The **atom probe** was introduced by Müller and Panitz in 1967 and is able to see He and Ne atoms (see Müller et al. 1969). The atom probe is however closely related to the field ion microscope developed by Müller (1951). This was the first instrument capable of atomic resolution.

Atom probe systems are unlike electron microscopes. ATP magnification is due to the highly curved applied electric field, rather than by the magnification from electron beam paths. Since ions are removed from the surface of the sample nano-tip, the method is destructive. Imaging of these ions is performed together with their identification. The coupling of this magnification method with time of flight mass spectrometry allows analysis of the ions sputtered by electric pulses applied in the nano-electrode specimen. The investigated volumes range usually from 10,000 to 10,000,000 nm^3. The freed ions (no cluster ions) can have their mass-to-charge ratio analyzed. Details on the principle and applications of APT can be found in the work of Miller and Forbes (2014).

The samples are required to have a nano-needle shape. They are produced by electropolishing or focused ion beam methods (see Sect. 2.2.2). Since 2006, commercial systems with laser pulsing have become available. These APT's have expanded applications from metallic specimens to semiconducting and insulating materials such as ceramics, and even geological materials. Preparation is done, often by hand, to manufacture a tip radius of the order of 100 nm, sufficient to induce a high electric field. Modern APT uses a position-sensitive detector to deduce the lateral location of atoms. This allows 3-D reconstructions to be generated. The Tomographic Atom Probe (TAP) introduced a multichannel timing system and multianode array (see Bunton 2006). Since then, there have been many refinements to increase the field of view, mass and position resolution, and data acquisition rate of the instrument. In 2005, the pulsed laser atom probe (PLAP) expanded the routes of research from highly conductive materials (metals) to poor conductors and insulating materials (see Seidman 2007).

The **ions released** from the sample are accelerated by electrostatic force, acquiring their energy within a distance of a few tip-radii. Subsequently, the accelerative force F on any ion is driven by the electrostatic equation:

$$F = ne\nabla\phi \tag{3.77}$$

where n is the ionization state of the ion, e is the fundamental electric charge and ϕ the electric field. Relativistic effects are ignored, as ion speeds are only a small fraction of the speed of light. Assuming that the ions are accelerated over a short interval, the ions are assumed to travel at constant velocity. As the ion travel from the tip at fixed voltage, the speed at which the ion is travelling can be estimated. The **magnification** M is due to the projection of ions radially away from the tip, which for a spherical emitter is inversely proportional to the radius of the tip r_{tip}, with projection onto a spherical screen of radius r_{scr}. M can be:

$$M = \frac{r_{\text{scr}}}{r_{\text{tip}}} \tag{3.78}$$

Practical tip to screen distances may range from several centimeters to several meters. Most models for reconstruction assume that the tip is a spherical object, and utilize empirical corrections to stereographic projection to convert detector positions back to a 2D surface embedded in R^3. By sweeping this surface through R^3 as a function of the ion sequence input data, such as via ion-ordering, a volume is generated onto which positions the 2D detector positions can be computed and placed in 3D space. Detailed information on the atom-probe may be found in the monograph by Müller (1970)

The minimum **equipment for** an atom probe includes a:

- vacuum system for maintaining the low pressures ($\sim 10^{-8}$ to 10^{-10} Pa) required
- system for manipulation of samples inside vacuum, including diagnostic systems
- cooling system to reduce temperature between room temperature to 15 K
- cooling system to reduce atomic motion, such as a helium refrigeration circuit
- high-voltage system to raise the sample voltage
- detection system for single ions for field evaporation.

Optionally, an atom probe may also include laser-optical systems for laser beam targeting and pulsing, if using laser-evaporation methods. Staged vacuum systems are regularly employed to ensure that the system vacuum conditions remain stable. In situ reaction systems may also be employed for some studies.

Performances are linked to the collectable ion volumes. They used to be limited to several thousand—or tens of thousands of—ionic events. Subsequent electronics development has increased the rate of data accumulation, with datasets of hundreds of million atoms (dataset volumes of 10^7 nm^3). Experiments take from a few minutes, to many hours to complete. Actually, atom probe has been used in the analysis of the chemical composition of a wide range of alloys.

Limitations are due to challenges, issues and conditions that restrict the use of this technique:

- Ions may be lost between release from the tip to the screen (may be 50%).
- Materials' shape control achievable spatial resolution.
- Uncontrolled specimen induces distortions into computer-generated 3D dataset.
- Features of interest might evaporate differently vs. bulk, altering reconstruction.
- Site specific preparations using FIB may be used to bypass such limitations.
- Ion overlap in some samples results in ambiguous analyzed species.
- Low molecular weight gases may be difficult to be purged from analysis chamber.
- Results may be contingent on the parameters used to convert the 2D into 3D data.

More information on the ATP may be found in the work of Gault et al. (2012).

The nuclear material example treated here deals with reactor pressure vessel (RPV) steel. When submitted to neutron irradiations followed by thermal annealing, RPV steel revealed that Cu and Ni impurities can form nanoclusters. In the unirradiated sample and in the only-irradiated sample no significant clustering is detected. In all irradiated and subsequently annealed samples increases of Cu and Ni atom densities are recorded. Since solute clustering is known to play a significant role in the embrittlement of RPV steels, specific analytical techniques have to

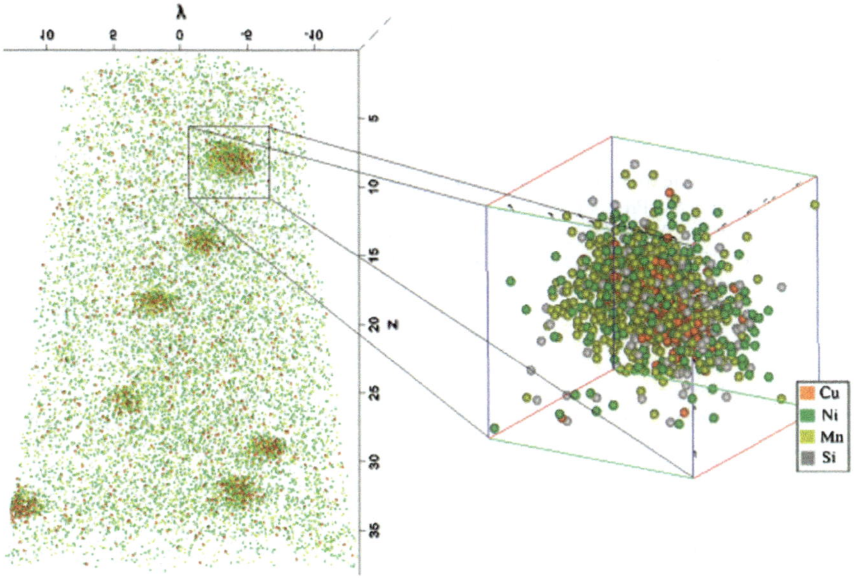

Fig. 3.90 AP map and APT of a 'high' copper RPV steel with copper-enriched precipitates. This steel has been thermally aged for 10,000 h at 365 °C. Adapted from Zelenty et al. (2016), with permission © Elsevier

be applied to investigate the properties and behavior of these clusters. The density of Cu, Ni, Mn and Si atoms (see Fig. 3.90) has been determined by AP mapping and APT after annealing treatment Zelenty et al. (2016).

References

Abdalla, A. M., Ashraf, O., Rammah, Y. S., Ashry, A. H., Eisa, M., & Tsuruta, T. (2015). Fast neutron detection in CR-39 and DAM–ADC nuclear track detectors. *Radiation Physics and Chemistry, 108,* 24–28.

Agarwal, B. K. (1991). X-Ray spectroscopy: An introduction. In *Springer series in optical sciences* (2nd ed.).

Aggarwal, S. K. (2016a). Alpha-particle spectrometry for the determination of alpha emitting isotopes in nuclear, environmental and biological samples: Past, present and future. *Analytical Methods.*

Aggarwal, S. K. (2016b). Thermal ionisation mass spectrometry (TIMS) in nuclear science and technology—a review. *Analytical Methods, 8,* 942–957.

Agostinelli, S., Allison, J., Amako, K., Apostolakis, J., Araujo, H., Arce, P., et al. (2003). Geant4 a simulation toolkit. *Nuclear Instruments and Methods in Physics Research Section A: Accelerators, Spectrometers, Detectors and Associated Equipment, 506,* 250–303.

Akopov, G. A., Krinitsyn, A. P., & Tikhonova, A. E. (1988). *Radiokhim, 30,* 578–583.

Alfassi, Z. B. (1990) *Activation analysis* (Vol. 1, pp. 97–109). Boca Raton, Fl: CRC Press.

Almqvist, N., Rubel, M., Fredriksson, S., Emmoth, B., Wienhold, P., & Ilyinsky, Lev S. (1995). AFM and STM characterization of surfaces exposed to high flux deuterium plasma. *Journal Nuclear Materials, 220,* 917–921.

Amemiya, S., Asawa, A., Tanaka, K., Tsurita, Y., Masuda, T., Katoh, T., et al. (1984). Application of PIXE to the study of nuclear fusion materials. *Nuclear Instruments and Methods in Physics Research Section B: Beam Interactions with Materials and Atoms, 3,* 549–552.

Apostol, A. I., Pantelica, A, Ortega-Feliu, I, Margineau, M., Sima, O., Straticiuc, M., et al. (2016). Ion beam analysis of elemental sugnatures in uranium dioxide samples: importance for nuclear forensics, *Journal of Radioanalytical and Nuclear Chemistry, 310,* 1–8.

Al Haj, O., Peres, V., Serris, E., Grosjean, F., Kittel, J., Ropital, F., & Cournil, M (2015). In situ high temperature oxidation analysis of Zircaloy-4 using acoustic emission coupled with thermogravimetry; Proc. Conference on Contribution of Materials Investigations and Operating Experience to LWRs' Safety; Avignon (France) 2014; INIS-FR–15-0411 Vol. 46

Allen, G. C., Butler, I. S., & Tuan, N. A. (1987). Characterisation of uranium oxides by micro-Raman spectroscopy. *Journal Nuclear Materials, 144,* 17–19.

Allen, G. C., Tucker, P. M., & Tyler, J. W. (1982). The behaviour of uranium oxides in low partial pressures of O_2 studied using X-ray photoelectron spectroscopy. *Vacuum, 32,* 481–486.

Ambe, F., & Ambe, S. (1973). Mössbauer spectroscopic verification of two different states of impurity ^{119}Sn atoms in Sb_2Te_3. *Physics Letters A, 43,* 399–400.

Ambe, S., & Ambe, F. (1975). Mössbauer emission spectrum of ^{121}Sb after the β^- decay of ^{121m} Sn in SnS_2: Nuclear decay synthesis of antimony (V) sulfide. *The Journal of Chemical Physics, 63,* 4077–4078.

Ambe, F., Ambe, S., Shoji, H., & Saito, N. (1974) Mössbauer emission spectra of ^{119}Sn after the EC decay of ^{119}Sb in metals, oxides, and chalcogenides of antimony and tellurium. *The Journal of Chemical Physics 60,* 3773–3778.

Ammar, M. R., Galy, N., Rouzaud, J. N., Toulhoat, N., Vaudey, C. E., Simon, P., et al. (2015). Characterizing various types of defects in nuclear graphite using Raman scattering: Heat treatment, ion irradiation and polishing. *Carbon, 95,* 364–373.

Anderson, I. S., Hurd, A. J., & McGreevy, R. (Eds). (2016). *Neutron scattering applications and techniques.* Springer series.

Anderson, I. S., McGreevy, R., & Bilheux, H. Z. (Eds.). (2009). *Neutron imaging and applications: A reference for the imaging community*. Berlin: Springer.

Anderson, J. A., Eberhard, C. D., Byrd, M. J., Carroll, J. J., Collins, C. B., Scarbrough, E. C., et al. (1989). Nuclear photoactivation cross sections for short-lived isomeric states excited with a 6 MeV linac. *Nuclear Instruments and Methods in Physics Research Section B: Beam Interactions with Materials and Atoms, 40,* 452–454.

Anderson, W., Kozak, D., Coleman, V. A., Jämting, Å. K., & Trau, M. (2013). A comparative study of submicron particle sizing platforms: Accuracy, precision and resolution analysis of polydisperse particle size distributions. *Journal of Colloid and Interface Science, 405,* 322–330.

Antonio, M. R., Soderholm, L., Williams, C. W., Blaudeau, J.-Ph., & Bursten, B. E. (2001). Neptunium redox speciation. *Radiochimica Acta, 89,* 17–25.

Apperley, D. C., Harris, R. K., & Hodgkinson, P. (2012). *Solid state NMR: Basic principles and practice*. Momentum: Press LLC.

Araki, T., Enomoto, S., Furuno, K., Gando, Y., Ichimura, K., Ikeda, H., et al. (2005). Experimental investigation of geologically produced antineutrinos with KamLAND. *Nature, 436,* 499–503.

Arinicheva, Y. (2016). *Private communication.*

Armstrong, D. E. J., Hardie, C. D., Gibson, J. S. K. L., Bushby, A. J., Edmondson, P. D., & Roberts, S. G. (2015). Small-scale characterisation of irradiated nuclear materials: Part II nanoindentation and micro-cantilever testing of ion irradiated nuclear material. *Journal of Nuclear Materials, 462,* 374–381.

Asner, D. M., Burns, K., Campbell, L. W., Greenfield, B., Kos, M. S., Orrell, J. L., et al. (2015). Method of fission product beta spectra measurements for predicting reactor anti-neutrino emission. *Nuclear Instruments and Methods in Physics Research Section A: Accelerators, Spectrometers, Detectors and Associated Equipment, 776,* 75–82.

Assenheim, H. M. (2014). *Introduction to electron spin resonance*. Berlin: Springer.

Aston, F. W. (1919). *Philosophical Magazine Series 6, 3*, 709.

Baechler, S., Materna, Th, Jolie, J., Cauwels, P., Crittin, M., Honkimaki, V., et al. (2001). Non-destructive analysis of a bulky sample from a natural fossil reactor. *Journal of Radioanalytical and Nuclear Chemistry, 250,* 39–45.

Barker, G. C. (1958). Pulse polarography. *Analytica Chimica Acta, 18,* 118–131.

Barker, G. C., & Gardner, A. W. (1958). *Pulse polarography* (pp. 79–83). Atomic Energy Research Establishment, AERE Harwell 2297.

Barkhausen, H. (1919). Zwei mit Hilfe der neuen Verstärker entdeckte Erscheinungen. *Physikalische Zeitschrift, 20,* 401–403.

Barnes, R. M. (1974). Chapter 21—Emission spectroscopy: Arc, spark, laser and plamas. *Systematic Materials Analysis*, 23–83.

Barzilov, A., & Novikov, I. (2015). Material classification by analysis of prompt photon spectra induced by 14-Mev neutrons. *Physics Procedia, 66,* 396–402.

Banerjee, S., Chen, Sh., Powers, N., Haden, D., Liu, C., Golovin, G., et al. (2015). Compact source of narrowband and tunable X-rays for radiography. *Nuclear Instruments and Methods in Physics Research Section B: Beam Interactions with Materials and Atoms, 350*, 106–111.

Basile, L. J., Ferraro, J. R., Mitchell, M. L., & Sullivan, J. C. (1978). The Raman scattering of actinide (VI) ions in carbonate media. *Applied Spectroscopy, 32,* 535–537.

Beaurepaire, E., Bulou, H., Scheurer, F., & Kappler, J. P. (Eds.). (2008). *Magnetism: A synchrotron radiation approach*. Berlin: Springer.

Becker, J. S. (2007). *Inorganic mass spectroscopy, principles and applications*. Hoboken: Wiley.

Beckhoff, B., Kanngießer, B., Langhoff, N., Wedell, R., & Wolff, H. (Eds.). (2006). Handbook of practical X-ray fluorescence analysis. Berlin: Springer.

Beer, A (1852). Bestimmung der Absorption des rothen Lichts in farbigen Flussigheiten [Determination of the absorption of red light in colored liquids], *Annual Review of Physical Chemistry, 86,* 78–88.

Beitz, J. V., & Hessler, J. P. (1980). Oxidation state specific detection of transuranic ions in solution. *Nuclear Technology, 51,* 169–177.

Beitz, J. V., Bowers, D. L., Doxater, M. M., Maroni, V. A., & Reed, D. T. (1988). Detection and speciation of transuranium elements in synthetic groundwater via pulsed-laser excitation. *Radiochimica Acta, 44*(45), 87–93.

Bellin, R. C., Strach, M., Truphémus, Th., Guéneau, Ch., Richaud, J.-Ch., & Rogez, J. (2015). In situ high temperature X-ray diffraction study of the phase equilibria in the UO_2–PuO_2–Pu_2O_3 system, *Journal of Nuclear Materials*, *465*, 407–417.

Benson, S., & Madey, J. M. J. (1985). Shot and quantum noise in free electron lasers. *Nuclear Instruments and Methods in Physics Research Section A: Accelerators, Spectrometers, Detectors and Associated Equipment, 237,* 55–60.

Benedetto, A., & Pajewski, L. (Eds.). (2015). *Civil engineering applications of ground penetrating radar*. Berlin: Springer.

Benninghoven, A., Colton, R. J., Simons, D. S., & Werner, H. W. (Eds.). (1986). *Secondary Ion Mass Spectrometry (SIMS) V, Proceedings of the fifth International Conference*.

Bertsch, P. M., Hunter, D. B., Sutton, S. R., Bajt, S., & Rivers, M. L. (1994). In situ chemical speciation of uranium in soils and sediments by micro x-ray absorption spectroscopy. *Environmental Science & Technology, 28,* 980–984.

Bickel, M. (1997). The Davies-Gray titration for the assay of uranium in nuclear materials: A performance study. *Journal of Nuclear Materials, 246,* 30–36.

Binnemans, K. (2015). Interpretation of europium(III) spectra. *Coordination Chemistry Reviews, 295,* 1–45.

Bitea, C., Müller, R., Neck, V., Walther, C., & Kim, J. I. (2003). Study of the generation and stability of thorium(IV) colloids by LIBD combined with ultrafiltration. *Colloids and Surfaces A: Physicochemical and Engineering Aspects, 217,* 63–70.

Blümich, B., Casanova, F., & Appelt, S. (2009). NMR at low magnetic field. *Chemical Physics Letters, 477,* 231–240.

Boolchand, P., Bresser, W., & Ehrhart, G. J. (1981). ^{129}I nuclear quadrupole interaction in trigonal Te and the role of oxygen contamination. *Physical Review B, 23*, 3669–3672.

Bragg, W. H. (1908). The nature of γ- and X-rays. *Nature, 77*(1995), 270.

Bogé, M., Blaise, A., Bonnisseau, D., Fournier, J. M., Thérond, P. G., Poirot, I., et al. (1986). ^{237}Np Mössbauer spectroscopy on neptunium doped borosilicate glasses. *Hyperfine Interactions, 28*, 765–767.

Bogomolova, L. D., Jachkin, V. A., Prushinsky, S. A., Dmitriev, S. A., Stefanovsky, S. V., Teplyakov, Yu G, et al. (1997). Paramagnetic defects induced by ion implantation in oxide glasses. *Journal of Non-crystalline Solids, 210,* 101–118.

Bogomolova, L. D., Pavlushkina, T. K., Stefanovskii, S. V., Teplyakov, Yu. G., & Trul, O. A. (1993). ESR and IR spectroscopic studies of sodium and aluminosilicophosphate glasses. *Glass Physics and Chemistry, 19*, 222–227.

Bogomolova, L. D., Stefanovsky, S. V., Troole, A. Y., & Vance, E. R. (2001). EPR spectra of V (IV) in zirconolite-rich ceramics. *Journal of Materials Science, 36,* 1213–1217.

Boizot, B., Petite, G., Ghaleb, D., & Calas, G. (1998). Radiation induced paramagnetic centres in nuclear glasses by EPR spectroscopy. *Nuclear Instruments and Methods in Physics Research Section B: Beam Interactions with Materials and Atoms, 141,* 580–584.

Bojanowski, R., Holm, E., & Whitehead, N. E. (1987). Determination of ^{227}Ac in environmental samples by ion-exchange and alpha spectrometry. *Journal of Radioanalytical and Nuclear Chemistry, 115,* 23–37.

Borger, J. J., Hashemi-Nezhad, S. R., Alexiev, D., Brandt, R., Westmeier, W., Thomauske, B., et al. (2012). Spatial distribution of thorium fission rate in a fast spallation and fission neutron field: An experimental and Monte Carlo study. *Nuclear Instruments and Methods in Physics Research Section A: Accelerators, Spectrometers, Detectors and Associated Equipment, 664,* 103–110.

Bond, A. M., & O'Halloran, R. J. (1976). Fundamental and second harmonic alternative current cyclic voltametric theory and experimental results for simple electrode reactions involving solution-soluble redox couples. *Analytical Chemistry, 48,* 872–883.

Briggs, A. (Ed.). (1995). *Advances in acoustic microscopy.* Berlin: Springer.

Bourgès, G., Lambertin, D., Rochefort, S., Delpech, S., & Picard, G. (2007). Electrochemical studies on plutonium in molten salts. *Journal of Alloys and Compounds, 444–445,* 404–409.

Boutchko, R., Rayz, V. L., Vandehey, N. T., O'Neil, J. P., Budinger, T. F., Nico, P. S., et al. (2012). Imaging and modeling of flow in porous media using clinical nuclear emission tomography systems and computational fluid dynamics. *Journal of Applied Geophysics, 76,* 74–81.

Buck, E. C., & Fortner, J. A. (1997). Detecting low levels of transuranics with electron energy loss spectroscopy. *Ultramicroscopy, 67,* 69–75.

Buckau, G., Stumpe, R., & Kim, J. I. (1986). Americium colloid generation in groundwaters and its speciation by laser-induced photoacoustic spectroscopy. *Journal of the Less Common Metals, 122,* 555–562.

Burchell, T. D., Rose, A. P. G., & McEnaney, B. (1986). Acoustic emission from irradiated nuclear graphite. *Journal of Nuclear Materials, 140,* 11–18.

Burghartz, M., Ledergerber, G., Ingold, F., Heimgartner, P., & Degueldre, C. (2001). X-ray diffraction and data interpretation of stabilised zirconia inert matrix fuel doped with plutonium. *Progress in Nuclear Energy, 38,* 247–250.

Bundschuh, T., Knopp, R., Müller, R., Kim, J. I., Neck, V., & Fanghänel, Th. (2000). Application of LIBD to the determination of the solubility product of thorium(IV)-colloids. *Radiochimica Acta, 88,* 625–629.

Bunton, J., Lenz, D., Olson, J., Thompson, K., Ulfig, R., Larson, D., et al. (2006). Instrumentation developments in atom probe tomography: Applications in semiconductor research. *Microscopy and Microanalysis, 12,* 1730–1731.

Bünzli, J.-C. G. (1989). *Luminescent probes in lanthanide probes in life.* Elsevier, Amsterdam: Chemical and Earth Sciences.

Byrne, A. R. (1986). Determination of ^{237}Np in Cumbrian (UK) sediments by neutron activation analysis: Preliminary results. *Journal of Environmental Radioactivity, 4,* 133–144.

Byrne, A. R. (1993). Review of neutron activation analysis in the standardization and study of reference materials. *Fresenius' Journal of Analytical Chemistry, 345,* 144–151.

Byrne, A. R., & Benedik, L. (1999). Applications of neutron activation analysis in determination of natural and man-made radionuclides including Pa-231. *Czechoslovak Journal of Physics, 49,* 263–270.

Cai, Z., & Liu, Sh. (Eds). (2013). *Applications of MALDI-TOF spectroscopy.* Berlin: Springer.

Cao, J., & Luk, K.-B. (2016). An overview of the Daya Bay reactor neutrino experiment. *Nuclear Physics B, 908,* 62–73.

Cameron, L. T., Lang, M., Zhang, F., Park, S., Palomares, R. I., & Ewing, R. C. (2017). Review of recent experimental results on the behavior of actinide-bearing oxides and related materials in extreme environments. *Progress in Nuclear Energy* (in Press).

Cammann, K. (1979). *Working with ion-selective electrodes. Chemical laboratory practice.* Berlin: Springer.

Canizarès, A., Guimbretière, G., Tobon, Y. A., Raimboux, N., Omnée, R., Perdicakis, M., et al. (2012). In situ Raman monitoring of materials under irradiation: Study of uranium dioxide alteration by water radiolysis. *Journal of Raman Spectroscopy, 43,* 1492–1497.

Casacuberta, N., Masqué, P., Henderson, G., Rutgers van-der-Loeff, M., Bauch, D., Vockenhuber, C., et al. (2016). First ^{236}U data from the Artic Ocean and use of $^{236}U/^{238}U$ and $^{129}I/^{236}U$ as a new dual tracor. *Earth and Planetary Science Letters, 440,* 127–134.

Capella, B., & Dietler, G. (1999). Force distance curves by atom force microscopy. *Surface Science Reports, 4,* 1–104.

Capote, R., Chen, Y.-J., Hambsch, F.-J., Kornilov, N. V., Lestone, J. P., Litaize, O., et al. (2016). Prompt fission neutron spectra of actinides. *Nuclear Data Sheets, 131,* 1–106.

Carretta, P., & Lascialfari, A. (Eds.). (2007). *NMR-MRI, μSR and mössbauer spectroscopies in molecular magnets.* Berlin: Springer

Chadwick, J. (1932). The existence of a neutron. In *Proceedings of the Royal Society 136,* 692–708.

Chatni, M. R., Maier, D. E., & Porterfield, D. M. (2009). Evaluation of microparticle materials for enhancing the performance of fluorescence lifetime based optrodes. *Sensors and Actuators B, 141,* 471–477.

Chapman, C. H. (2004). *Fundamentals of seismic wave propagation.* Cambridge: Cambridge University Press.

Chakarvarti, S. K., Lal, N., & Nagpaul, K. K. (1980). Uranium trace analysis of some materials using solid state nuclear track detectors. *Solid State Nuclear Track Detectors*, 701–715.

Cherniak, D. J., & Lanford, W. A. (1992). 11.4—NRA nuclear reaction analysis. *Encyclopedia of Materials Characterization,* 680–694.

Chen, J.-W., & Milnes, A. G. (1980). Energy levels in silicon. *Annual Review of Materials Science, 10,* 157–228.

Chen, Y., Wang, F., Zhao, Y.-G., Li, L.-L., Zhang, Y., Shen, Y., et al. (2015). An improved FT-TIMS method of measuring uranium isotope ratios in the uranium -bearing particles, *Radiation Measurements, 83*, 63–67.

Christl, M., Casacuberta, N., Lachner, J., Maxeiner, S., Vockenhuber, C., Synal, H.-A., et al. (2015). Status of ^{236}U analyses at ETH Zurich and the distribution of ^{236}U and ^{129}I in the North Sea in 2009. *Nuclear Instruments and Methods in Physics Research Section B: Beam Interactions with Materials and Atoms, 361*, 510–516.

Clough, R. B., & Wadley, H. N. G. (1982). Indentation loading studies of acoustic emission from temper and hydrogen embrittled A533B steel. *Metallurgical Transactions A, 13*(11), 1965–1975.

Citrin, P. H., Eisenberger, P., & Hewitt, R. C. (1979). SEXAFS studies of iodine adsorbed on single crystal substrates. *Surface Science, 89,* 28–40.

Claassen, A., & Vissen, J. (1946). Determination of uranium with 8-hydroxyquinoline (oxine). *Recueil des Travaux Chimiques des Pays-Bas, 65,* 211–215.

Clayton, C. G. (1990). Some comments on the development of radiation and radioisotope measurement applications in industry. *International Journal of Radiation Applications and Instrumentation. Part A. Applied Radiation and Isotopes, 41,* 917–934.

Close, F. (2012). Neutrino. Berlin: Springer.

Collaboration, B. (2010). Observation of geo-neutrinos. *Physics Letters B, 687,* 299–304.

Conradson, S. (1998). Application of X-ray absorption fine structure spectroscopy to materials and environmental science. *Applied Spectroscopy, 52,* 252A–279A.

Costarramone, N., Gleyzes, C., Castetbon, A., & Berger, M. (2000). Lithium ion selective electrodes: Application to the lithium measurements in PWR plants. In *Proceedings of the Water Chemistry of Nuclear Reactor Systems, 8 BNES, 1,* 259–261.

Cottrell, F. G. (1903). Der Reststrom bei galvanischer Polarisation, betrachtet als ein. Diffusionsproblem. *Zeitschrift für Physikalische Chemie, 42,* 385–431.

Couprie, M. E. (2014). New generation of light source: Present and future. *Journal of Electron Spectroscopy and Related Phenomena, 196,* 3–13.

Cox, S. F. J. (1987). Implanted muon studies in condensed matter science. *Journal of Physics C: Solid State Physics, 20,* 3187–3319.

Croft, W. L., Stone, J. A., & Pillinger, W. L. (1968). Mössbauer effect in ^{231}Pa. *Journal of Inorganic and Nuclear Chemistry, 30,* 3203–3208.

Das, S. K., Kedari, C. S., & Tripathi, S. C. (2010). Spectrophotometric determination of trace amount of uranium (VI) in different aqueous and organic streams of nuclear fuel processing using 2-(5-bromo-2-pyridylazo-5-diethylaminophenol). *Journal of Radioanalytical and Nuclear Chemistry, 285,* 675–681.

Davis, J. L., & Annan, A. P. (1989). Ground penetrating radar for high resolution mapping of soil and rock stratigraphy. *Geophysical Prospecting, 37,* 531–551.

Davies, W., & Gray, W. (1964). A rapid and specific titrimetric method for the precise determination of uranium using iron(II) sulphate as reductant. *Talanta, 11,* 1203–1211.

David, A., Kögel, G., Sperr, P., & Triftshäuser, W. (2001). Lifetime measurements with a scanning positron microscope. *Physical Review Letters, 87*, 067402-1–4.

Davies, W., Gray, W., & McLeod, K. C. (1970). Coulometric determination of uranium with a platinum working electrode. *Talanta, 17,* 937–944.

Davidovits, P., & Egger, M. D. (1969). Scanning laser microscope. *Nature, 223,* 831.
de Broglie, L.-V. (1923). Radiation—Waves and quanta [*Translated from Comptes rendus*] (Vol. 177, pp. 507–510).
Debye, P., & Hückel, E. (1924). Bemerkungen zu einem Satze über die kataphoretische Wanderungsgeschwindigkeit suspendierter Teilchen. *Physikalische Zeitschrift, 25,* 49–52.
De Geuser, F., & Deschamps, A. (2012). Precipitate characterisation in metallic system by small angle X-rays or neutron scattering. *Comptes Rendus Physique, 13,* 246–256.
Degueldre, C. (1978). *Constant-current coulometric determination of uranium in the pure metal (Thesis).* Université de Liège, Belgium.
Degueldre, C., & Alekseev, E. V. (2015). Uranium trioxide behavior during electron energy loss spectroscopy analysis. *Radiation Physics and Chemistry, 108,* 7–12.
Degueldre, C., & Favarger, P.-Y. (2003). Colloid analysis by single particle inductively coupled plasma-mass spectroscopy: A feasibility study. *Colloids and Surfaces A: Physicochemical and Engineering Aspects, 217,* 137–142.
Degueldre, C., & Favarger, P.-Y. (2004). Thorium colloid analysis by single particle inductively coupled plasma-mass spectrometry. *Talanta, 62,* 1051–1054.
Degueldre, C., & Fiorina, C. (2016). The proto-Earth geo-reactor: Reassessing the hypotheses. *Solid Earth Sciences, 1,* 49–63.
Degueldre, C., & Hellwig, Ch. (2003). Study of a zirconia based inert matrix fuel under irradiation. *Journal of Nuclear Materials, 320,* 96–105.
Degueldre, C., & Laaksoharju, M. (2014). Ground water colloid properties from the Bangombé system. *Applied Geochemistry, 45,* 130–143.
Degueldre, C. A., & Meklati, M. (1983). Polarographic behaviour of uranium (VI) in organic solutions. *Journal of Radioanalytical Chemistry, 77,* 87–96.
Degueldre, C. A., & Meklati, M. (1984a). Polarographic behaviour of uranium (VI) in tributyl phosphate organic solutions. *Radiochimica Acta, 35,* 53–56.
Degueldre, C. A., & Meklati, M. (1984b). Differential pulse polarographic behaviour of hexavalent uranium and molybdenum from its trioctylphosphine oxide extract. *Journal of the Less Common Metals, 97,* 11–20.
Degueldre, C., Bertsch, J., & Martin, M. (2016). Post irradiation examination of nuclear fuel: Toward a complete analysis. *Progress in Nuclear Energy, 92,* 242–253.
Degueldre, C., Borca, C., & Cozzo, C. (2013a). Curium analysis in plutonium uranium mixed oxide by x-ray fluorescence and absorption fine structure spectroscopy. *Talanta, 115,* 986–991.
Degueldre, C., Buckley, D., Dran, J. C., & Schenker, E. (1998b). Study of the oxide layer formed on stainless steel exposed to boiling water reactor conditions by ion beam techniques. *Journal of Nuclear Materials, 252,* 22–27.
Degueldre, C., Cozzo, C., Martin, M., Grolimund, D., & Mieszczynski, C. (2013b). Americium characterization by X-ray fluorescence and absorption spectroscopy in plutonium uranium mixed oxide. *Journal of Solid State Chemistry, 202,* 315–319.
Degueldre, C., Conradson, St., Amato, A., & Campitelli, E. (2006a). Feeling defects in Zircaloy by extended X-ray absorption fine structure and muon spin relaxation analyses. *Journal of Nuclear Materials, 352,* 126–135.
Degueldre, C., Favarger, P.-Y., & Rossé, S. (2006b). Wold, uranium colloid analysis by single particle inductively coupled plasma-mass spectrometry. *Talanta, 68,* 623–628.
Degueldre, C., Fuks, L., & Schenker, E. (1998a). In-line measurement of the oxide layer build-up on Zircaloy under Boiling Water Reactor conditions using diffuse reflection spectroscopy. *Measurement Science and Technology, 9,* 809–815.
Degueldre, C., Rocchiccioli, F., & Laube, A. (1999). Accelerated measurement of groundwater redox potentials: Method and application. *Analytica Chimica Acta, 396,* 23–31.
Degueldre, C., Kastoryano, M., & Dardenne, K. (2007). Variable incidence angle—X-ray absorption spectroscopy for the study of Zircaloy corrosion layers. *Journal of Nuclear Materials, 362,* 316–326.

Degueldre, C., Kuri, G., Borca, C. N., & Grolimund, D. (2009). X-ray micro- fluorescence, diffraction and absorption spectroscopy for local structure investigation of a radioactive zinc ferrite deposit. *Corrosion Science, 51,* 1690–1695.

Degueldre, C., Kuri, G., Martin, M., Froideval, A., Cammelli, S., Orlov, A., et al. (2010). Nuclear material investigations by advanced analytical techniques. *Nuclear Instruments and Methods in Physics Research Section B: Beam Interactions with Materials and Atoms, 268,* 3364–3370.

Degueldre, C., Martin, M., Kuri, G., Grolimund, D., & Borca, C. (2011a). Plutonium—uranium mixed oxide characterization by coupling micro-X-ray diffraction and absorption investigations. *Journal of Nuclear Materials, 416,* 142–150.

Degueldre, C., Schaeublin, R., Krbanjevic, J., & Minikus, E. (2013c). Electron energy loss spectroscopy investigation through a nano ablated uranium dioxide sample. *Talanta, 106,* 408–413.

Degueldre, C., Schenker, E., & Nhuis-Weda, H. (1996a). Investigation of colloid characteristics in the water of boiling water reactors. In *Proceedings of the Water Chemistry of Reactor Systems, 7 BNES*.

Degueldre, C., O'Prey, S., & Francioni, W. (1996c, October). An in-line diffuse reflection spectroscopy study of the oxidation of stainless steel under boiling water reactor conditions. *Corrosion Science, 38,*(10), 1763–1782.

Degueldre, C., Pleinert, H., Maguire, P., Lehman, E., Missimer, J., Hammer, J., Leenders, K., Böck, H., & Townsend, D. (1996d, May). Porosity and pathway determination in crystalline rock by positron emission tomography and neutron radiography. *Earth and Planetary Science Letters, 140,*(1–4), 213–225.

Degueldre, C., & Taibi, K. (1996). Polarographic behaviour and determination of uranium(VI) in alcoholic solutions from organic extraction phases. *Analytica Chimica Acta, 321,* 201–207.

Degueldre, C., Triay, I., Kim, J.-I., Vilks, P., Laaksoharju, M., & Miekeley, N. (2000). Groundwater colloids properties: A global approach. *Applied Geochemistry, 15,* 1043–1051.

Degueldre, C., Pouchon, M., Streit, M., Zaharko, O., & Di Michiel, M. (2001). Analysis of porous features in zirconia based inert matrix, impact on the material qualification. *Progress in Nuclear Energy, 38,* 241–246.

Degueldre, C., Reed, D., Kroft, A. J., & Mertz, C. (2004). XAFS study of americium sorbed onto groundwater colloids. *Journal of Synchrotron Radiation, 11,* 198–203.

Degueldre, C., Ulrich, H. J., & Silbi, H. (1994). Sorption of ^{241}Am onto Montmorillonite, illite and hematite colloids. *Radiochimica Acta, 65,* 173–179.

Derjaguin, B. V., Muller, V. M., & Toporov, Y. P. (1975). Effect of contact deformations on the adhesion of particles. *Journal of Colloid and Interface, 53,* 314–325.

De Menibus, A. H, Auzoux, Q., Dieye, O., Macdonald, V., Besson, J., Crépin, J. (2012). Hydride blisters formation, characterization and effect on the fracture of Zircaloy-4 cladding tubes under reactivity initiated accident conditions. In *21st International conference Nuclear Energy for New Europe*, September 2012, Ljubljana, Slovenia (hal-01057285).

Denecke, M., Dardenne, K., & Macquard, Ch. (2004). Np(IV)/Np(V) valence determination from Np L3 edge XANES/EXAFS. *Talanta, 65,* 1008–1014.

Dodge, C. J., Francis, A. J., & Clayton, C. R. (1995). Application of synchrotron radiation techniques in industrial chemical and materials science. *Emerging Technologies in Hazardous Waste Management ACS, CONF, 9509139,* 1352–1377.

Drescher, M., & Jeschke, G. (2012). *EPR spectroscopy: Applications in chemistry and biology.* Berlin: Springer.

Drot, R., Simoni, E., Alnot, M., & Ehrhart, J. J. (1998). Structural environment of uranium (VI) and europium (III) species sorbed onto phosphate surfaces: XPS and optical spectroscopy studies. *Journal of Colloid and Interface Science, 205,* 410–416.

Dev, B., & Jain, B. D. (1961). Gravimetric determination of uranium and thorium with 1-hydroxy-xanthone. *Proceedings of the Indian Academy of Sciences—Section A, 54,* 341–344.

Diggens, A. A., Lichtenstein, S., Synnott, J. C., & West, S. J. (1981). High-purity water quality monitoring based on ion-selective electrode technology. Power plant instrumentation for measurement of high-purity water. *Lane Otten EDRS, American Society for Testing Materials, ASTM STP, 742,* 131–138.

Doyama, M., Kogure, Y., Inoue, M., Kurihara, T., Cao, X., Nishiyama, K., et al. (2009). Comparison between muon and positron images using imaging plates. *Nuclear Instruments and Methods in Physics Research Section A: Accelerators, Spectrometers, Detectors and Associated Equipment, 600,* 60–63.

Dutt, D. A., Higby, P. L., & Griscom, D. L. (1991). An electron spin resonance study of X-irradiated calcium aluminosilicate glasses. *Journal of Non-crystalline Solids, 130,* 41–51.

Dworak, V., Augustin, S., & Gebbers, R. (2011). Application of terahertz radiation to soil measurements: Initial results. *Sensors, 11,* 9973–9988.

Dye, St. T. (2007). Neutrino geophysics. In *Proceedings of Neutrino Sciences 2005.* Berlin: Springer.

Edelson, M. C., & Fassel, V. A. (1981). Isotopic abundance determinations by inductively coupled plasma atomic emission atomic emission spectrometry. *Analytical Chemistry, 53,* 2345–2347.

Edelstein, N. M., Klenze, R., Fanghänel, T., & Hubert, S. (2006). Optical properties of Cm(III) in crystals and solutions and their application to Cm(III) speciation. *Coordination Chemistry Reviews, 250,* 948–973.

Egerton, R. (2011). *Electron energy-loss spectroscopy in the electron microscope.* Berlin: Springer.

Egger, W., Kögel, G., Sperr, P., Triftshäuser, W., Rödling, S., Bär, J., et al. (2002). Vacancy clusters close to a fatigue crack observed with the München scanning positron microscope. *Applied Surface Science, 194,* 214–217.

Ehmann, W. D., & Vance D. E. (1991). *Radiochemistry and nuclear methods of analysis.* Hoboken: Wiley Interscience.

Einstein, A. (1905a). Über einen die Erzeugung und Verwandlung des Lichtes betreffenden heuristischen Gesichtspunkt. *Annalen der Physik, 322,* 132–148.

Einstein, A. (1905b). Über die von der molekularkinetischen Theorie der Wärme geforderte Bewegung von in ruhenden Flüssigkeiten suspendierten Teilchen. *Annalen der Physik, 322,* 549–560.

Einstein, A. (1917). Quantentheorie der Strahlung [On the quantum theory of radiation]. *Physikalische Zeitschrift, 18,* 121–128.

Eldrup, M., Ligthbody, D., & Sherwood, J. N. (1981). The temperature dependence of positron lifetime in solid pivalic acid. *Journal of Chemical Physics, 63,* 51–58.

Elliman, R. G., Timmers, H., Palmer, G. R., & Ophel, T. R. (1998). Limitations to depth resolution in high-energy, heavy-ion elastic recoil detection analysis. *Nuclear Instruments and Methods in Physics Research Section B: Beam Interactions with Materials and Atoms, 136–138,* 649–653.

Erdmann, N., Nunnemann, M., Eberhardt, K., Herrmann, G., Huber, G., Köhler, S., et al. (1998). Determination of the first ionization potential of nine actinide elements by resonance ionization mass spectroscopy (RIMS). *Journal of Alloys and Compounds, 271*(3), 837–840.

Erdoğan, H., & Fessler, J. A. (1999). Ordered subsets algorithms for transmission tomography. *Physics in Medicine and Biology, 44,* 2835–2851.

Faisal, N. H., Ahmed, R., & Reuben, R. L. (2013). Indentation testing and its acoustic emission response: applications and emerging trends. *International Materials Reviews, 56*(2), 98–142.

Feldhaus, J., Arthur, J., & Hastings, J. B. (2005). X-ray free-electron lasers. *Journal of Physics B: Atomic, Molecular and Optical Physics, 38,* S799–S819.

Fessler, J. A., & Hero, A. O. (1994). Space-alternating generalized expectation-maximization algorithm. *IEEE Transaction on signal processing, 42,* 2664–2677.

Fitch, F. R., & Rees, L. V. C. (1981). Mössbauer emission studies of zeolite A. Part 1—Effect of dehydration on $^{57}Co^{2+}$ doped Na^{+} and Co^{2+} exchanged zeolite A. *Zeolites, 1,* 19–29.

Fortner, J. A., Buck, E. C., Ellison, A. J. G., & Bates, J. K. (1997). EELS analysis of redox in glasses for plutonium immobilization. *Ultramicroscopy, 67,* 77–81.

Frégeau, M. O., Oberstedt, S., Brys', T., Gamboni, Th., Geerts, W., Hambsch, F.-J., et al. (2015). First use of single-crystal diamonds as fission-fragment detector. *Nuclear Instruments and Methods in Physics Research Section A: Accelerators, Spectrometers, Detectors and Associated Equipment, 791,* 58–64.

Friedlander, G., Kennedy, J. W., Macias, E. S., & Miller. J. M. (1981). *Nuclear and radiochemistry* (3rd ed.) Hoboken: Wiley Interscience.
Freiser, H. (1980). *Ion-selective electrodes in analytical chemistry.* Berlin: Springer.
Fritz, J. S., & Johnson-Richard, M. (1959). Colorimetric uranium determination with arsenazo. *Analytica Chimica Acta, 20,* 164–171.
Fritzsche, H., Huot, J., & Fruchart, D. (Eds.). (2016). *Neutron scattering and other nuclear techniques for hydrogen in materials.* Berlin: Springer.
Froideval, A., Badillo, A., Bertsch, J., Churakov, S., Dähn, R., Degueldre, C., et al. (2011). Towards possible opportunities in nuclear materials science and technology at an X-ray free electron laser research facility. *Journal of Nuclear Materials, 416,* 242–251.
Froideval, A., Iglesias, R., Samaras, M., Schuppler, S., Nagel, P., Grolimund, D., et al. (2007). Magnetic and structural properties of FeCr alloys. *Physical Review Letters, 99,* 237201–237204.
Fultz, B., & Howe, J. (2013). *Transmission electron microscopy and diffractometry of materials.* Berlin: Springer.
Fukugita, M., & Yanagida, T. (2003). *Physics of neutrinos.* Berlin: Springer.
Fukutani, K. (2002). Below-surface behavior of hydrogen studied by nuclear reaction analysis. *Current Opinion in Solid State and Materials Science, 6,* 153–161.
Gando, A., Gando, Y., Hanakago, H., Ikeda, H., Inoue, K., Ishidoshiro, K., et al. (2013). Reactor on-off antineutrino measurement with KamLAND. *Physical Review D, 88,* 033001–033010.
Gault, B., Moody, M. P., Cairney, J. M., Ringer, & S. P. (2012). Atom probe microscopy. *Springer series in materials science* (Vol. 160). New York: Springer.
Gaunt, J. (1956). The analysis of heavy water by infra-red spectroscopy. *Spectrochimica Acta, 8,* 57–65.
Gauthier, R., Ilmstädter, V., & Lieser, K. H. (1983). Simultaneous quantitative determination of the various oxidation states of neptunium at low concentrations by spectrophotometry. *Radiochimica Acta, 33,* 35–39.
Glatter, O., & Kratky, O. (Eds.). (1982). *Small angle X-ray scattering.* London: Academic Press.
Gehrels, N., Crannell, C. J., Forrest, D. J., Lin, R. P., Orwig, L. E., & Starr, R. (1988). Hard X-ray and low-energy gamma-ray spectrometers. *Solar Physics, 118,* 233–268.
Geipel, G., Bernhard, G., Brendle, V., & Nitsche, H. (1998). Complex formation between UO_2^{2+} and CO_3^{2-}: Studied by Laser-Induced Photoacoustic Spectroscopy (LIPAS). *Radiochimica Acta, 82,* 59–62.
Geipel, G., Reich, T., Brendler, V., Bernard, G., & Nitsche, H. (1997). Laser and X-ray spectroscopic studies of uranium-calcite interface phenomena. *Journal of Nuclear Materials, 248,* 408–411.
George, E. P. (1955). Cosmic rays measure overburden of tunnel. *Commonwealth Engineer, 1,* 455–457.
Guardincerri, E., Bacon, J., Borozdin, K., Durham, J. M., Fabritius, J., II, Hecht, A., et al. (2015). Detecting special nuclear material using muon-induced neutron emission. *Nuclear Instruments and Methods in Physics Research Section A: Accelerators, Spectrometers, Detectors and Associated Equipment, 789,* 109–113.
Griscom, D. (1984). Electron spin resonance studies of trapped hole centers in irradiated alkali silicate glasses: A critical comment on current models for HC1 and HC2. *Journal of Non-crystalline Solids, 64,* 229–247.
Griscom, D. L. (1974). E.S.R. studies of radiation damage and structure in oxide glasses not containing transition group ions: A contemporary overview with illustrations from the alkali borate system. *Journal of Non-crystalline Solids, 13,* 251–285.
Grosse, Ch. U. & Ohtsu, M. (Eds.). (2008). *Acoustic emission testing.* Berlin: Springer.
Grosse, M. K., Stuckert, J., Steinbrück, M., Kaestner, A. P., & Hartmann, S. (2013). Neutron radiography and Tomography Investigations of the Secondary Hydriding of Zircaloy-4 during Simulated Loss of Coolant Nuclear Accidents. *Physics Procedia, 43,* 294–306.

Ghosh, A., Patel, K. S., & Mishra, R. K. (1991). Extraction-spectrophotometric determination of uranium(VI) with PAR and N-octylacetamide. *Journal of Radioanalytical and Nuclear Chemistry, 152,* 243–249.

Kearley, G. J., & Peterson, V. K. (Eds.). (2015). *Neutron applications in materials for energy.* Berlin: Springer.

Glans, P.-A., Szigethy, G., Demoin, D., Tyliszczak T., Xu, J., Guo, J., et al. (2010). Actinide science with a soft X-ray Scanning Transmission X-ray Microscope (STXM). In *Materials research society symposium proceedings* (Vol. 1264, pp. 137–149).

Goldstein, J. I., Newbury, D. E., Echlin, P., Joy, D.C., Romig, A. D., Lyman, C. E. Jr, et al. (1992). *Scanning electron microscopy and X-ray microanalysis* (p. 820). New York: Plenum Press.

Gouder, T. (1998). Thin layers in actinide research. *Journal of Alloys and Compounds, 271*(3), 841–845.

Grigoriev, M. S., Fedoseev, M., Gelis, A. V., Budantseva, N. A., Shilov, V. P., Perminov, V. P., et al. (2001). Study of the interaction of Pu(IV) and Np(IV, V, VI) with Fe hydroxides to predict the behavior of actinides in environmental media. *Radiochimica Acta, 89,* 95–100.

Gross, J. H. (2004). *Mass spectrometry*. Berlin: Springer.

Guillot, L. (2001). Extraction of full absorption peaks in airborne gamma-spectrometry by filtering techniques coupled with a study of the derivatives. *Journal of Environmental Radioactivity, 53,* 381–398.

Günther-Leopold, I., Krois, M., Kobler Waldis, J., Linder, H., & Abolhassani, S. (2012). Investigation of fuel crud by means of ICP-MS and TEM. *Procedia Chemistry, 7,* 673–678.

Gurbic, A. F. (2014). Evaluation of cross-sections for particle induced gamma-ray emission (PIGE) spectroscopy. *Nuclear Instruments and Methods in Physics Research Section B: Beam Interactions with Materials and Atoms, 331,* 31–33.

Gutmacher, R. G., Cremers, D. A., Wachter, Z. (1987). Laser-induced breakdown spectroscopy: A new technique for nondestructive analysis of solutions *Transactions of the American Nuclear Society Journal, 55,* 19–20.

Hadley, J. H., Hsu, F. H., Jr., Vance, E. R., & Begg, B. D. (2005). Positron annihilation lifetime spectroscopy of air-fired $Ca_{(1-x)}(La)_xTiO_3$ Perovskites. *Journal of the American Ceramic Society, 88,* 246–248.

Hamstad, M. A. (1986). A review: Acoustic emission, a tool for composite-materials studies. *Experimental Mechanics, 26,* 7–13.

Haschke, M. (2014). *Laboratory micro X-ray fluorescence spectroscopy instrumentation and applications*. Berlin: Springer.

Havenith, M. H. (2002). *An introduction to intermolecular forces*. Berlin: Springer.

Hébert, C., Schattschneider, P., Rubino, S., Novak, P., Rusz, J., & Stöger-Pollach, M. (2008). Magnetic circular dichroism in electron energy loss spectrometry. *Ultramicroscopy, 108,* 277–284.

Hecht, F., & Reich-Rohrwig, W. (1929). Über die Bestimmung von Uran und Thorium mittels 8-Oxychinolins [On the determination of uranium and thorium using 8-oxyquinone]. *Monatshefte Fur Chemie, 53,* 596–606.

Hen, A., Magnani, N., Griveau, J.-C., Eloirdi, R., Colineau, E., Sanchez, J.-P., et al. (2015). Site-selective magnetic order of neptunium in Np_2Ni_{17}. *Physical Review B, 92,* 024410.

Hein, Ch., Sonder, J. M., & Kautenburger, R. (2017). New approach of a transient ICPMS measurement method for samples with high salinity. *Talanta, 164,* 477–482.

Heise, H., & Matthews, S. (Eds.). (2013). *Modern NMR methodology.* Berlin: Springer

Herman, G. T. & Frank, J. (Eds.). (2014). *Computational methods for three-dimensional microscopy reconstruction*. Berlin: Springer.

Henze, G. (2001). *Polarographie und voltammetrie*. Heidelberg: Springer-verlag Berlin.

Hess, N. J., Felmy, A. R., Rai, D., & Conradson, S. D. (1997). Characterization of Th carbonate solutions using XAS and implications for thermodynamic modeling. In *Materials Research Society Symposium Proceedings, 465,* 729–734.

Hess, P., & Pelzl, J. (Eds.). (1988). Photoacoustic and photothermal phenomena. In *Proceedings of the 5th International Topical Meeting*. Berlin: Springer.

Heyrovský, J. (1922). Elektrolysa se rtuťovou kapkovou kathodou. *Chemicke Listy, 16,* 256–264.

Hiraoka, K. (Ed.). (2013). *Fundamentals of mass spectrometry*. Berlin: Springer.

Hirose, M., Miyake, C., & Iida, M. (1993). "The third phase" of extraction processes in fuel reprocessing, (III) ^{31}P-NMR study of coordination behavior of Zirconium Dibutylphosphates. *Journal of Nuclear Science and Technology 30*, 232–238.

Hofmann, S. (2013). *Auger- and X- photoelectron spectroscopy in materials science*. Berlin: Springer.

Ho, C. S., Lam, C. W. K., Chan, M. H. M., Cheung, R. C. K., Law, L. K., Lit, L. C. W., et al. (2003). Electrospray ionisation mass spectrometry: Principles and clinical applications. *Clinical Biochemist Reviews, 24,* 3–12.

Holliday, K. S., Babelot, C., Walther, C., Neumeier, S., Bosbach, D., & Stumpf, Th. (2012). Site selective time resolved laser fluorescence spectroscopy of Eu and Cm dopted $LaPO_4$. *Radiochimica Acta, 100,* 189–195.

Hong-Yan, G., Chang-Chun, G., Min, X., Li-Ping, G., Ji-Hong, C., & Qing-Zhi, Y. (2015). Effect of helium implantation on SiC and graphite. *Chinese Physics B, 24,* 037803.

Horn, I., Rudnick, R. L., & McDonough, W. F. (2000). Precise elemental and isotopic ratio determination by combined solution nebulisation and laser ablation ICP-MS: Application to U/Pb geochronology. *Chemical Geology, 164,* 283–301.

Horvath, M., Guillong, M., Izmer, A., Kivel, N., Restani, R., Günther-Leopold, I., et al. (2007). Analysis of xenon gas inclusions in nuclear fuel using laser ablation ICP-MS. *Journal of Analytical Atomic Spectrometry, 22,* 1266–1274.

Hori, F., Takenaka, M., Kuramoto, E., & Aono, U. (1993). Positron annihilation study of electron-irradiation FeCu and FeCuc alloys. *Scripta Metallurgica, 29,* 243–248.

Hosten, E., & Rohwer, H. E. (1997). Complexation reactions of uranyl with arsenazo III. *Analytica Chimica Acta, 355,* 95–100.

Hotchkis, M., Child, D., & Tumlz, C. (2002). Application of accelerator mass spectrometry for ^{236}U analysis. *Journal of Nuclear Science and Technology, 39*, 532–536.

Hrdlička, A., Zaorálková, L., Galiová, M., Čtvrtníčková, T., Kanický, V., Otruba, V., et al. (2009). Correlation of acoustic and optical emission signals produced at 1064 and 532 nm laser-induced breakdown spectroscopy (LIBS) of glazed wall tiles. *Spectrochimica Acta Part B: Atomic Spectroscopy, 64,* 74–78.

Huang, J., Li, Z., Liaw, B. Y., & Zhang, J. (2016). Graphical analysis of electrochemical impedance spectroscopy data in Bode and Nyquist representations. *Journal of Power Sources, 309,* 82–98.

Hues, A. D., Henicksman, A. L., Ashley, W. H., & Romero, D. (1977). Fluorometric determination of uranium in natural waters. In *Inorganic, organic, physical and analytical chemistry (B1100) Los Alamos report 15 p; LA–6683-MS; GJBX–24(77).*

Hunter, D. B., & Bertsch, P. M. (1998). In situ examination of uranium contaminated soil particles by micro-X-ray absorption and micro-fluorescence spectroscopies. *Journal of Radioanalytical and Nuclear Chemistry, 234,* 237–242.

Ice, G. E., & Specht, E. D. (2012). Microbeam, timing and signal-resolved studies of nuclear materials with synchrotron X-ray sources. *Journal of Nuclear Materials, 425,* 233–237.

Ilkovič, D. (1934). Polarographic studies with the dropping mercury kathode. Part XLIV. The dependence of limiting currents on the diffusion constant, on the rate of dropping and on the size of drops. *Collection of Czechoslovak Chemical Communications, 6,* 498–513.

Iizuka, M., Inoue, T., Shirai, O., Iwai, T., & Arai, Y. (2001). Application of normal pulse voltammetry to on-line monitoring of actinide concentrations in molten salt electrolyte. *Journal of Nuclear Materials, 297,* 43–51.

Ionescu, S., Uţă, O., Pârvan, M., & Ohâi, D. (2009). Pressurized heavy water reactor fuel behaviour in power ramp conditions. *Journal of Nuclear Materials, 385,* 387–391.

Imam, M., Gaul, K., Stegmüller, A., Höglund, C., Jensen, J., Hultman, L., et al. (2015). Gas phase chemical vapor deposition chemistry of triethylboron probed by boron–carbon thin film deposition and quantum chemical calculations. *Journal of Materials Chemistry C, 3,* 10898–10906.

Ireland, T. R. (2014). *15.21—Ion microscopes and microprobes, reference module in earth systems and environmental sciences, from treatise on geochemistry* (2nd ed.), *15*, 385–409.

Issa, I. M., Issa, R. M., & Ahmed, Y. Z. (1978). Complexometric and spectrophotometric determination of thorium(IV), cerium(III), and uranium(VI), using quinizarin sulphonic acid after separation with ion-exchange resins. *Microchemical Journal, 18*, 569–576.

Ishigure, K. (1996). State of the art of water chemistry of Japanese BWRs. *Nuclear Engineering and Design, 160,* 171–183.

Ivanova, B., & Spiteller, M. (2014). Uranyl-water containing complexes: solid-state UV-MALDI mass spectrometric and IR spectroscopic approachs for selective quantitation. *Environmental Science and Pollution Research, 21*, 1548–1563.

Jaeschke, E. J., Khan, S., Schneider, J. R., & Hastings, J. B. (Eds.). (2016). *Synchrotron light sources and free-electron lasers, accelerator physicsinstrumentation and science applications.* Berlin: Springer.

Jaiswal, D. D., Dang, V. R., Pullat, H. S., & Sharma, R. C. (1994). Ultra-Trace Analytical Techniques for Internal Dosimetry of Actinides: An Appraisal. *Bulletin of Radiation Protection, 17,* 44–47.

Jarvis, K. E. (1992). *Handbook of inductively coupled plasma mass spectrometry*. Berlin: Springer.

James, D., Pandey, A. K., Naidu, G. R. K., & Rao, T. P. (2008). Design of two-dimensional biomimetic uranyl optrode and its application to the analysis of natural waters. *Talanta, 74,* 1420–1427.

Jameson, C. J. (2004). Xe Chemical Shift Tensor in Silicalite and SSZ-24. *Journal of the American Chemical Society, 126,* 10450–10456.

Jeong, Y. H., & Hwang, S. S. (2013). Materials management strategies for pressurized water reactors (PWRs). *Materials Ageing and Degradation in Light Water Reactors*, 315–334.

Jia, G., Belli, M., Sansone, U., Rosamilia, S., Ocone, R., & Gaudino, S. (2002). Determination of uranium isotopes in environmental samples by alpha-spectrometry. *Journal of Radioanalytical and Nuclear Chemistry, 253,* 395–406.

Jiang, C., Müller-Petke, M., Lin, J., & Yaramanci, U. (2015). Imaging shallow three dimensional water-bearing structures using magnetic resonance tomography. *Journal of Applied Geophysics, 116,* 17–27.

Johansson, S. A., Campbell, J. L., & Malmqvist, K. G. (1995). *Particle Induced X-ray Emission Spectroscopy (PIXE).* Hoboken: Wiley.

Johnson, D. A., & Florence, T. M. (1971). Spectrophotometric determination of uranium (vi) with 2-(5-bromo-2-pyridylazo)-5-diethylaminophenol. *Analytica Chimica Acta, 53,* 73–79.

Johnson, S. G., & Feary, B. L. (1993). Spectroscopic study of thorium using continuous-wave resonance ionization mass spectrometry with ultraviolet ionization. *Spectrochimica Acta, 48B,* 1065–1077.

Jonkmans, G., Anghel, V. N. P., Jewett, C., & Thompson, M. (2013). Nuclear waste imaging and spent fuel verification by muon tomography. *Annals of Nuclear Energy, 53,* 267–273.

Jol, H. M. (1995). Ground penetrating antennae frequencies and transmitter powers compared for penetration depth, resolution and reflection continuity. *Geophysical Prospecting, 43,* 693–709.

Jones, R. B., Younes, C. M., Heard, P. J., Wild, R. K., & Flewitt, P. E. J. (2002). The effect of the microscale distribution of boron on the yield strength of C-Mn steels subjected to neutron irradiation. *Acta Materialia, 50,* 4395–4417.

Juhojuntti, N., Wood, G., Juhlin, C., O'Dowd, C., & Cosma, C. (2012). 3D seismic survey at the Millennium uranium deposit, Saskatchewan, Canada: Mapping depth to basement and imaging post-Athabasca structure near the orebody. *Geophysics, 77,* 245–258.

Kalinin, S. V., & Gruverman, A. (2007). *Scanning probe microscopy electrical and electromechanical phenomena at the nanoscale*. Berlin: Springer.

Kalmykov, St. N., Aliev, R. A., Sapozhnikov, D. Yu., Sapozhnikov, Yu. A., & Afinogenov, A. M. (2004). Determination of ^{237}Np by radiochemical neutron activation analysis combined with extraction chromatography. *Applied Radiation and Isotopes, 60*, 595–599.

Kamat, R. V., Sawant, R. M., Mhatre, H. R., Chaudhuri, N. K., & Vaidya, V. N. (1998). Development of a pH titration method for the simultaneous determination of uranium, nitrate and free-acid in the feed solution of the sol-gel process of nuclear fuel fabrication. *Journal of Radioanalytical and Nuclear Chemistry, 238,* 33–41.

Kansy, J. (1996). Microcomputer program for analysis of positron annihilation lifetime spectra. *Nuclear Instruments and Methods A, 374,* 235–244.

Kato, Y., & Takahashi, M. (1976). Determination of uranium and plutonium by sequential potentiometric titration. *Bunseki Kagaku, 25,* 841–846.

Keil, R. (1979). Hochselektive spektralphotometrische Spurenbestimmung von Uran(VI) mit Arsenazo III nach Extraktionstrennung. *Fresenius' Zeitschrift für analytische Chemie, 297,* 384–387.

Keil, R. (1981). Selektive spektralphotometrische Spurenbestimmung von Uran(VI) mit Arsenazo III nach Extraktionstrennung. *'Fresenius' Zeitschrift für analytische Chemie, 305,* 374–378.

Kerbelov, L. M., & Rangelov, R. (1997). Airborne gamma ray spectroscopy—An efficient method for finding and mapping pollution with radioactive elements from uranium extraction and ore mining. In *Uranium exploration data and techniques applied to the preparation of radioelement maps. IAEA TECDOC* (Vol. 980, pp. 299–304).

Keeler, J. (2010). *Understanding NMR spectroscopy*. Hoboken: Wiley.

Keil, R. (1978). Highly selective trace determination of uranium by differential pulse polarography (catalytic nitrate reduction) following extraction separation. *Fresenius' Zeitschrift für analytische Chemie, 292,* 13–19.

Khan, M. H., Warwick, P., & Evans, N. (2006). Spectrophotometric determination of uranium with arsenazo-III in perchloric acid. *Chemosphere, 63,* 1165–1169.

Kikawa, T. (2016). *Measurement of neutrino interactions and three flavor neutrino oscillations in the T2K experiment*. Berlin: Springer.

Kim, J. I., Buckau, G., & Klenze, R. (1987). Natural colloids and generation of actinide pseudocolloids in groundwater. In B. Come & N. Chapman (Eds.), *Natural analogues in radioactive waste disposal*. London: Graham & Trottman.

Kim, B. C., Chang, K. O., Choi, S.-P., & Lee, S. L. (1997). Nondestructive evaluation techniques on the radiation damage of reactor pressure vessel steel due to neutron irradiation. *Journal of the Korean Society for Nondestructive Testing, 17*, 31–40.

Kimura, T., Serrano, J., Nakayama, S., Takahashi, K., & Takeishi, H. (1992). Speciation of uranium in aqueous solutions and in precipitates by photoacoustic spectroscopy. *Radiochimica Acta, 58*(9), 173–178.

Kissinger, P. T., & Heineman, W. R. (1983). Cyclic voltammetry. *Journal of Chemical Education, 60,* 702–706.

Klenze, R., & Kim, J. I. (1988). A direct speciation of transuranium elements in natural aquatic systems by laser-induced photoacoustic spectroscopy. *Radiochimica Acta, 44,*(5), 77–85.

Klenze, R., Kim, J. I., & Wimmer, H. (1991). Speciation of aquatic actinide ions by pulsed laser spectroscopy. *Radiochimica Acta, 52*(3), 97–103.

Klett, A. (1999). Plutonium detection with a new fission neutron survey meter. *IEEE Transactions on Nuclear Science, 46,* 877–879.

Knyazev, O. A., & Stefanovsky, S. V. (1997). EPR of paramagnetic ions in synthetic zirconolite. In *Proceedings of Sixth International Conference on Radioactive Waste Management and Environmental Restoration ASME-1997*, October 12–16, 1997 (pp. 333–335). Singapore.

Krachler, M., Alvarez-Sarandes, R., Souček, P., & Carbol, P. (2014). High resolution ICP-OES analysis of neptunium-237 in samples from pyrochemical treatment of spent nuclear fuel. *Microchemical Journal, 117,* 225–232.

Krachler, M., Alvarez-Sarandes, R., & Van Winckel, St. (2015). Challenges in the quality assurance of elemental and isotopic analyses in the nuclear domain benefitting from high resolution ICP-OES and sector field ICP-MS. *Journal of Radioanalytical and Nuclear Chemistry, 304,* 1201–1209.

Kohli, R. (2012). Chapter 5—*Developments in imaging and analysis techniques for micro- and nanosize particles and surface features. Developments in surface contamination and cleaning* (pp. 215–306).

Korpel, A., & Kessler, L. W. (1971). Comparison of methods of acoustic microscopy. In A. F. Metherell (Ed.), *Acoustical Holography, 3* (pp. 23–43). New York: Plenum.

Koopmans, Tj. (1934). Über die Zuordnung von Wellenfunktionen und Eigenwerten zu den einzelnen Elektronen eines Atoms. *Physica, 1*, 104–113.

Kordas, G., Camara, B., & Oel, H. J. (1982). Electron spin resonance studies of radiation damage in silicate glasses. *Journal of Non-crystalline Solids, 50,* 79–95.

Kosterev, A. A. (2014). *Photo-acoustic spectroscopy laser spectroscopy for sensing* (pp. 208–234).

Kubelka, P., & Munk, Fr. (1931). Ein Beitrag zur Optik der Farbanstriche. *Zeitschrift für technische Physik, 12*, 593–601.

Kubota, N., Kondo, K., Ochiai, K., & Nishitani, T. (2007). Neutron elastic recoil detection for hydrogen isotope analysis in fusion materials. *Journal of Nuclear Materials, 367–370,* 1596–1600.

Kubota, N., Ochiai, K., Kutsukake, C., Kondo, K., Shu, W., Nishi, M., et al. (2006). Ion and neutron beam analyses of hydrogen isotopes. *Fusion Engineering and Design, 81,* 227–231.

Kuhn, E., & Hartmut, I. K. (1944) Neutron image convertor, *Patent US 2344042A*

Kulenkampff, J., Gründig, M., Richter, M., & Enzmann, F. (2008). Evaluation of positron-emission-tomography for visualisation of migration processes in geomaterials. *Physics and Chemistry of the Earth, Parts A/B/C, 33,* 937–942.

Kumar, Ch. S. S. R. (Ed.). (2013). *UV-VIS and photoluminescence spectroscopy for nanomaterials characterization*. Berlin: Springer.

Kuperman, A. Ya., Smirnov, Yu. A., Fedotov, S. N., Nikol'skaya, T. L., & Efimova, N. S. (1989). Potentiostatic voltammetric method for determination of submicrogram quantities of neptunium and plutonium. *Soviet Radiochemistry, 30*, 750–755.

Laborda, F., Bolea, E., & Jiménez-Lamana, J. (2016). Post hoc interlaboratory comparison of single particle ICP-MS size. *Trend in Environemental Analytical Chemistry, 9,* 15–23.

Landman, U., & Adams, D. L. (1976). Extended x-ray-absorption fine structure—Auger process for surface structure analysis: Theoretical considerations of a proposed experiment In *Proceedings of the National Academy of Sciences*, USA (Vol. 73, pp. 2550–2553).

LaBreque, J. J. (1994). Distribution of ^{137}Cs, ^{40}K, ^{238}U and ^{232}Th in soils from Northern Venezuela. *Journal of Radioanalytical and Nuclear Chemistry, 178*, 327–336.

Lachner, J., Christl, M., Vockenhuber, C., & Synal, H.-A. (2013). Detection of UH^{3+} and ThH^{3+} molecules and ^{236}U background studies with low-energy AMS. *Nuclear Instruments and Methods in Physics Research Section B: Beam Interactions with Materials and Atoms, 294*, 364–368.

Laux, D., Baron, D., Despaux, G., Kellerbauer, A., & Kinoshita, M. (2012). Determination of high burn-up nuclear fuel elastic properties with acoustic microscopy. *Journal of Nuclear Materials, 420,* 94–100.

Le Guillou, M., Toulhoat, N., Pipon, Y., Moncoffre, N., & Khodja, H. (2015). Deuterium migration in nuclear graphite: Consequences for the behavior of tritium in CO_2-cooled reactors and for the decontamination of irradiated graphite waste. *Journal of Nuclear Materials, 461,* 72–77.

Lenz, D. (1954). Zur stenung mittelschneller Elektronen in kleinste Winkel. *Zeitschrift für Naturforschung A, 9,* 185–204.

Li, B., Wang, M., Lu, B., & Wu, J. (1991). Determination of eight trace elements in U_3O_8 standard reference material by isotope dilution spark source mass spectrometry. *Atomic Energy Science and Technology 25*, 66–70.

Linge, K. L., & Jarvis, K. E. (2009). Quadrupole ICPMS introduction to instrumentation, measurement techniques and analytical capabilities. *Geostandards and Geoanalytical Research, 33,* 445–467.

Linsmeier, Ch., Fu, C.-C., Kaprolat, A., Nielsen, S. F., Mergia, K., Schäublin, R. et al. (2013). Advanced materials characterization and modeling using synchrotron, neutron, TEM, and novel micro-mechanical techniques—A European effort to accelerate fusion materials development. *Journal of Nuclear Materials*, *442*, S834–S845.

Litherland, A. E. (1980). Ultrasensitive mass spectrometry with accelerators. *Annual Review of Nuclear and Particle Science, 30*(1980), 437–473.
Long, G. G., Fischer, D. A., Kruger, J., Black, D. R., Tanaka, D. K., & Danko, G. A. (1989). Surface-extended x-ray-absorption fine-structure experiments at atmospheric pressure by means of a photocathode proportional counter with monolayer sensitivity. *Physical Review B, 39,* 10651–10655.
Lumpkin, G. R. (1999). Physical and chemical characteristics of baddeleyite (monoclinic zirconia) in natural environments: An overview and case study. *Journal of Nuclear Materials, 274,* 206–217.
Lunney, D., Audi, G. M., & Kluge, H. J. (Eds). (2001). Atomic physics at accelerators: Mass spectrometry. In *Proceedings of the APAC 2000*. Berlin: Springer.
Lund, A., & Shiotani, M. (Eds.). (2014). *Applications of EPR in radiation research.*
Lumpkin, G. R. (1999). Physical and chemical characteristics of baddeleyite (monoclinic zirconia) in natural environments: An overview and case study. *Journal of Nuclear Materials, 274,* 206–217.
Lyons, P. C., Hercules, D. M., Morelli, J. J., Sellers, G. A., Mattern, D., Thomson Rizer, C. L., et al. (1987). Application of laser microprobe (LAMMA 1000) to "fingerprinting" of coal constituents in bituminous coal. *International Journal of Coal Geology, 7,* 185–194.
Macdonald, D. D., Scott, A. C., & Wentrcek, P. (1981). Redox potential measurements in high temperature aqueous systems. *Journal of the Electrochemical Society, 128,* 250–257.
Madey, J. M. J. (1971). Stimulated emission of bremsstrahlung in a periodic magnetic field. *Journal of Applied Physics, 42,* 1906–1913.
Maeda, N., Nakamura, N., Uchida, M., Ohta, Y., & Yoshida, K. (1996). Application of positron annihilation line-shape analysis to fatigue damage for nuclear plant materials. *Nuclear Engineering and Design, 167,* 169–174.
Magilin, D., Ponomarev, A., Rebrov, V., & Ponomarov, A. (2015). High-voltage scanning ion microscope: Beam optics and design. *Nuclear Instruments and Methods in Physics Research Section B: Beam Interactions with Materials and Atoms, 350*, 32–35.
Maheswari, M. A., & Subramanian, M. S. (2005). Selective extraction of actinides using EEBEHBA grafted polymer: A green process for nuclear reprocessing program. *Separation Science and Technology, 39,* 3621–3638.
Malchukova, E., & Boizot, B. (2010). Reduction of Eu^{3+} to Eu^{2+} in aluminoborosilicate glasses under ionizing radiation. *Materials Research Bulletin, 45,* 1299–1303.
Marinello, F., Passeri, D., & Savio, E. (Eds.). (2013). *Acoustic scanning probe microscopy*. Berlin: Springer.
Manning, D. L., & Mamantov, G. (1968). Disproportionation of electrochemically-generated uranium(V) in molten $LiF\text{-}BeF_2\text{-}ZrF_4$ at 500 °C. *Journal of Electroanalytical Chemistry and Interfacial Electrochemistry, 18,* 137–141.
Marks, N. (1995). Synchrotron radiation sources. *Radiation Physics and Chemistry, 45,* 315–331.
Marquis, E. A., Hyde, J. M., Saxey, D. W., Lozano-Perez, S., de Castro, V., Hudson, D., et al. (2009). Nuclear reactor materials at the atomic scale. *Materials Today, 12,* 30–37.
Martin, P., Hancock, G. J., Paulka, & Akber, S. R. A. (1995). Determination of ^{227}Ac by α-particle spectrometry. *Applied Radiation and Isotopes, 46*, 1065–1070.
Marschall, R. (1990). *Aspects of seismic reflection data processing*. Berlin: Springer.
Marschall, P., & Lunati, I. (Eds.). (2011). *GAM—Gas Migration Experiments in a heterogeneous shear zone of the Grimsel Test Site* (*Nagra NTB 03-11*) (p. 135).
Mauchien, P., Pailloux, A., & Vercouter, T. (2014). 17—Applications of laser spectroscopy in nuclear research and industry. *Laser Spectroscopy for Sensing*, 522–543.
Mavko, G., Mukerji, T., & Dvorkin, J. (1998). *The rock physics handbook—tool for seismic analysis in porous media.* Cambridge: Cambridge University Press.
May, L. (1971). *An introduction to Mössbauer spectroscopy*. Berlin: Springer.
Maya, L., & Begun, G. M. (1981). A Raman spectroscopic study of hydroxo and carbonato species of the uranyl (VI) ion. *Journal of Inorganic and Nuclear Chemistry, 43,* 2827–2832.
McDonald, R. S. (1986). Review: infrared spectrometry. *Analytical Chemistry, 58,* 1906–1925.

Mennecart, Th., Cachoir, C., Lemmens, K., Govers, K., Dobney, A., & Adriaensen, L. (2016). Repartition of the uranium isotopes within the Belgian UOX spent fuel. In *Proceedings of MRS symposium. SBNWM*, 2016.

Merciny, E., Pattyn-fauville, G., Swennen, L., & Duyckaerts, G. (1981). Constant-current coulometric determination of uranium in the pure metal. *Analytica Chimica Acta, 129*, 113–124.

Mieszczynski, C., Degueldre, C., Kuri, G., Bertsch, J., & Borca, C. N. (2012). Investigation of irradiated uranium-plutonium mixed oxide fuel by synchrotron based micro X-ray diffraction. *Progress in Nuclear Energy, 57*, 130–137.

Miller, M. K., & Forbes, R. G. (2014). *Atom-probe tomography, The local electrode atom probe*. Berlin: Springer.

Miller, K. M., Shebell, P., & Klemic, G. A. (1994). In situ gamma-ray spectrometry for the measurement of uranium in surface soils. *Health Physics, 67*, 140–150.

Misaelides, P, Samara, C., Noli, F., Kouimtzis, Th., & Anousis, I. (1993). Toxic element concentrations in airborne particulate matter in the area of Thessaloniki, Greece. *Science of the Total Environment, 130–131*, 139–146.

Miziolek, A., Palleschi, V., & Schechter, I. (2006). *Laser induced breakdown spectroscopy*. Cambridge: Cambridge University Press.

Mogensen, O. E. (2004). Positron annihilation in chemistry. Berlin: Springer.

Moharram, B. M., Lamaze, G., Elfiki, M., & Khalil, N. (2002). Neutron-based analysis of fission rates and ultra-trace concentrations of 235U using gamma spectrometry and CR-39 (plastic track detector). *Radiation Measurements, 35*, 113–117.

Molina, A., & González, J. (2007). *Pulse voltammetry in physical electrochemistry and electroanalysis theory and applications*. Berlin: Springer

Monteith, J. L., & Unsworth, M. H. (2013). *Transport of radiant energy. Principles of Environmental Physics—Chapter 4* (4th ed., pp. 37–48).

Morita, S., Giessibl, F. J., Meyer, E., & Wiesendanger, R. (Eds.). (2015). *Noncontact atomic force microscopy* (Vol. 3). Berlin: Springer.

Morse, D. H., Antolak, A. J., & Doyle, B. L. (2007). Photofission in uranium by nuclear reaction gamma-rays. *Nuclear Instruments and Methods in Physics Research Section B: Beam Interactions with Materials and Atoms, 261*, 378–381.

Moseley, H. G. J. (1913). The high frequency spectra of the elements. *Philosophical Magazine Series 6, 26*, 1024–1034.

Mössbauer, R. L. (1958). Kernresonanzfluoreszenz von Gammastrahlung in Ir^{191}. *Zeitschrift für Physik, 151*, 124–143.

Mössbauer, R. L., Seelbach, H. E., Persson, B., Bent, M., & Longworth, G. (1968). Self-inversion of gamma lines. *Physics Letters A, 28*, 94–95.

Moulin, C., Charron, N., Plancque, G., & Virelizier, H. (2000). Speciation of uranium by ES-MS: comparison with TRLIF. *Applied Spectroscopy, 54*, 843–848.

Moulin, C., Amekraz, B., Hubert, S., & Moulin, V. (2001). Study of thorium hydrolysis species by electrospray-ionization mass spectrometry. *Analytica Chimica Acta, 21*, 1–11.

Moulin, C., Decambox, P., Mauchien, P., Moulin, V., & Theyssier, M. (1991). On the use of laser-induced time-resolved Spectrofluorometry: Application to curium. *Radiochimica Acta, 52/3*, 119–125.

Moulin, C., Decambox, P., & Mauchien, P. (1997). State-of-the-art in time resolved laser fluorescence for actinides analysis applications and trends. *Journal of Radioanalytical and Nuclear Chemistry, 226*, 135–138.

Moulin, C., Decambox, P., Moulin, V., & Decaillon, J. G. (1995). Uranium speciation in solution by time- resolved laser-induced fluorescence. *Analytical Chemistry, 67*, 348–353.

Musazzi, S., & Perini, U. (Eds.). (2014). *Laser-induced breakdown spectroscopy theory and applications*. Berlin: Springer.

Müller, E. W. (1951). Das Feldionenmikroskop. *Zeitschrift für Physik, 131*, 136

Müller, E. W. (1970). The atom-probe field ion microscope. *Naturwissenschaften, 5*, 222–230.

Müller, E. W., MacLane, S. B., Panitz, J. A. (1969) Field adsorption and desorption of helium and neon. *Surface Science, 17,* 430–438.

Müller, W., Worrack, M., & Zapara, M. (2011). Analysis of nanoidentation experiment by mean of atom force microscopy. In *Proceedings in Applied Mathematics and Mechanics* (Vol. 11, pp. 413–414).

Mundschau, M. (1991). Photoelectron emission microscopy. *Synchrotron Radiation News, 4,* 29–34.

Mwenifumbo, C. J., & Kjarsgaard, B. A. (1999). Gamma-ray logging and radioelement distribution in the Fort à la Corne kimberlite pipe 169. *Exploration and Mining Geology, 8,* 137–147.

Nagra. (1991) *Sondierbohrung Leuggern* (Technischer Bericht 88-10, Beilage Band 5.16). Wettingen, Switzerland.

Nakada, M., Saeki, M., Masaki, N. M., & Tsutsui, S. (1998). Mössbauer spectroscopy of ^{237}Np. *Journal of Radioanalytical and Nuclear Chemistry, 232,* 201–207.

Nakashima, S. (1992). Complexation and reduction of uranium by lignite. *Science of the Total Environment, 117*(8), 425–437.

Nasu, S. (2012). Chapter: General introduction to mössbauer spectroscopy. In *Mössbauer spectroscopy* (pp. 1–22). Berlin: Springer.

Néher-Neumann, E. (2009). *Advanced potentiometr, potentiometric titrations and their systematic errors.* Berlin: Springer.

Neudecker, D., Taddeucci, T. N., Haight, R. C., Lee, H. Y., White, M. C., & Rising, M. E. (2016). The need for precise and well-documented experimental data on prompt fission neutron spectra from neutron -induced fission of ^{239}Pu. *Nuclear Data Sheets, 131,* 289–318.

Nernst, W. (1888) zur Kinetik der in Lösung befindlichen Körper 1.Theorie der Diffusion, *Zeitschrift für Physikalische Chemie, 2,* 613–637; Ueber freie lonen. (1889). *ibid, 3,* 120–130.

Neu, M., Hoffman, D., Roberts, K. E., Nitsche, H., & Silva, R. J. (1994). Comparision of extractions and laser photoacoustic spectroscopy for the determination of plutonium species in carbonate solution IV. *Radiochim.ica Acta, 66*(7), 251–258.

Neuville, D. R., Cormier, L., Boizot, B., & Flank, A. M. (2003). Structure of β-irradiated glasses studied by X-ray absorption and Raman spectroscopies. *Journal of Non-crystalline Solids, 323,* 207–213.

Nicholson, R. S., & Shain, I. (1964). Theory of stationary electrode polarography. *Analytical Chemistry, 36,* 706–723.

Nitsche, H. (1995). Synchrotron X-ray absorption spectroscopy: A new tool for actinide and lanthanide speciation in solids and solution. *Journal of Alloys and Compounds, 223,* 274–279.

Noller, B. N., & Hart, B. T. (1993). Uranium in sediments from the Magela Creek catchment, northern territory. *Environmental Technology, 14,* 649–656.

Ohkubo, Y., Kobayashi, Y., Harasawa, K., Ambe, S., Okada, T., Ambe, F., et al. (1995). Time-differential perturbed-angular-correlation and emission Moessbauer studies on 99Ru dispersed in $YBa_2Cu_3O_{6.8}$ and $YBa_2Cu_3O_6$. *The Journal of Physical Chemistry, 99,* 10629–10634.

Ohtani, T. (1986). ^{129}Te Mössbauer emission spectroscopic study of the V-Te system: V_3Te_4, V_5Te_8 and VTe_2 phases. *Solid State Communication, 57,* 81–83.

Onsager, L. (1927). Zur Theorie der Elektrolyte. II. *Physikalische Zeitschrift, 28,* 277–298.

Orlov, A. V., Restani, R., Kuri, G., Degueldre, C., & Valizadeh, S. (2010). Investigation on a corrosion product deposit layer on a boiling water reactor fuel cladding. *Nuclear Instruments and Methods in Physics Research Section B: Beam Interactions with Materials and Atoms, 268,* 297–305.

Ortega Paredes, V. I., Neyra Astudillo, M. R., Nu–ez, N., Ruzzante, J. E., Lopez Pumarega, M. I., Gomez, M. P., et al. (2012). Analysis of barkhausen noise signals of samples Fe–1 wt % Cu. *Procedia Materials Science, 1,* 651–658.

Okajima, S., Reed, D. T., Beitz, J. V., Sabau, C. A., & Bowers, D. L. (1991). Speciation of Pu (VI) in near-neutral solutions via laser photoacoustic spectroscopy. *Radiochimica Acta, 52*(3), 111–117.

Onsager, L. (1926). Zur Theorie der Elektrolyte. *Physikalische Zeitschrift, 27,* 388–392.

Opilik, L., Schmid, Th, & Zenobi, R. (2013). Modern Raman imaging: Vibrational spectroscopy on the micrometer and nanometer scales. *Annual Review of Analytical Chemistry, 6,* 379–398.

Page, A. G., Godbole, S. V., Kulkarni, Madhuri J., Porwal, N. K., Shelar, S. S., & Joshi, B. D. (1983). Trace metal assay of U_3O_8 powder by electrothermal AAS. *Talanta, 30,* 783–786.

Papadopulos, N. N., & Tsagas, N. F. (1994). Rapid nondestructive isotopic uranium analysis by neutron activation delayed neutron counting. *Journal of Radioanalytical and Nuclear Chemistry, 179,* 35–43.

Pareto, V. (1965). La courbe de répartition de la richesse. In G. Busino (Ed.), *Œuvres completes de Vilfredo Pareto*. Geneva: Librairie Droz (Originally published in 1896).

Park, J. S., Zhang, X., Sharma, H., Kenesei, P., Hoelzer, D., Li, M., et al. (2015). High-energy synchrotron x-ray techniques for studying irradiated materials. *Journal of Materials Research, 30,* 1380–1391.

Parker, H. M., & Joyce, M. J. (2015). The use of ionising radiation to image nuclear fuel: A review. *Progress in Nuclear Energy, 85,* 297–318.

Parry, E. P., & Osteryoung, R. A. (1965). Evaluation of analytical pulse polarography. *Analytical Chemistry, 37,* 1634–1637.

Pedersen, H., Höglund, C., Birch, J., Jensen, J., & Henry, A. (2012). *Chemical Vapor Deposition, 18,* 221–224.

Perkampus, H. H. (1992). *UV-VIS spectroscopy and its applications*. Berlin: Springer.

Perkins, W. T., Pearce, N. J. G., & Jefferies, T. E. (1993). Laser ablation inductively coupled plasma mass spectrometry: A new technique for the determination of trace and ultra-trace elements in silicates *Geochim. Cosmochimica Acta, 57,* 475–482.

Petersen, T. B. (2009). Acoustic emission from impact on rigid body, *Journal of Acoustic Emission*, *27*, 98–113.

Pfennig, G., Klewe-Nebenius, H., & Seelmann-Eggebert, W. (1995). *Chart of the nuclides, Forschungszentrum Karlsruhe.*

Pike, E. R. & Abbiss, J. B. (Eds.). (1997). *Light scattering and photon correlation spectroscopy*. Berlin: Springer.

Pleniceany, M., Isvoranu, M., & Spinu, C. (2005). Liquid membrane ion-selectve electrodes for potentiometric dosage of coper and nickel. *Journal of the Serbian Chemical Society, 70,* 269–276.

Pollard, P. M., Liezers, M., McMillan, J. W., Phillips, G., Thomason, H. P., & Ewart, F. T. (1988). Some actinide speciation using laser induced photoacoustic spectroscopy. *Radiochimica Acta, 44*(5), 95–101.

Portier, S., Degueldre, C., & Kivel, N. (2012). Solving isobaric interferences in secondary ion mass spectrometry: The case of Am and Pu in irradiated thorium-based fuel. *Spectrochimica Acta B, 73,* 35–38.

Pouchon, M. A., Chen, J., & Hoffelner, W. (2009). He implantation induced microstructure- and hardness-modification of the intermetallic γ-TiAl. *Nuclear Instruments and Methods in Physics Research Section B: Beam Interactions with Materials and Atoms, 267,* 1500–1504.

Pozsgai, S. (2007). Mass thickness determination and microanalysis of thin films in the TEM-revisited. *Ultramicroscopy, 107,* 191–195.

Probst, T., Zeh, P., & Kim, J. I. (1995). Multielement determinations in ground water. *Fresenius' Journal of Analytical Chemistry, 351,* 745–751.

Quaim, S. M., Bisinger, T., Hilgers, K., Nayak, D., & Coenen, H. H. (2007). Positron emission intensities in the decay of ^{64}Cu, ^{76}Br and ^{124}I. *Radiochimica Acta, 95,* 67–73.

Raman, C. V. (1928). A new radiation. *Indian Journal of Physics, 2,* 387–398.

Randles, J. E. B. (1948). A cathode ray polarograph. *Transactions of the Faraday Society, 44,* 322–327.

Rapkin, E. (1964). Liquid scintillation counting 1957–1963: A review. *The International Journal of Applied Radiation and Isotopes, 15,* 69–87.

Rao, T. S., Shriwastwa, B. B., Dubey, J. N., Patil, B. P., Chandrasekharan, K. N., Pandey, V. D., et al. (2003). Quantitative estimation of plutonium-rich areas in thorium-based MOX fuels using alpha autoradiography technique. *Radiation Measurements, 36,* 747–750.

Ray, A. K. (1963). 3-Oximinomethylsalicylic acid as a reagent for the gravimetric determination of thoriumAnalytica Chimica Acta, 28, 580–583.

Rehr, J. J., & Albers, R. C. (2000). Theoretical approaches to X-ray absorption fine structure. *Reviews of Modern Physics, 72,* 621–654.

Reimer, L. (1998). *Scanning electron microscopy, physics of image formation and microanalysis.* Berlin: Springer.

Restani, R., Martin, M., Kivel, N., & Gavillet, D. (2009). Analytical investigations of irradiated inert matrix fuel. *Journal of Nuclear Materials, 385,* 435–442.

Riggi, S., Antonuccio-Delogu, V., Bandieramonte, M., Becciani, U., & Costa, A. (2013). Muon tomography imaging algorithms for nuclear threat detection inside large volume containers with the Muon Portal detector. *Nuclear Instruments and Methods in Physics Research Section A: Accelerators, Spectrometers, Detectors and Associated Equipment, 728,* 59–68.

Riley, K. J., & Harling, O. K. (1998). An improved prompt gamma neutron activation analysis facility using a focused diffracted neutron beam. *Nuclear Instruments and Methods in Physics Research Section B: Beam Interactions with Materials and Atoms, 143,* 414–421.

Rietveld, H. M. (1969). A profile refinement method for nuclear and magnetic structures. *Journal of applied Crystallography, 2,* 65–71.

Robertson, C., Panigrahi, B. K., Balaji, S., Kataria, S., Serruys, Y., Mathon, M.-H., et al. (2012). Particle stability in model ODS steel irradiated up to 100 dpa at 600 & #xB0;C: TEM and nano-indentation investigation. *Journal of Nuclear Materials, 426,* 240–246.

Rogers, A. W. (1979). *Techniques of autoradiography* (3rd ed.). New York: Elsevier North Holland.

Roncal-Herrero, J. D., Rodríguez-Blanco, E. H., Oelkers, L. G., & Benning, L. G. J. (2011). The direct precipitation of rhabdophane ($REEPO_4 \cdot nH_2O$) nano-rods from acidic aqueous solutions at 5–100 °C. *Journal of Nanoparticle Research, 13,* 4049–4062.

Ronchi, C., Ottaviani, J. P., Degueldre, C., & Calabrese, R. (2003). Thermophysical properties of inert matrix fuels for actinide transmutation. *Journal of Nuclear Materials, 320,* 54–65.

Rothchild, S. (1963). Advances in tracer methodology (Vol. 1). Berlin: LSC, Springer.

Rudenko, A., & Rolles, D. (2015). Time-resolved studies with FELs. *Journal of Electron Spectroscopy and Related Phenomena, 204,* 228–236.

Sah, D. N., Viswanathan, U. K., Ramadasan, E., Unnikrishnan, K., & Anantharaman, S. (2008). Post irradiation examination of thermal reactor fuels. *Journal of Nuclear Materials, 383,* 45–53.

Sahoo, P., Mallika, C., Ananthanarayanan, R., Lawrence, F., Murali, N., & Kamachi, U. (2012). Mudali, Potentiometric titration in a low volume of solution for rapid assay of uranium. Application to quantitative electro-reduction of uranium(VI). *Journal of Radioanalytical and Nuclear Chemistry, 292,* 1401–1409.

Sarma, D. V. N., & Raghava Rao, Bh S V. (1955). Alizarin-s, a reagent for thorium. A gravimetric, colorimetric, and spectrophotometric study. *Analytica Chimica Acta, 13,* 142–149.

Sarott, F. A. (2005). Water chemistry in boiling water reactor—a Leibstadt—specific overview. *CHIMIA International Journal for Chemistry, 59,* 923–928.

Savvin S. B. (1961). Analytical use of arsenazoIII, determination of thorium, zirconium, uranium and rare earth elements. *Talanta, 8,* 673–585.

Ševčík, A. (1948). Oscillographic polarography with periodical triangular voltage. *Collection of Czechoslovak Chemical, 13,* 349–377.

Sharma, A., & Schulman, St. G. (1999). *Introduction to fluorescence spectroscopy*. Hoboken: Wiley.

Shamsipur, M., Ghiasvand, A. R., & Yamini, Y. (1999). Solid-phase extraction of ultratrace uranium(VI) in natural waters using octadecyl silica membrane disks modified by tri-*n*-octylphosphine oxide and its spectrophotometric determination with dibenzoylmethane. *Analytical Chemistry, 71,* 4892–4895.

Sheng, Z., Zhao, Y., & Gu, D. (1985). NAA determination of the $^{235}U/^{238}U$ ratio of uranium in geological samples. *Geochimica, 18*, 188–195.

Shimizu, I. (2013, September). In *Proceeding of 13th Conference Astroparticle & Underground Physics*.

Scholz, F. (2010). *Electroanalytical methods*. Berlin: Springer.

Schlemmer, G. (1999). *Analytical graphite furnace atomic absorption spectrometry a laboratory guide*. Berlin: Springer.

Schlichting, I., & Miao, J. (2012). Emerging opportunities in structural biology with X-ray free-electron lasers. *Current Opinion in Structural Biology, 22,* 613–626.

Schoonover, J. R., & Havrilla, G. J. (1999). Combining X-ray fluorescence spectrometry and vibrational microscopy to assess highly heterogeneous, actinide-contaminated materials. *Applied Spectroscopy, 53,* 257–265.

Scott, D. B. (2013). Internal inspection of reinforced concrete for nuclear structures using shear wave tomography. *Energy Conversion and Management, 74,* 582–586.

Schmüser, P., Dohlus, M., & Rossbach, J. (2009). *Ultraviolet and soft x-ray free-electron lasers introduction to physical principles, experimental results, technological challenges, springer tracts in modern physics* (p. 229).

Scott, M. C., Chen, C. C., Mecklenburg, M., Zhu, C., Xu, R., Ercius, P., et al. (2012). Electron tomography at 2.4-ångström resolution. *Nature, 483,* 444–447.

Seidman, D. N. (2007). Three-dimensional atom-probe tomography: Advances and applications. *Annual Review of Materials Research, 37,* 127–158.

Sekine, T., Yoshihara, K., Lakosi, L., Németh, Z., & Veres, A. (1991). Integral cross section of the ^{99}Tc(γ, γ′)^{99m}Tc reaction in the 15–50 MeV energy region. *International Journal of Radiation Applications and Instrumentation. A., 42,* 149–153.

Sheriff, R. E., & Geldart, L. P. (1995). *Exploration seismology*. Cambridge: Cambridge University Press (2nd ed).

Silverman, L., Billings, C. E., & First, M. W. (1971). 5—Automatic particle counting and sizing instruments. *Particle Size Analysis in Industrial Hygiene*, 196–234.

Singh, J. P., Gautam, S., Srivastava, R. C., Asokan, K., & Chae, K. H. (2015). XAS and XMCD investigation of zinc ferrite nanoparticles irradiated with 100 MeV O beam. In *Proceedings of Magnetics Conference (INTERMAG), 2015 IEEE.*

Siren, T., Hakala, M., Valli, J., Kantia, P., Hudson, J. A., & Johansson, E. (2015). In situ strength and failure mechanisms of migmatitic gneiss and pegmatitic granite at the nuclear waste disposal site in Olkiluoto, Western Finland. *International Journal of Rock Mechanics and Mining Sciences, 79,* 135–148.

Sim, C.-M., Park, S.-S., Song, Y.-Y., Park, D.-G., & Chang, K.-O. (2000). Harmonic frequency analysis of acoustic Barkhausen noise on neutron irradiated material. In *Springer 15th World Conference on Non-Destructive Testing 15–21 October 2000*, Rome.

Slichter, Ch. P. (1996). *Principles of magnetic resonance, springer series in solid-state sciences.*

Simeone, D., Mallet, C., Dubuisson, P., Baldinozzi, G., Gervais, C., & Maquet, J. (2000). Study of boron carbide evolution under neutron irradiation by Raman spectroscopy. *Journal of Nuclear Materials, 277,* 1–10.

Song, D., Wang, Z., & Zhu, J. (2015). Effect of the asymmetry of dynamical electron diffraction on intensity of acquired EMCD signals. *Ultramicroscopy, 148,* 42–51.

Soto-Guerrero, J., Gajdosova, D., & Havel, J. (2001). Uranium oxide clusters by laser desorption ionization during MALDI-TOF MS analysis of uranium(VI). *Journal of Radioanalytical and Nuclear Chemistry, 249,* 139–143.

Smedley, S. (1980). *The interpretation of ionic conductivity in liquids*. Berlin: Springer.

Smith, A. L., Raison, P. E., Hen, A., Bykov, D., Colineau, E., Sanchez, J.-P., et al. (2015). Structural investigation of Na_3NpO_4 and Na_3PuO_4 using X-ray diffraction and ^{237}Np Mössbauer spectroscopy. *Dalton Transactions, 44,* 18370–18377.

Stefanovsky, S. V., & Stefanovskaya, O. I. (2014). EPR of radiation-induced paramagnetic centers in irradiated sodium silicate glass for immobilization of solid radioactive wastes synthesized in various conditions. *Inorganic Materials: Applied Research, 3,* 271–277.

Stefanovsky, S. V., Stefanovsky, O. I., Kadyko, M. I., Zhachkin, V. A., & Bogomolova, L. D. (2016). The effect of electron irradiation on the structure of sodium aluminum-iron phosphate glasses. *Materials Research Society Symposium: the Japanese Society for Non-Destructive Inspection. 1*, 4227–4234.

Steinhausen, Ch., Dobler, M., Glückler, H., & Weidinger, A. (1994). Elastic recoil detection analysis with heavy ions. *Nuclear Instruments and Methods in Physics Research Section B: Beam Interactions with Materials and Atoms, 89,* 131–139.

Suekane, F., & Junqueira de Castro Bezerra, Th. (2016). Double Chooz and a history of reactor θ_{13} experiments. *Nuclear Physics B, 908,* 74–93.

Szöghy, I. M., Simon, J., & Kish, L. (1981). Si(Li) detector efficiency in standard X-ray fluorescence geometries. *X-Ray Spectrometry, 10,* 168–170.

Szymanski, H. (1970). *Raman spectroscopy—Theory and practice*. Berlin: Springer.

Tao, S. J. (1972). Positronium annihilation in molecular substances. *The Journal of Chemical Physics, 56,* 5499–5510.

Taylor, J. E. L., Hall, G. N., & Mummery, P. M. (2016). Investigating the effects of stress on the pore structures of nuclear grade graphites. *Journal of Nuclear Materials, 470,* 216–228.

Takamatsu, K., Takegami, H., Ito, C., Suzuki, K., Ohnuma, H., Hino, R., et al. (2015). Cosmic-ray muon radiography for reactor core observation. *Annals of Nuclear Energy, 78,* 166–175.

Taurines,T., & Boizot, B. (2011). Synthesis of powellite-rich glasses for high level waste immobilization. *Journal of Non-crystalline Solids, 357,* 2723–2725.

Teixeira, L. S. G., Costa, A. C. S., Ferreira, S. L. C., Freitas, M. D. L., & Carvalho, M. S. D. (1999). Spectrophotometric determination of uranium using 2-(2-Thiazolylazo)-p-cresol (TAC) in the presence of surfactants. *Journal of the Brazilian Chemical Society, 10* (On-line). ISSN 1678-4790.

Teo, B.-K. (1986). *EXAFS: Basic principles and data analysis*. Berlin: Springer

Theisen, R. (1965). *Quantitative electron microprobe analysis*. Berlin: Springer.

The Japanese, Society for Non-Destructive Inspection. (Ed.). (2016). *Practical acoustic emission testing*. Berlin: Springer.

Theophanides. (Ed). (1984). *Fourier transform infrared spectroscopy industrial chemical and biochemical applications*. Berlin: Springer.

Theuwissen, A. J. P. (1995). *Solid-state imaging with charge-coupled device*. Dordrecht: Kluwer Academic Publisher.

Thieme, J., Schmahl, G., Rudolph, D. & Umbach, E. (Eds). (1998). *X-Ray microscopy and spectromicroscopy—status report from the Fifth International Conference*, Wüzburg, August 19–23, 1996. Berlin: Springer.

Thomson, J. J. (1923). The electron in chemistry. *Journal of the Franklin Institute, 195,* 737–785.

Thompson, M. (Ed). (1989). *Handbook of inductively coupled plasma spectrometry* (2nd ed.). Berlin: Springer.

Trautmann, N., Peuser, P., Rimke, H., Sattelberger, P., Herrmann, G., Ames, F., et al. (1986). A resonance ionization mass spectrometer as an analytical instrument for trace analysis. *Journal of the Less Common Metals, 122,* 533–538.

Triay, I., Hobart, D. E., Mitchell, A. J., Newton, T. W., Ott, M. A., Palmer, P. D., et al. (1991). Size distribution of plutonium colloids using autocorrelation photon spectroscopy. *Radiochimica Acta, 52*(53), 127–131.

Tycko, R. (Ed.). (1994). Nuclear magnetic resonance probes of molecular dynamics. Berlin: Springer.

Tylka, M. M., Willit, J. L., Prakash, J., & Williamson, M. A. (2015a). Method development for quantitative analysis of actinides in molten salts. *Journal of the Electrochemical Society, 162,* H1. doi:10.1149/2.0401509jes

Tylka, M. M., Willit, J. L., Prakash, J., & Williamson, M. A. (2015b). Application of voltammetry for quantitative analysis of actinides in molten salts. *Journal of The Electrochemical Society, 162,* H852–H859.

Uchida, S., Satoh, T., & Tsukada, T. (2008). High temperature water chemistry sensors related to nuclear power plant. *Preprint ICPWS XV* (p. 10).

Upadhyay, S. K. (2004). *Seismic reflection processing, with special reference to anisotropy*. Berlin: Springer.

Usman, S. M., Jocher, G. R., Dye, S. T., McDonough, W. F., & Learned J. G. (2015a). Corrigendum: AGM2015: Antineutrino global map 2015. Scientific Reports, 5, 15308.
Usman, S. M., Jocher, G. R., Dye, S. T., McDonough, W. F., & Learned, J. G. (2015b). Corrigendum: AGM2015: Antineutrino global map 2015. Scientific Reports, 5, 15308.
Valeur, B., & Brochon, J.-C. (Eds.). (2001). *New Trends in fluorescence spectroscopy applications to chemical and life sciences*. Berlin: Springer.
Vegard, L. (1921). Die Konstitution der Mischkristalle und die Raumfüllung der Atome. *Zeitschrift für Physik, 5,* 17–26.
Verbeeck, J., Hébert, C., Rubino, S., Novak, P., Rusz, J., et al. (2008). Optimal aperture sizes and positions for EMCD experiments. *Ultramicroscopy, 108,* 865–872.
Veleva, L., Schäublin, R., Plocinski, T., Walter, M., & Baluc, N. (2011). Processing and characterization of a W–2Y material for fusion power reactors. *Fusion Engineering and Design, 86,* 2450–2453.
Veselsky, J. C., & Degueldre, C. A. (1986). Quenching behaviour of lanthanides on the ultraviolet fluorescence of uranium(VI). *Analyst, 111,* 535–538.
Veselsky, J. C., & Ratsimandresy, Y. (1979). An investigation of quenching effects in the direct fluorimetric determination of uranium in minerals. *Analytica Chimica Acta, 104,* 345–353.
Vizkelethy, G. (2003). Nuclear reaction analysis and proton-induced gamma ray emission. In *Characterisation of materials*. Hoboken: Wiley.
von Laue, M. (1913). Röntgenstrahlinterferenzen. *Physikalische Zeitschrift, 14,* 1075–1079.
Xu, H., & Wang, Y. (1999). Electron energy-loss spectroscopy (EELS) study of oxidation states of Ce and U in pyrochlore and uraninite—natural analogues for Pu- and U-bearing waste forms. *Journal of Nuclear Materials, 265,* 117–123.
Xu, J., & Shi, S.-Q. (2004). Investigation of mechanical properties of ε-zirconium hydride using micro- and nano-indentation techniques. *Journal of Nuclear Materials, 327,* 165–170.
Walker, C. T., Bremier, S., Portier, S., Hasnaoui, R., & Goll, W. (2009). SIMS analysis of an UO_2 fuel irradiated at low temperature to 65 MWd/kgHM. *Journal of Nuclear Materials, 393,* 212–223.
Walstedt, R. E., Tokunaga, Y., & Kambe, Sh. (2014). NMR studies of actinide oxides—A review. *Comptes Rendus Physique, 15,* 563–572.
Wanga, Zh, Morrisa, C. L., Makelaa, M. F., Bacona, J. D., Baera, E. E., Brockwella, M. I., et al. (2009). Inexpensive and practical sealed drift-tube neutron detector. *Nuclear Instruments and Methods in Physics Research Section A: Accelerators, Spectrometers, Detectors and Associated Equipment, 605,* 430–432.
Ward, J. D., Bowden, M., Resch, C. T., Smith, St., McNamara, Br. K., Buck, & E. C. (2015). Identification of uranyl minerals using oxygen K-edge X-ray absorption spectroscopy. *Geostandards and Geoanalytical Research, 02.* doi:10.1111/j.1751-908X.2015.00337.x
Waseda, Y., Matsubara, E., & Shinoda, K. (2011). *X-Ray diffraction crystallography. introduction, examples and solved problems*. Berlin: Springer.
Weeks, R. A. (1994). The many varieties of E′ centers: A review. *Journal of Non-crystalline Solids, 179,* 1–9.
Weisenburger, S., & Sandoghdar, V. (2015). Light Microscopy: An ongoing contemporary revolution. *Contemporary Physics, 52,* 123–143.
Weiland, E., Springuel-Huet, M.-A., Nossov, A., & Gédéon, A. (2016). ^{129}Xenon NMR: Review of recent insights into porous materials. *Microporous and Mesoporous Materials, 225,* 41–65.
Wevers, M. (1997). Listening to the sound of materials: Acoustic emission for the analysis of material behaviour. *NDT & E International, 30,* 99–106.
Weyer, G. (1976). Applications of parallel-plate avalanche counters in Mössbauer spectroscopy, chapter. In *Mössbauer effect methodology* (pp. 301–319). Berlin: Springer Publisher
Wilbraham, R. J., Boxall, C., Goddard, D. T., Taylor, R. J., & Woodbury, S. E. (2015). The effect of hydrogen peroxide on uranium oxide films on 316L stainless steel. *Journal of Nuclear Materials, 464,* 86–96.

Winkelmann, I., Thomas, M., & Vogl, K. (2001). Aerial measurements on uranium ore mining, milling and processing areas in Germany. *Journal of Environmental Radioactivity, 53,* 301–311.

Whitehouse, A. I., Young, J., Botheroyd, I. M., Lawson, S., Evans, C. P., & Wright, J. (2001). Remote material analysis of nuclear power station steam generator tubes by laser-induced breakdown spectroscopy. *Spectrochimica Acta B, 56,* 821–830.

Xiong, Q., Baychev, T. G., & Jivkov, A. P. (2016). Review of porenetwork modeling of porous media: Experimental characterizations, network constructions and application to reactive transport. *Journal of Contaminant Hydrology, 192,* 101–117.

Yakshin, V. V., & Krokhin, M. N. (2011). Determination of small amounts of uranium by redox potentiometric titration. *Radiochemistry, 53,* 327–331.

Yang, Ch.-H, & Huang, M.-F. (2004). Characterization of hydrogen concentration in Zircaloy claddings using a low-frequency acoustic microscope with a PVDF/LFB transducer. *Journal of Nuclear Materials, 335,* 359–365.

Yang, F., & Li, J. C. M. (Eds.). (2008). *Micro and nano mechanical testing of materials and devices*. Berlin: Springer.

Yang, K. J., Wang, T. S., Zhang, G. F., Peng, H. B., Chen, L., Zhang, L. M., et al. (2013). Study of irradiation damage in borosilicate glass induced by He ions and electrons. *Nuclear Instruments and Methods in Physics Research Section B: Beam Interactions with Materials and Atoms, 307,* 541–544.

Yasuoka, H., Koutroulakis, G., Chudo, H., Richmond, S., Veirs, D. K., Smith, A. I., et al. (2012). Observation of ^{239}Pu nuclear magnetic resonance. *Science, 336,* 901–904.

Yoe, J. H., Will, F., III, & Black, R. A. (1953). Colorimetric determination of uranium with dibenzoylmethane. *Analytical Chemistry, 25,* 1200–1204.

Yoshida, Y., & Langouche, G. (Eds.). (2013). *Mössbauer spectroscopy, tutorial book springer*. Berlin: Springer.

Yu-fu, Y., Salbu, B., Bjørnstad, H. E., & Lien, H. (1990). Improvement for α-energy resolution in determination of low level plutonium by liquid scintillation counting. *Journal of Radioanalytical and Nuclear Chemistry, 145,* 345–353.

Zeeman, P. (1897). On the influence of magnetism on the nature of light emitted by a substance. *Philosophical Magazine, 43*, 226.

Zelenty, J., Smith, G. D. W., Wilford, K., Hyde, J. M., & Moody, M. P. (2016). Secondary precipitation within the cementite phase of reactor pressure vessel steels. *Scripta Materialia, 115,* 118–122.

Zhang, F., & Juhlin, Ch. (2014). Full waveform inversion of seismic reflection data from the Forsmark planned repository for spent nuclear fuel, eastern central Sweden. *Geophys. J. Int., 192,* 1106–1122.

Zhang, W., Yi, J., Mekarski, P., Ungar, K., Hauck, B., & Kramer, G. H. (2011). Quantification of ^{235}U and ^{238}U activity concentrations for undeclared nuclear materials by a digital gamma—gamma coincidence spectroscopy. *Applied Radiation and Isotopes, 69,* 904–907.

Zheng, Y., Maev, R Gr, & Solodov, I Yu. (2000). Review/Sythèse nonlinear acoustic applications for material characterization: A review. *Canadian Journal of Physics, 77,* 927–967.

Ziebold, Th O. (1967). Precision and sensitivity in electron microprobe analysis. *Analytical Chemistry, 39,* 858–861.

Zink, F. E. (1997). X-ray tubes. *Radiographic, 17,* 1257–1268.

Zschornack, G. H. (2007). *Handbook of X-ray data*. Berlin: Springer.

Chapter 4
Comparing and Combining Analytical Techniques

The literature on nuclear material analysis includes a large number of studies combining analytical techniques. This chapter explores the advantage of combining complementary techniques to gain information from isotopic, element, chemical and structural nature on the nuclear material from the nano- to the macro- scale. However, the comparison of these technique performances in terms of detection limit and lateral resolution, and resolution in depth or in terms of interferences, is required to optimize the selection of the analytical techniques before combining them.

4.1 Comparing Analytical Techniques

When comparing the potential of analytical techniques to characterize nuclear materials or their environments, several questions need to be addressed. Do we have to:

- Analyze isotopes, elements, chemical species and/or structural issues?
- Characterize objects at the nano-, micro- or macro- scale?
- Apply a passive or an interactive technique?
- Consider waves or particles?
- Work at high or low energy?
- Work with nano- micro- macro- beam or quantity of reagents?

Consequently, the comparison of these technique performances in terms of detection limit, and lateral resolution, resolution in depth or in terms of interferences resolution, is mandatory.

Important issues are noted under the six bullet points above. They are not totally independent. The energy spectrum guides the level of investigation for targeting the isotope, element, molecular or structural level. The energy scale which goes from neV (e.g. Mössbauer spectroscopy) to MeV (nuclear spectroscopy) ranges over 15

C.A. Degueldre, *The Analysis of Nuclear Materials and Their Environments*,
DOI 10.1007/978-3-319-58006-7_4

orders of magnitude going from intermolecular to intranuclide issues. This should also be completed with other considerations such as time required for sample preparation and analysis, instrumentation needed and investigation costs.

In this frame, high-efficiency detectors produce spectra in less time than low-efficiency detectors. They can be used for a 1D scan along *x*, a 2D radiography on the *x-y* plan, a 3D tomography in the volume *x-y-z* or along plans over time: *x-y-t*, or, a 4D tomography along a volume over time: *x-y-z-t*. In all cases detection limit as well as resolutions (lateral and in depth) need to be known. Finally, the time and costs limit also the quantity of investigations to be carried out to avoid a use and abuse case.

4.1.1 Comparing Passive Techniques

In the case of passive techniques, limitations are inherent to the activity (radiochemical, chemical) of the sample, the geometry of the analytical system as well as the performance or response of the detector. The ratio between the experimentally detected activity, A_{exp}, by the real activity, A_{real}, of the investigated sample may be calculated to quantify the efficiency *Eff* of the counting:

$$Eff = A_{exp}/A_{real} \tag{4.1}$$

In Sect. 3.1 the efficiency for αS, βS and γS have been evaluated for the actinides (e.g. as DL in Tables 3.2, 3.4 and 3.6). In all counting techniques (e.g. LSC, αS, βS and γS), the counting yield is driven by the ionization properties of the ionizing particles, hence for similar detector geometries the following inequations are observed:

$$Eff_{\alpha} > Eff_{\beta} > Eff_{\gamma} \tag{4.2}$$

Consequently and for comparable blank counting rate, the detection limits, *DL* (mol), gives:

$$\mathrm{DL}(\gamma\mathrm{S}) > \mathrm{DL}(\alpha\mathrm{S}) \tag{4.3}$$

This is valid for solid detector but also for liquid detector such as LSC. Detection limit comparison for actinides is to be found in Fig. 4.3 and more especially 4.3b. The detection limit in terms of mass or amount in the case of radionuclides is a function of their specific activity or of their decay constant (see Eq. 3.2) or according to Eq. 1.2 the inverse of their half-life. This issue has been quantified in Tables 3.2 and 3.6 for both α and γ tracers respectively.

The comparison of detection limits is also a function of the technique performances in terms of resolution and sample size. In this frame, alpha autoradiography and gamma autoradiography may be compared in the case of a nuclear fuel. An autoradiograph on a nuclear emulsion reveals a pattern of decay emissions from the

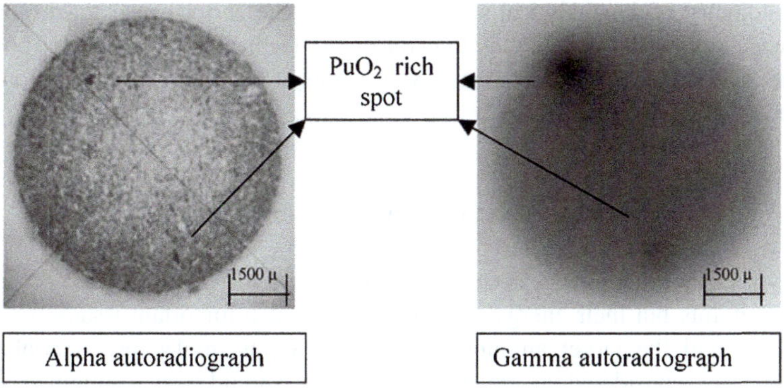

Fig. 4.1 Alpha autoradiograph and gamma autoradiograph of MOX fuel pellet (ThO_2 –4 wt% PuO_2) showing PuO_2-rich spots. Rao et al. (2003), with permission © Elsevier

distribution of radioisotopes on a nuclear material section. Methodology is as described in Sect. 3.1.2. In Fig. 4.1, both alpha and gamma autoradiographs of a (ThO_2 –4 wt% PuO_2) MOX fuel pellet are compared, (see Rao et al. 2003). Both autoradiograms are showing PuO_2-rich spots. However the resolution of the alpha autoradiography is of the order of the 20 µm while the resolution of the gamma autoradiography is 50 times larger. Thanks to his better resolution and efficiency of alpha particle absorption, the PuO_2 rich spots are better detected showing the performances of the alpha technique versus the gamma technique.

4.1.2 Comparing Interactive Techniques

In the case of the interactive techniques, limitations in terms of performance may also be due to the activity (radiochemical, chemical, physical) of the analyte or components in the sample. However, the intensity of interaction is induced by the added reagent, the incident particle or wave beam into the sample (see Fig. 4.2).

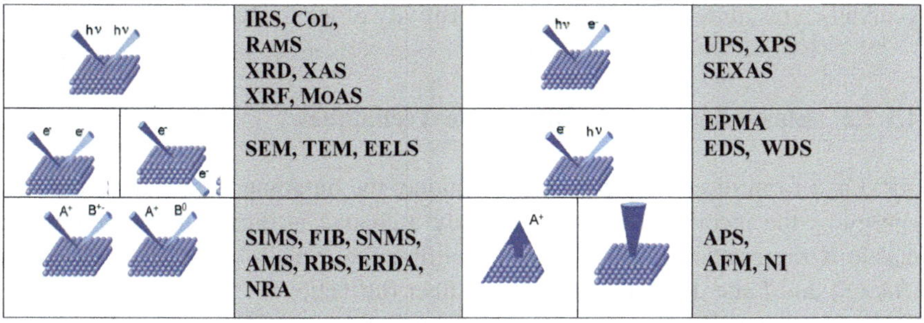

Fig. 4.2 Comparing the scheme of photon, electron and ion-induced techniques

The geometry of the analytical system as well as the performances or responses of the detector are also key issues that make the technique more or less powerful. Figure 4.2 compares in a schematic way some of the interactive techniques discussed in this work.

4.1.2.1 Interaction-Detection: Options and Limitations

The types of particle (mass, speed, charge) or wave (wavelength) used for analysis are key points but their energy and intensity are also important issues concerning the scale and the target amount as nuclides, atoms, molecules, nanophases or macrophases considered in the interaction. One should also distinguish surface, sub-surface and bulk and again lateral resolution as well as depth resolution. A comparison of the potential of macro- micro- nano- analysis is required.

The comparison may start with the performances in terms of resolution comparing the optical microscope (~ 1000 nm), the confocal microscope (~ 100 nm), the SEM (~ 10 nm), the TEM (~ 1 nm), then the promising muon and ion microscope with a resolution function of the mass of the particle and their wavelength. The lateral resolution may result from the fact that photons are not charged and electrons or ions are charged allowing the beam to be better focused by electromagnetic lenses, while photons are focused by optical systems limited by their wavelength.

The wavelength of the electrons in a 10 kV SEM is then 12.2 pm while in a 200 kV TEM the wavelength is 2.5 pm. In comparison the wavelength of X-rays usually used in X-ray diffraction is in the order of 100 pm (e.g. Cu Kα: λ = 154 pm). The X-ray photons may subsequently be used for microscopy, e.g. STXM, making this technique very powerful for nano-spectroscopy (see Sect. 3.2.3.12).

On the other side, the ion microscope also has potential. The possibilities and limitations for microstructure characterization using backscattered particles from a sharply focused helium ion beam was demonstrated by van Gastel et al. (2015). The interaction of helium ions with matter enables the imaging, spectroscopic characterization, as well as the nanometer scale modification of samples. The contrast that is seen in helium ion microscopy (HIM) images differs from that in SEM and is generally a result of the higher surface sensitivity of the method (see Sect. 3.2.18.5).

4.1.2.2 Elastic Interaction: Diffraction Techniques

XRD is a form of elastic scattering technique; the outgoing X-rays have the same energy as the incoming X-rays, with only a change in direction. Other forms of elastic X-ray scattering include powder diffraction, Small-Angle X-ray Scattering (SAXS) and Laue diffraction and X-ray fiber diffraction.

Since all particles are associated to waves, they can be used for probing the crystallographic properties. Several scattering techniques may be applied in an elastic diffraction mode, namely: XRD (photon scattering), eD (electron scattering) and ND (neutron scattering). These particles may be used to produce useful diffraction patterns. Although electron, neutron, and X-ray scattering are based on different physical processes, the resulting diffraction patterns are analyzed using the same coherent diffraction imaging techniques.

Comparison of XRD and ND has been performed on zirconia-based inert matrix fuel (IMF). Clearly the advantage of ND was noted with a form factor revealing sharp peaks even at large angles, allowing a more sensitive analysis of the lattice parameters of the solid solution, (see Degueldre et al. 2001).

As discussed in Sect. 3.2.3.13, the electron density within the crystal and the diffraction patterns are related by a simple mathematical method, the Fourier transformation, which allows the density to be calculated relatively easily from the patterns. However, this works only if the scattering is weak, i.e. if the scattered beams are much less intense than the incoming beam.

Neutron diffraction is an excellent method for structure determination, e.g. Reiche et al. (2016), although it has been difficult to obtain intense, monochromatic beams of neutrons in sufficient quantities. Being uncharged, neutrons scatter much more readily from the atomic nuclei rather than from the electrons. Therefore, neutron scattering is very useful for observing the positions of light atoms with few electrons, especially hydrogen, e.g. Wang et al. (2016), which is essentially invisible in the X-ray diffraction. Neutron scattering also has the property that solvent can be made invisible by substituting light water, H_2O, by heavy water, D_2O.

In the same frame, application of SAXS and SANS in evaluation of porosity, pore size distribution and surface area of material, as well as investigations of early stage precipitation in a tungsten-rich nickel-base superalloy, show that the use of SAXS and SANS may be complementary, (see Gilles et al. 2014).

By contrast, electron beams may produce strong secondary scattering even for relatively thin crystals (>100 nm). Moreover, the strong interaction of electrons with matter (about 1000 times stronger than for X-rays) allows determination of the atomic structure of extremely small volumes (see Sect. 3.2.8.2). The field of applications for electron crystallography includes the complex structures of (nanocrystalline $Th_2Ni_{10}Al_{15}$) intermetallic compounds, e.g. Yaniv and Meshi (2016), and zeolites, e.g. Su et al. (2014).

4.1.2.3 Inelastic Interaction: Absorption/Fluorescence Techniques

Inelastic scattering takes place when energy is exchanged from the incident X-ray to the crystal by exciting an electron inner-shell, for example. Since it deals with wavelength changes, inelastic scattering is treated in the RamS, PhoS, XRF and XAS Sections for the photons.

XAS and other X-ray techniques such as XRF, XAFS, EXAFS, XANES … belong to the so called inelastic scattering techniques (except transmission XAFS). The lateral resolution is dictated by the incident beam properties (i.e. beam size).

Since these techniques used high brilliance sources, degradation of the sample may be anticipated. A priori no degradation occurs on metals and ceramics. However, in solution some reactions are possible: the ionizing nature of the X-rays allows oxidation of species to occur. This is the case for Np(IV) species that can undergo photo-oxidation to Np(V) when investigated by the X-ray beam, (see Denecke et al. 2004).

Considering the electrons, EELS as daughter technique of XAFS is also an inelastic scattering technique. The analysis of Extended Electron Energy Loss Fine Structure (ExEELFS) on the Ti K line in pure titanium has been reported by using EELS, (see Blanche et al. 1993). Although the Ti K edge at 4966 eV is well beyond the usual EELS domain (e.g. 5–500 eV), results of the analysis can be competitively compared to the equivalent extended X-ray absorption fine structures (EXAFS).

EELS has a potential similar to XAFS for the atomic environment or the structural analysis; however, the interaction potential between electron (monoenergetic) and matter is larger than the photon (monochromatic) and matter. Recently Degueldre and Alekseev (2015) recorded EEL spectra of presumed UO_3 small grains attached to a TEM grid (FIBed or as delivered). The EEL spectrum was studied through a range of thicknesses going from $\sim$60 to $\sim$260 nm in the low energy range 0–350 eV. The EEL spectra recorded for UO_3 were compared with those recorded for UO_2. The reduction of UO_3 into U_4O_9 and/or UO_2 are readily observed apparently during the TEM investigations and as confirmed by electron diffraction (ED). This redox effect is similar to that known for other redox-sensitive oxides. Recommendations were suggested by Degueldre and Alekseev (2015) to avoid sample decomposition.

4.1.3 Comparing Passive and Interactive Methods

The interactive techniques are considered for characterization or imaging using ions, neutrons, electrons or photons.

4.1.3.1 Comparing Techniques for Characterization

Passive techniques are limited by the intrinsic activity of the sample and the reasonable time required for collecting the data. The detection limit is dictated by Eqs. 1.3, 1.4, 1.5 and 3.2.

The detection limit on the number of nuclide *i*, N_{DL} is linked to its activity *A* by:

$$N_{DL}(t) = A_{DL}(t)/\lambda_i \tag{4.4}$$

where λ_i is the decay constant of nuclide *i*. Clearly the larger λ_i (the shorter the half-life) the better the detection limit (see Fig. 4.3).

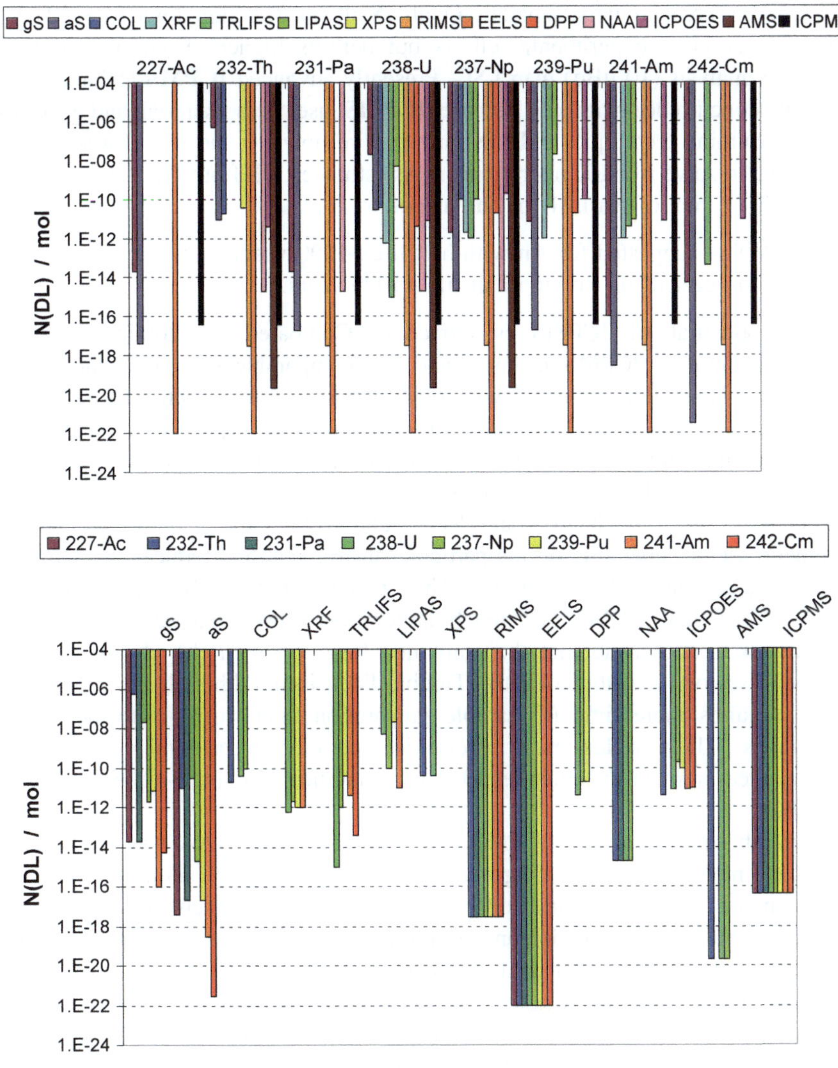

Fig. 4.3 Indicative comparison of technique detection limits for actinide isotope ^{A}An. Conditions, data see Sects. 3.1 and 3.2: Data from techniques such as XAFS (comparable to XRF), APT (comparable to EELS) or LAICPMS (comparable to AMS) are not included for reading

Interactive techniques are not limited by this intrinsic condition; however, quantity of reagent or intensity of the incident particles or waves flux can be increased to enlarge the product or the output signal. The later parameter is intrinsic to the sensitivity κ(i) of the analytical protocol for i (see Eq. 1.3). The second option valid for passive or interactive technique is to reduce the standards deviation on the background or limiting noise of the analytical signal (blank). Figure 4.3 presents some estimated detection limits for both passive and interactive techniques.

Actinides are a emitters; separation with **αS** allows detection down to ~10 μBq. **αS** requires more experimental efforts but permits higher resolution and more sensitive analysis (10 μBq) than LSC. Presently advanced **ICPMS** techniques are used in concurrence with **αS** because **ICPMS** uses less material and is a less time-consuming technique. **ICPMS** may require however the use of a separation technique to solve isobaric interferences. As estimated DL from Eq. 4.4 for ICPMS is 5×10^{-17} mol of An in 1 ml, the use of

- **ICPMS** is suggested for An with half-life >1000 years, and
- **αS** is suggested for An with half-life <1000 years.

The potential of ICPMS is attractive. Compared to atomic absorption (AAS) techniques, ICPMS is faster and has comparable precision, and greater sensitivity. However, compared with other types of mass spectrometry, such as TIMS and GDMS, ICPMS introduces a lot of interfering species: argon from the plasma, component gasses from air that leaks through the cone orifices, and contaminations from glassware and the cones.

For high resolution ICPMS, an electrostatic sector may be before or after the magnetic sector depending on the particular instrument, and reduces the spread in kinetic energy caused by the plasma. This spread is particularly large for ICPMS, being larger than GDMS and much larger than TIMS. The geometry of the instrument is chosen so that the combined focal point of the electrostatic and magnetic sectors is at the collector (Double Focusing). Although the TIMS technique requires more time for sample preparation than single–collector ICPMS techniques, TIMS yields more precise isotope ratio measurements leading to smaller measurement uncertainties (0.5–1%) than those of single-collector ICPMS (~10%).

Until the introduction of ICPMS, ICPOES and PIXE, NAA was the standard analytical method for performing multi-element analyses with minimum detection limits in the sub-ppm range. Today NAA has lost its attractiveness because it requires preparation, irradiation and counting time. In addition it requires a neutron source and the sample after analysis needs to be treated as a radioactive waste. Fluorescence techniques are also attractive but their detection limits—such as for uranium, e.g. Rathore (2008)—are not better than that reached by ICPMS.

Table 4.1 provides a comparison of interactive techniques with the analyzed properties. Techniques are evaluated for their depth resolution (DR), lateral resolution (LR) and their detection limit (DL). For isotope analysis SIMS may be compared to ERDA with for both DR >2 nm, LR(SIMS)0.05–2< LR(ERDA) 20 μm, and DL(SIMS) < DL(ERDA); while potential interferences are noted for SIMS, SIMS is a priori better than ERDA. NERDA may be applied for depth profile of H and D with atomic density DL of $\sim 10^{20}$ cm^3. For element analysis comparison of TEM/EELS versus TEM/EDS reveals that EELS has better LR and DR as well as DL than EDS, however the interpretation is easier for EDS. For speciation,

Table 4.1 Comparison of interactive techniques with characteristic properties. Data from Sect. 3.2

Technique	Atomic C_{DL}	Depth resolution	Lateral resolution	Provided information
DRS	–	10 nm	10 μm	Film thickness
XPS	0.1–1%	1 nm	10 μm	Depth profile
TEM/EDS	~10^{21} cm^{-3}	Through sample e.g. 20–200 nm	1 nm–10 μm	Structure analysis, concentration
AFM	Variable	0.5 nm	0.5 nm	Cross sectional view
SIMS	10^{12}–10^{16} cm^{-3}	2–30 nm	50 nm–2 μm	Elemental depth profile
ERDA	~0.1 a%	2–5 nm	20 μm	Depth profile and elemental composition

TEM/EELS is more difficult to apply than XPS, the latter is very sensitive for speciation. Finally, for structures, TEM investigates the bulk of the sample while AFM focuses on the atoms at the sample surface.

4.1.3.2 Comparing Techniques for Imaging

System imaging goes from the km (SR, νTOM) to the m (μTom), to the mm and μm (NRAD, OM) to some nm (SEM, EPMA, TEM, IM) and to some pm (HR-TEM, AFM) covering 15 orders of magnitude. The lateral resolution follows the same trend going from some m to some pm. These techniques may be used to image in a 1D, 2D, 3D or 4D the nuclear materials or their environments.

Parker and Joyce (2015) recently reviewed the use of radiation to image nuclear fuel in a passive or interactive way. Nuclear fuel radioisotope or density imaging has been carried out for physical inventory verification, e.g. safeguard and for quality assurance. Fuels investigated are fresh or spent fuels, in various physical forms from the assembly to the element, e.g. pellet. The review included the use of γ, X-ray, β, α, μ and neutrons. The discussion spans over four decades, describing how imaging quality has improved. Imaging started by radiography (2D) and turned to tomography (3D) with for example the 0.5 mm microspheres of UO_2 as dispersed in carbon pebble. With the increasing computer technology and detector sensitivity, an increase of the quality of image is expected.

Comparing n- and X-ray radiography, the analysis procedures on digital radiographic images and the corrections that have to be included to obtain the highest measurement accuracy have recently been reviewed by Rogers et al. (2015). They refer to dimensional measurements in determining particle movements in fluid flow and in comparisons of dimensional changes or movements in operating machinery or test objects. Emphasis was given to accuracy of the measurements.

Neutron radiography (NRAD) has been performed as a non-destructive method to acquire post-irradiation examination (PIE) data (e.g. fuel integrity and hydrogen uptake in cladding) from safety and integrity tests of nuclear fuels performed within the research reactor, e.g. Jenssen et al. (2014). The method was later applied under refabrication and instrumentation operations of experimental nuclear fuel rods prior to in reactor testing, and for a variety of PIE projects, e.g. reactor power ramp testing, PCI failure detection and fuel degradation experiments. Neutron radiography has also proved to be a very useful tool for examination of nuclear fuels irradiated, e.g. for high burn-up nuclear fuels. Hydrogen uptake of the fuel cladding, fuel pellet-clad bonding condition, fuel fragmentation, particle size distributions, and other features obtained from neutron tomography data are quite relevant for reactor core safety impact study.

Neutron radiography (NR) has been compared to PET focusing on lateral resolution, e.g. Degueldre et al. (1996), while NMRI may also be noted for its versatile potential of 4D imaging distribution of ^1H molecules such as water in host rock samples near potential repository, (see Waggoner and Fukushima 1996; Cheng et al. 2015).

The ERT reconstructions are compared with neutron radiographs, which provide high resolution information on the 2D distribution of the moisture. The results indicate that ERT is able to detect the moisture movement and to show approximately the shape and position of the water front even if the flow is not uniform (see Hallaji et al. 2015).

4.1.3.3 Comparing Isotope and Element Analysis

The two ICP techniques—namely optical emission and mass spectroscopy—may be compared to cover this comparison issue on element and isotope analysis. Comparison of ICPOES and ICPMS has been reported for the analysis of trace elements in graphite by Cruz et al. (2015). The following is noted.

Compared to ICPOES, ICPMS is faster and has greater precision and sensitivity. However, ICPMS introduces a lot of interfering species: argon from the plasma, component gasses of air that leak through the cone orifices, and contamination from glassware and the cones that may require HR-ICP-MS, which is less flexible for use.

The second issue on ICPMS is that it may use a combination of parallel plates, rings, quadrupoles, hexapoles and octopoles to optimize the shape and focus the beam to produce symmetrical peaks. An electrostatic sector may be used together with the magnetic sector to reduce in kinetic energy enlargement due to the plasma. The geometry of the instrument is fixed to combine focal point of the electrostatic and magnetic sectors at the collector (Table 4.2).

Finally, comparison of ICPMS with other MS techniques confirms the versatility of the ICP method. Although the TIMS technique requires more time for sample preparation than single-collector ICPMS techniques, **TIMS** yields more precise isotope **ratio** measurements leading to smaller measurement uncertainties (0.5–1%)

Table 4.2 Comparing the performances of ICPOES and ICPMS Sample types: Liquids, Metals, Alloys, Plastics, Polymers, Other Industrial Applications: Steels, Sputter Targets, Fuel

Properties	ICPOES	ICPMS
Sample mass	0.1–1 g	0.01–0.1 g
Optimal detection	50 ppm wt–50% wt	0.1 ppm wt–1% wt
Uncertainty	0.5–3% relative to reported value	3–5% relative to reported value
Element range	Li–U, except atmospheric species (C, H, O, N) and noble gases (spectral interferences)	Li–U, except S, atmospheric species (C, H, O, N) and noble gases (isobar interferences)
Typical applications	Precise determinations of known constituents (i.e. alloy composition and identification, dopants/additives)	Trace and ultra-trace determinations of deleterious elements, impurity surveys for unknown/unexpected species

than those of single-collector ICPMS ($\sim$10%). **AMS** is less precise; however, its **mass resolution** is outstanding.

4.1.4 Comparing the Detetion Limits and Relevant Issues

One of the first performances to be known for applying an analytical technique is the detection limit which allows evaluation of whether the technique can be applied or not. Indicative detection limits are compared for actinide analysis in Fig. 4.3. They focus only on actinides that could be used as examples for other elements. The following points may be sorted out:

- Actinides are α emitters, separation with **αS** or **LSC** allows detection down to $\sim$10 mBq. **αS** requires more experimental efforts but may permits higher resolution and more sensitive analysis (10 μBq).
- Alternative to **αS** may be **NAA** for which limitations of low counting rates should be improved by activation. **NAA** has been found to be prone to interference, e.g. from uranium nuclides.
- Today, advanced **ICPMS** techniques are used instead of **αS** because **ICPMS** uses less material and is a less time-consuming technique. **ICPMS** may require the use of a separation technique, e.g. HPLC, which may be performed on-line.
- Recommendation:

  use **αS** for An with half-life <1000 years, and
  use **ICPMS** for An with half-life >1000 years.

- Better standards, reference materials and benchmarks are required. For some techniques standards are even unavailable, e.g. for **SIMS**, **EMPA**, **LAICPMS** and **XRF**.

- **RBS**, **AMS** and **EELS** under vacuum and **XAS** in solution, require internal calibrations, the sensitivity of the latter is expected to be improved (from minor component to traces) in future.
- Advanced analytical techniques provide data on nuclear materials.
- Layers are characterized by combining **μXRF**, **μXRD**, **μXAFS**, **DRS**, **RBS** …
- Layer species are determined by using **VIAXAFS** or **μXAFS**, **TEM/EELS** ….
- Defects and vacancies in Zircaloy are identified by **MuSR**, **XAFS**, **STXM**, **TEM**…
- Fuel porosity and defects are characterized by **SANS**, **ToM**, **XAFS**, **NR** …

Techniques required to perform the characterization of actinides are needed that will have better performance in terms of capacity, realization, precision, and detection limit. This will remain a challenge for this century. Since most of the actinides are α emitters, development of separation techniques coupled with highly sensitive α spectroscopy (αS) or liquid scintillation counting (LSC) allows their detection below the 10 mBq level. LSC allows a fast analysis while αS requires more experimental efforts but permits higher resolution and more sensitive analysis (10 μBq). An alternative to conventional alpha spectrometry may be neutron activation analysis (NAA), for which limitations of low counting rates on the non-activated actinide-bearing samples should be improved by their activation; the background is, however, also enhanced. Neutron activation analysis (NAA) has been found to be particularly prone to interference, especially from uranium nuclides.

The trend today is to use advanced ICPMS techniques instead of separation, electroplating, and α spectroscopy because ICPMS uses less material and is a less time-consuming technique. ICPMS may, however, require the use of a separation technique, which may advantageously be performed on-line because ICPMS is fast enough, compared to NAA or αS, to analyze the eluant on-line. Evaluation of NAA and ICPMS methods for the assay of neptunium and other long-lived actinides in environmental matrices has been discussed. ICPMS offers suitable sensitivity, precision and accuracy compared to other techniques, with considerably faster sample throughput relative to radiometric and activation approaches. Added advantages of ICPMS include the abilities to determine concentrations of several long-lived actinides simultaneously and to quantify ^{239}Pu–^{240}Pu ratios.

ICPMS, NAA and αS are compared for the determination of ^{237}Np in environmental samples. ICPMS accuracy and precision was assessed by comparison with those obtained by NAA and αS. Results are in good agreement at a relative deviation of 2–9%. The routine analysis detection limit is 0.02 mBq mL^{-1} or 4×10^{-15} mol calculated with the specific activity of 26 Bq per ng of ^{237}Np.

Plutonium is somewhat difficult to analyze quantitatively by γ-ray spectroscopy because all the commonly encountered isotopes ^{236}Pu, ^{238}Pu, ^{239}Pu, ^{240}Pu, ^{242}Pu and ^{244}Pu have very low gamma-ray abundance emissions, making conventional 4π α-γ coincidence-counting techniques impractical. ^{241}Pu is a low-energy β emitter with low (20 keV) end point energy and no associated γ-ray. ^{241}Pu is best-analyzed using liquid scintillation counting by calibration with ^{3}H with some corrections for differing end-point energies and β spectrum shapes.

4.2 Combining Techniques

Based on the comparison of the analytical performance of the techniques described in Sect. 4.1, the combination of these techniques can be discussed and carried out to produce the data needed to understand the behavior of nuclear materials and their environments at the level of the material nuclides, atoms, molecules, phases and sample. The combination of the techniques may be based on their classification as described in Table 3.1, combining passive techniques, interactive techniques or both of them. For the interactive techniques combination, the photon-photon, then lepton-photon, lepton-lepton, neutron-photon, ion-photon until ion-ion are possible together with ternary combinations.

A large number of techniques are today available for quantifying various properties of the nuclear materials investigated. However, most of these techniques are restricted to the analysis of one parameter only. It is difficult to find a technique that allows element, isotope and species analysis. This can however be done by combining specific techniques. Particles in suspension such as in the water of nuclear reactor primary are also analyzed by combining techniques since a "colloidograph" that would analyze colloid size, concentration and nature is not yet available. Again, all that can be done is to combine techniques including in situ analysis, e.g. SPC for size distribution analysis, and ultra-filtration followed by SEM/EDS analysis of the colloidal particles loaded on a filter or by TEM/EELS analysis on a TEM grid after centrifugation on the sample carrier (see Perret et al. 1991).

Finally, since no technique allows a zoom from the m to the pm, a combination of various microscopies and spectroscopies such as optical macroscopic, optical microscopy, electronic microscopy, XRD and XAS is needed. The combination of techniques is just sketched in this chapter because full description of the case would be at least several times larger than Chap. 3.

4.2.1 Combining Passive Techniques

The combination of passive techniques is very attractive because it does not require excitation of the sample and can be applied in-line (see discussion in Sect. 2.1.2). The combination of radioanalytical techniques is usually performed to complete the radioisotope analysis.

However, the combination of radioanalytical techniques with passive chemical analysis may be gained using passive electrochemical techniques.

This combination has been applied on-line for the analysis of reactor water (see Fig. 3.7). This combined the evolution of the redox potential measured in the feed water and the mean dose rate increase in the main steam line as reported by Sarott (2005).

Other passive techniques such as gravimetry may be combined to yield specific analytical data.

As an example, particles and colloids in suspension in reactor water may be analyzed by combining membrane filtration and investigation on the colloid cake produced by gravimetry, γS and αS. Combinations of these techniques were tested during the 'Grimsel Colloid Exercise' (see Degueldre et al. 1990).

Several combinations of techniques may be applied during tracer breakthrough experiments through porous samples. PET may be used to monitor breakthrough of β^+ tracer; ISE has been combined to follow pM in the eluates and to compare the breakthrough of the PET tracer together with other dissolved metals.

γ counting, radiography or tomography has been combined with ISE to follow breakthroughs. Finally the redox species of Fe could be measured by MoAS, e.g. after breakthrough and correlated with EHE data measured in situ.

4.2.2 *Combining Interactive Techniques*

Here, the potential of combination is great. It concerns not only the combination of the photon-photon, lepton-lepton, neutron/ion-neutron/ion techniques and their mixed wave/particle interaction techniques, but also for each of them the typical methods for the energy range used for the analysis that makes the combination nearly infinite.

4.2.2.1 Combining Photon-Photon Interactive Techniques

In this section the technique combination deals with methods described in Sect. 3.2.3. This is for example the combination of the use of GPR in formation interfaces and optrodes used for fluid analysis or XRF of rock specimen analysis, the combination of DRS of a metal corrosion sample in hot water through a window and the use of Raman spectroscopy or XAFS to complete the characterization, or for particle or colloid characterization in water the combination of SPC or PCS and after filtration on a membrane RAMS and/or XAFS, STXM or MoAS analysis.

The concept of photon-photon interactive technique combination is illustrated below. The combination of μXRF, μXRD and μXAFS is quite easy because these techniques can be utilized successively on the same beam line optical bank. Corrosion product deposits on BWR fuel rods were for example analyzed by combining μXRF, μXRD and μXAFS (see Degueldre et al. 2009). The deposit material is called CRUD (see Sect. 3.2.3.11, Fig. 3.33). BWR CRUD's are mainly made of iron oxide and nickel spinel, traditionally. However, the addition of soluble Zn in the primary water proposed for reduction of ^{60}Co build-up in the corrosion layers induced unexpected problems including precipitation of magnetic-sensitive phases which were identified to falsify Eddy current measurements used to quantify the thickness of the Zircaloy corrosion layer. Initially, EPMA measurements were carried out to identify the element associated with the CRUD layer and it was found that Zn was associated with Fe and O. It was provisionally deduced that the material was made of a zinc ferritic spinel. Additional studies were needed to better understand the nature of the material. It was consequently required to combine µXRF, µXRD and µXAFS to gain information on the element distribution, crystallographic phase distribution and the local description of the chemical environment at the sub-nano-scale level. Figure 4.4 shows µXRF Zn and Fe scans along and across a segment of the CRUD layer confirming occurrence of deposits rich in Zn and Fe.

The XRF scans were completed by XRD investigations. The semi-circular pattern as recorded for the marked (arrows in µXRF maps) on the CCD camera, and, angular diffraction data as retrieved using FIT2D, showed that the film was composed of zinc ferrite deposited on zirconia substrate. Peaks of ZrO_2(m) substrate, deposit $ZnFe_2O_4$(c) and additional α-Fe_2O_3(h) phase formed. In addition, the µXAFS spectra recorded for Zn were found to be comparable with that recorded for $ZnFe_2O_4$ (Franklinite); see Degueldre et al. (2009).

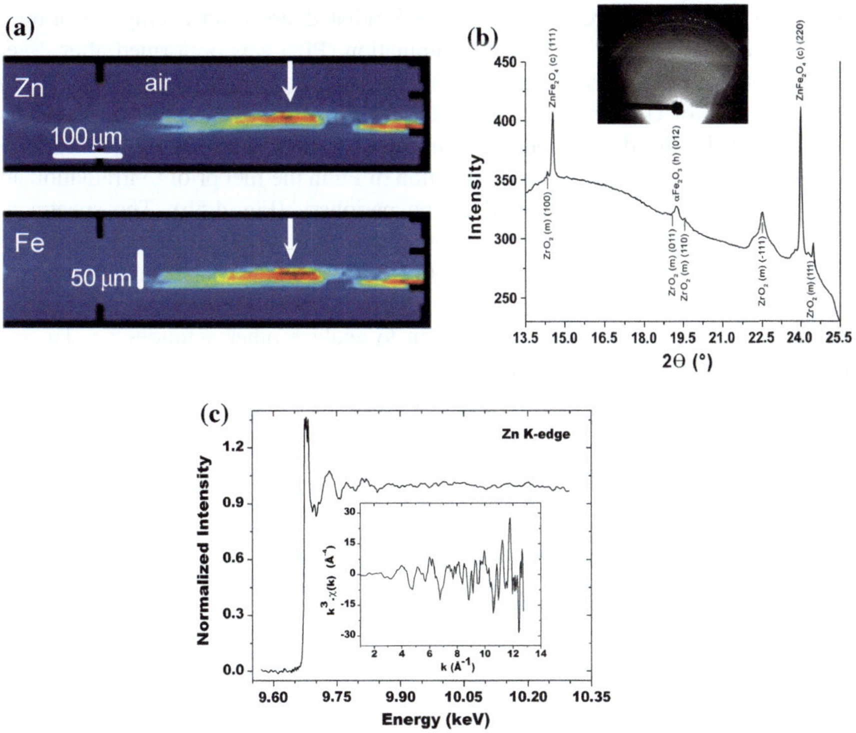

Fig. 4.4 µXRF, µXRD and µXAFS of Zn rich spot in corrosion product deposit layer on a BWR cladding. Degueldre et al. (2009) with permission © Elsevier

Finally, the characterization of fluids may be carried out in combining specific techniques. Evaluation of nanometer-scale droplets in a ternary oil/water microemulsion using SAXS and ^{129}Xe NMR were for example carried out by Kataoka et al. (2007). This could be used to study phase decantation during liquid-liquid extraction during reprocessing.

4.2.2.2 Combining Photon-Photon and Electron-Photon Interactive Techniques

In this section the technique combination deals with methods described in Sects. 3.2.7 and 3.2.8. This is for example the combination of the use of SEM/EDS or EPMA/WDS with other techniques as widely used in material science. This can be extended using TEM/EDS/ED/EELS in a combined way. The concept of photon and electron-photon and electron interactive technique combination is illustrated below. Recently, irradiated plutonium uranium mixed oxide MOX analysis has been reported to have been performed by combining EPMA and μXRF/μXAFS (Degueldre et al. 2011a, b). The MOX fuel was obtained by internal gelation and produced as microspheres. The fuel was irradiated up to an average burn-up of ~60 MW d kg^{-1}. Post irradiation, examination (PIE) was performed after 2-year cooling. The EPMA performed on a fuel cross-section (Fig. 4.5a) shows that the radial fraction of U decreases towards the border of the fuel, corresponding to an increase of Pu in the periphery, confirming higher burn-up in this location. The XAFS allows chemical characterization of Pu in the fuel prior to irradiation and after irradiation in the center and in the periphery (Fig. 4.5b). The spectra are similar and it can be concluded that Pu is apparently Pu(IV) and Pu state is likely to remain constant during burning. No Pu(VI) (<5%) was found, which is important for safety concerning its solubility in reactor or in repository groundwaters. Additional investigations were carried out to analyze other actinides i.e. Th, Am, Cm in the fuel periphery mainly by XAFS (e.g. Fig. 3.34).

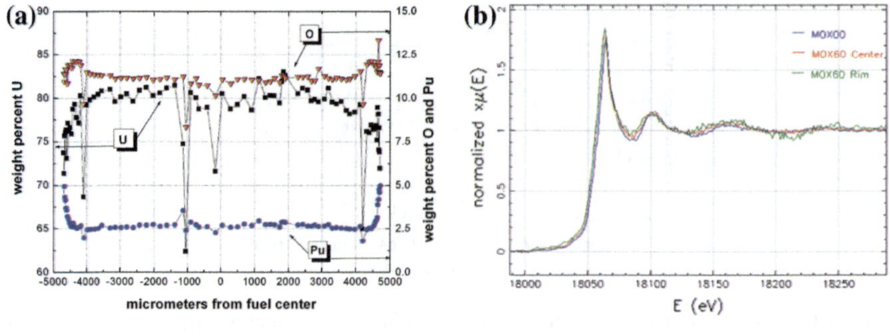

Fig. 4.5 MOX analysis by (**a**) EPMA and (**b**) XAFS. Conditions: homogeneous fuel obtained by internal gelation, burn-up 60 MW d kg^{-1}. Degueldre et al. (2011a) with permission © Elsevier

4.2.2.3 Combining Other Interactive Techniques

In this section the technique combination deals with methods described in Sect. 3.2. This is for example not only the combination of methods treated in Sects. 4.2.2.1 and 4.2.2.2 but also those including neutrons and/or ions implying combination of the use of NRad or NTom and γTom, μTom and ETom such as for imaging or NAA, ICPOES, ICPMS, Con, NIR-Vis-UVS and Vol for bulk analysis in situ or in leachates. Honestly, the number of dual combination is very large, and the number of triple combination enormous.

For imaging, Casalta et al. (2003) combine digital image analysis of X-ray and neutron radiography for the inspection and monitoring of nuclear materials. Parker and Joyce (2015) show examples where neutron and photon imaging can be combined for gaining in situ quality of investigations. The combination of elastic scattering techniques such as XRD and ND may also be pointed out as each technique has its inherent limitations (see Sects. 3.2.3.13 and 3.2.13.3 respectively.) In this frame, the combination of inelastic scattering techniques such as XRF (Sect. 3.2.3.11) and NAA (Sect. 3.2.11) may also be mentioned.

Combining nuclear fuel post irradiation examination (PIE) techniques for the characterization of isotopes, elements and species/phases is very useful. For bulk analysis after sample preparation by leaching, the combination of high-performance ion chromatography (HPLC) and inductively coupled plasma quadrupole mass spectrometry has to be suggested for determining among other things the burn-up of a nuclear fuel using lanthanide or actinide isotopes as indicators. The determination of actinides (U, Np, Pu, Am and Cm) is very sensitive and accurate by combining HPLC and ICPMS (e.g. Röllin et al. 1996). This separation technique followed by mass spectroscopic detection allows the elimination of isobaric interferences such as between actinide elements, e.g. ^{243}Am and ^{243}Cm, or from cluster ions. e.g. $^{155}Gd^{40}Ar_2^+$ or $^{207}Pb^{16}O^{12}C^+$ for $^{235}U^+$, $^{238}U^1H^+$ for $^{239}Pu^+$ e.g. Moreland et al. (1970), or $^{232}Th^{12}C^+$ for $^{244}Cm^+$ (e.g. from electrothermal vaporization [ETV]) or $^{232}Th^{16}O^+$ for $^{248}Cm^+$ (see Becker et al. 1999).

Combining fuel PIE at the microscopic level is also very useful. Lamontagne et al. (2006) combine SEM, EPMA and SIMS for the characterization of fission gas "bubbles" in irradiated UO_2 fuel. These apparent "bubbles" were characterized by LAICPMS (Fig. 3.46) and by XAFS (Degueldre et al. 2014) in UO_2 fuel with a burn-up of ~40 MW d kg^{-1}. The results suggest the high density of the fission gases, Xe and Kr atoms ~20 nm^{-3} or ~5 g cm^{-3} which corresponds to a solid phase at room temperature. This result completes earlier TEM/eD analyses from Nogita and Une (1995, 1998), performed on ~50 MW d kg^{-1} fuel specimens, showing fission Xe atoms as precipitates. Actually, the combination of these data gives a more accurate vision of the real structure of these nano-phase inclusions rather than "bubbles".

The combination of STM and AFM studies of $^2H^+$ irradiated graphite and graphite-silicon mixtures has been reported by Almqvist et al. (1995). The microscopes were used for studying surface topography and for measuring the surface roughness. The substrates were exposed at various temperatures (60 and 700 °C) to different doses of deuterium ions in simulators of plasma-surface interactions and in a tokamak. Also NRA and RBS were applied for the qualitative and quantitative determination of surface composition.

The combination of TEM/EELS and ion back-scattering (e.g. RBS/ERDA during ion implantation) has also be performed on various installations around the world; e.g. Jannus, University of Orsay. Finally, as mentioned in Sect. 3.2.18.1, most of the volumetric techniques are usually performed in combination with spectroscopic or electroanalytic techniques to detect the titration factor and the end point.

4.2.3 Combining All Analytical Techniques

In this section, the technique combination deals with methods described in Chap. 3. Again, the number of dual combination is very large, and the number of triple combination enormous. Some examples are given below.

Concerning nuclear material diagnostic, the possibilities of techniques to detect certain types of damages can only be proven by using complementary techniques (replica techniques, radiography, ultrasonics, OM and SEM, SANS, SAXS. PALS) in a combined way. Talking about diagnostics, Nowak et al. (2015) suggest the diagnostic of boron carbide coatings for neutron detection to be performed by X-rays, ions, and neutrons to determine thin film quality. This combination of techniques allows improvement of the accuracy on the analytical results.

The release of contaminants such as radiocesium has been of great concern following reactor accidents. Since the fuel matrix is considered as the first barrier toward the environment, the retention of Cs in UO_2 fuel (UOx) as well as MOX is the topic of safety-relevant investigations in numerous organizations. Fuel cross-sections have been specifically analyzed by SIMS (Fig. 4.6a) EPMA, (Fig. 4.6b) and β/γRAD (Fig. 4.6c). The Cs isotope distribution recorded for ^{137}Cs in Fig. 3.4 can be confirmed with a better resolution by recording the radial concentration of ^{134}Cs, ^{135}Cs and ^{137}Cs obtained by SIMS (Fig. 4.6a). The ^{137}Cs signal ratio from periphery to centre is about 1.5 that could be compared with the ratio estimated from tomography, e.g. Caruso and Jatuff (2014). A non-symmetrical burn-up is followed by the Cs isotopes spatial distribution. A similar radial profile is found by EPMA of similar UOx fuel that underwent a faster burn-up rate that the former. Clearly, Cs and Xe are depleted in the center of the fuel and concentrate in the colder periphery, while Pu and Nd (both burn-up indicators) show that the burn-up increases in the periphery compared to the center. Even if EPMA does not distinguish between the isotopes, the information is safety-relevant. In addition, irradiated plutonium uranium mixed oxide MOX analysis has been reported to have been performed by combining β/γRAD, (see Degueldre et al. 2011a). The β/γRAD (Fig. 4.6c) shows larger fraction of fission product in the periphery, suggesting larger local burn-ups than in the centre. However, cesium—being volatile—may condense in the colder periphery.

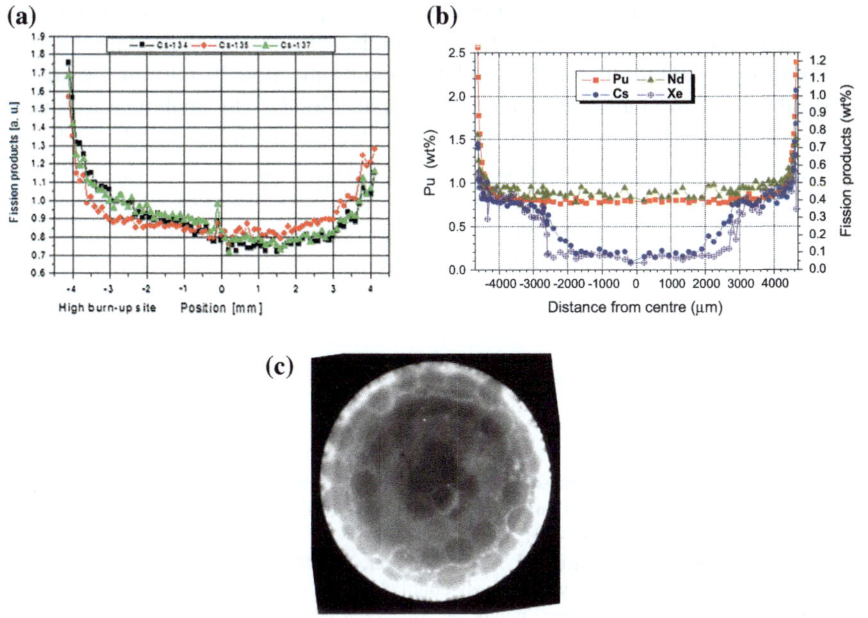

Fig. 4.6 Cs analysis in irradiated fuel. **a** SIMS of irradiated UOx fuel **b** EPMA and **c** MOX analysis by β/γRad, Degueldre et al. (2011a). SIMS ATOMICA 4000, source 25 keV Ga^+, 200 nA, EPMA Cameca unit, 20 keV, 200 nA, fuel periphery. Degueldre et al. (2016) with permission © Elsevier

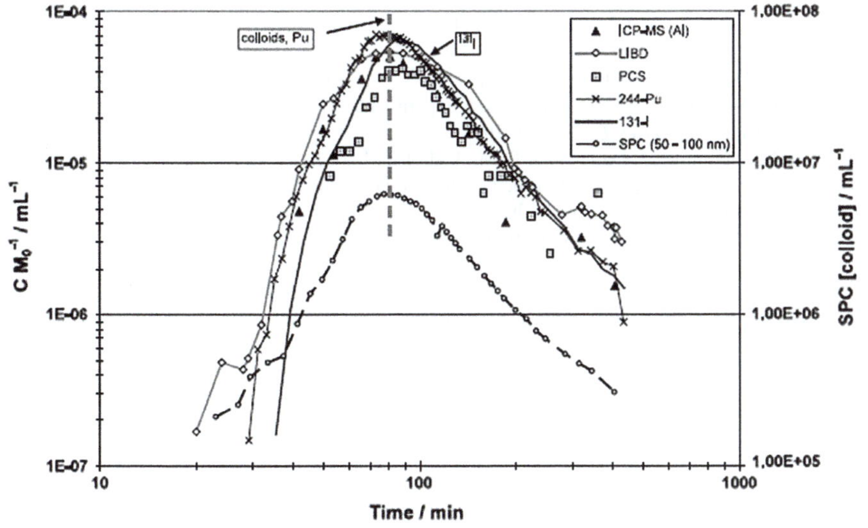

Fig. 4.7 Colloid breakthrough through a dipole in comparison to breakthrough curve recorded by LIBD, PCS, SPC, ICPMS (Al) and the ^{131}I and ^{244}Pu breakthrough. Al is the surrogate element for clay colloid. Möri et al. (2003) with permission © Elsevier

Finally **concerning environmental** studies, a quite spectacular combination of analytical techniques has been obtained by following the breakthrough of labeled colloid suspension through a 2.23 m long dipole (injection 10 ml min^{-1} extraction 150 ml min^{-1}) in granitic formation Möri et al. (2003). The analytical techniques included ICPMS of Al (as surrogate element for clay colloids), LIBD, PCS, SPC (for direct colloid number or concentration per size analysis), ^{244}Pu αS (4.589 meV α tracer associated to clay colloids) as well as ^{131}I γS (364 keV) as conservative (non-sorbing tracer). All breakthrough curves presented a maximum after 80 min (see Fig. 4.7), showing that colloids can migrate as fast as a non-conservative or non-sorbing tracer.

References

Almqvist, N., Rubel, M., Fredriksson, S., Emmoth, B., Wienhold, P., & Ilyinsky, Lev S. (1995). AFM and STM characterization of surfaces exposed to high flux deuterium plasma. *Journal Nuclear Materials, 220,* 917–921.

Becker, J. S., Dietze, H.-J., McLean, J. A., & Montaser, A. (1999). Ultratrace and isotope analysis of long-lived radionuclides by inductively coupled plasma quadrupole mass spectrometry using a direct injection high efficiency nebulizer. *Analytical Chemistry, 71,* 3077–3084.

Blanche, G., Hug, G., Jaouen, M., & Flank, A. M. (1993). Comparison of the TiK extended fine structure obtained from electron energy loss spectroscopy and x-ray absorption spectroscopy. *Ultramicroscopy, 50,* 141–145.

Caruso, S., & Jatuff, F. (2014). Design, development and utilisation of a tomography station for γ-ray emission and transmission analyses of light water reactor spent fuel rods. *Progress in Nuclear Energy, 72,* 49–54.

Casalta, S., Daquino, G. G., Metten, L., Oudaert, J., & Van de Sande, A. (2003). Digital image analysis of X-ray and neutron radiography for the inspection and the monitoring of nuclear materials. *NDT & E International, 36,* 349–355.

Cheng, Y., Di, Q., Gu, C., Feng, Y. E., Hua, S., & Yang, P. (2015). Visualization study on fluid distribution and end effects in core flow experiments with low-field MRI method. *Journal of Hydrodynamics, Series B, 27,* 187–194.

Cruz, S. M., Schmidt, L., Dalla Nora, F. M., Pedrotti, M. F., Bizzi, C. A., Barin, J. S., et al. (2015). Microwave-induced combustion method for the determination of trace and ultratrace element impurities in graphite samples by ICP-OES and ICP-MS. *Microchemical Journal, 123,* 28–32.

Degueldre, C., & Alekseev, E. V. (2015). Uranium trioxide behavior during electron energy loss spectroscopy analysis. *Radiation Physics and Chemistry, 108,* 7–12.

Degueldre, C., Bertsch, J., Kuri, G., & Martin, M. (2011b). Nuclear fuel in generation II and III reactors: Research issues related to high burn-up. *Energy & Environmental Science, 4,* 1651–1661.

Degueldre, C., Bertsch, J., & Martin, M. (2016). Post irradiation examination of nuclear fuel: Toward a complete analysis. *Progress in Nuclear Energy, 92,* 242–253.

Degueldre, C., Kuri, G., Borca, C. N., & Grolimund, D. (2009). X-ray micro- fluorescence, diffraction and absorption spectroscopy for local structure investigation of a radioactive zinc ferrite deposit. *Corrosion Science, 51,* 1690–1695.

Degueldre, C., Longworth, G., Moulin, V., Vilks, P. et al (1990). *Grimsel colloid exercise* (CEC report and Nagra report NTB-90-01).

Degueldre, C., Martin, M., Kuri, G., Grolimund, D., & Borca, C. (2011a). Plutonium—uranium mixed oxide characterization by coupling micro-X-ray diffraction and absorption investigations. *Journal of Nuclear Materials, 416,* 142–150.

Degueldre, C., Mieszczynski, C., Borca, C., Grolimund, D., Martin, M., & Bertsch, J. (2014). X-ray fluorescence and absorption analysis of krypton in irradiated nuclear fuel. *Nuclear Instruments and Methods in Physics Research Section B: Beam Interactions with Materials and Atoms, 336,* 116–122.

Degueldre, C., Pleinert, H., Maguire, P., Lehman, E., Missimer, J., Hammer, J., et al. (1996). Porosity and pathway determination in crystalline rock by positron emission tomography and neutron radiography. *Earth and Planetary Science Letters, 140,* 213–225.

Degueldre, C., Pouchon, M., Streit, M., Zaharko, O., & Di Michiel, M. (2001). Analysis of porous features in zirconia based inert matrix, impact on the material qualification. *Progress in Nuclear Energy, 38,* 241–246.

Denecke, M., Dardenne, K., & Macquard, Ch. (2004). Np(IV)/Np(V) valence determination from Np L3 edge XANES/EXAFS. *Talanta, 65,* 1008–1014.

Gilles, R., Mukherji, D., Eckerlebe, H., Karge, L., Staron, P., Strunz, P., et al. (2014). Investigations of early stage precipitation in a tungsten-rich nickel-base superalloy using SAXS and SANS. *Journal of Alloys and Compounds, 612,* 90–97.

Hallaji, M., Seppänen, A., & Pour-Ghaz, M. (2015). Electrical resistance tomography to monitor unsaturated moisture flow in cementitious materials. *Cement and Concrete Research, 69,* 10–18.

Jenssen, H. K., Oberländer, B. C., Beenhouwer, J. D., & Sijbers, J., Verwerft, M. (2014). Neutron radiography and tomography applied to fuel degradation during ramp tests and loss of coolant accident tests in a research reactor. *Progress in Nuclear Energy, 72,* 55–62.

Kataoka, H., Ueda, T., Ichimei, D., Miyakubo, K., Eguchi, T., Takeichi, N., et al. (2007). Evaluation of nanometer-scale droplets in a ternary oil/water microemulsion using SAXS and ^{129}Xe NMR. *Chemical Physics Letters, 441,* 109–114.

Lamontagne, J., Desgranges, L., Valot, C., Noirot, J., Blay, T., Roure, I., et al. (2006). Fission gas bubbles characterisation in irradiated UO_2 Fuel by SEM, EPMA and SIMS. *Microchimica Acta, 155,* 183–187.

Moreland, P. E., Rokop, D. J., & Stevens, C. M. (1970). Mass-spectrometric observations of uranium and plutonium monohydrides formed by ion—molecule reaction. *International Journal of Mass Spectrometry and Ion Physics, 5,* 127–136.

Möri, A., Alexander, W. R., Geckeis, H., Hauser, W., Schäfer, T., Eikenberg, J., et al. (2003). The colloid and radionuclide retardation experiment at the Grimsel Test Site: Influence of bentonite colloids on radionuclide migration in a fractured rock. *Colloids and Surfaces A, 217,* 33–47.

Nogita, K., & Une, K. (1995). Irradiation-induced recrystallization in high burnup, UO_2 fuel. *Journal of Nuclear Materials, 226,* 302–310.

Nogita, K., & Une, K. (1998). High resolution TEM observation and density estimation of Xe bubbles in high burnup UO_2 fuels. *Nuclear Instruments and Methods in Physics Research Section B: Beam Interactions with Materials and Atoms, 141,* 481–486.

Nowak, G., Störmer, M., Horstmann, C., Kampmann, R., & Hoeche, D. (2015). Boron carbide coatings for neutron detection probed by x-rays, ions, and neutrons to determine thin film quality. *Journal of Applied Physics, 117*(034901), 1–14.

Parker, H. M., & Joyce, M. J. (2015). The use of ionising radiation to image nuclear fuel: A review. *Progress in Nuclear Energy, 85,* 297–318.

Perret, D., Leppard, G. G., Müller, M., Belzile, N., De Vitre, R., & Buffle, J. (1991). Electron microscopy of aquatic colloids: Non-perturbing preparation of specimens in the field. *Water Research, 25,* 1333–1343.

Rao, T. S., Shriwastwa, B. B., Dubey, J. N., Patil, B. P., Chandrasekharan, K. N., Pandey, V. D., et al. (2003). Quantitative estimation of plutonium-rich areas in thorium-based MOX fuels using alpha autoradiography technique. *Radiation Measurements, 36,* 747–750.

Rathore, D. P. S. (2008). Advances in technologies for the measurement of uranium in diverse matrices. *Talanta, 77,* 9–20.

Reiche, H. M., Vogel, S. C., & Tang, M. (2016). In situ synthesis and characterization of uranium carbide using high temperature neutron diffraction. *Journal of Nuclear Materials, 471,* 308–316.

Rogers, J., Amaral-Rogers, A., & Christodoulou, M. (2015). Evaluation of measurement accuracy in neutron and X-ray radiography. *Physics Procedia,* 366–373.

Röllin, S., Kopajtic, Z., Wernli, B., & Magyar, B. (1996). Determination of lanthanides and actinides in uranium materials by HPLC with. ICP-MS detection. *Journal of Chromatography A, 739,* 139–149.

Sarott, F. A. (2005). Water chemistry in boiling water reactor—a Leibstadt—specific overview. *CHIMIA International Journal for Chemistry, 59,* 923–928.

Su, J., Kapaca, E., Liu, L., Georgieva, V., Wan, W., Sun, J., et al. (2014). Structure analysis of zeolites by rotation electron diffraction (RED). *Microporous and Mesoporous Materials, 189,* 115–125.

van Gastel, R., Hlawacek, G., Dutta, S., & Poelsema, B. (2015). Backscattered helium spectroscopy in the helium ion microscope: Principles, resolution and applications. *Nuclear Instruments and Methods in Physics Research Section B: Beam Interactions with Materials and Atoms, 344,* 44–49.

Waggoner, R. A., & Fukushima, E. (1996). Velocity distribution of slow fluid flows in bentheimer sandstone: An NMRI and propagator study. *Magnetic Resonance Imaging, 14,* 1085–1091.

Wang, Z., Steuwer, A., Liu, N., Maimaitiyili, T., Avdeev, M., Blomqvist, J., et al. (2016). Observations of temperature stability of γ-zirconium hydride by high-resolution neutron powder diffraction. *Journal of Alloys and Compounds, 661,* 55–61.

Yaniv, G., & Meshi, L. (2016). Crystal structure of the $Th_2Ni_{10}Al_{15}$ phase solved using electron diffraction tomography. *Journal of Alloys and Compounds, 660,* 496–502.

Chapter 5
Concluding Summary

The analysis of nuclear materials and their environments is required before and during their installation, utilization or operation, as well as after service, during various phases of decommissioning or disposition. Quality control, safety requirement, safeguard measures, security issues are demanded by the vendors, operators or regulators concerning not only the current nuclear materials of the nuclear fuel cycles as they are deployed but also as they should be foreseen for the planned nuclear installations or systems.

Prior analysis, sampling, in a general sense, and sample treatment, with physical or chemical separation if needed, must be carried out when the analytical technique is not applied in situ, in a non-invasive way, or in an in-line or on-line mode. The analytical procedure typically includes sampling or sample preparation (e.g. decomposition), separation and/or enrichment prior to analysis in either a passive or interactive way. Separation of elements of interest, which are later analyzed by different analytical techniques, is an important prerequisite of any analytical method.

Analysis may be performed in-line, or on-line, within a flow bypass with direct detection of activity, for example, or at-line with intermediate samples, or off-line with transfer of the samples in the laboratory. The analysis may be carried out in situ, for example using a remote system, or in an underground rock laboratory in the considered phase, or ex situ with transfer of the sample and separation. To complete the picture it must be mentioned that separation techniques such as filtration, centrifugation, diffusion, electrodiffusion, electroplating, partitioning (liquid-liquid, or solid-liquid) may also be applied, making the analysis more specific or efficient.

The analyte concentration or amount may be determined up to an experimental limit called detection limit (DL). It is usually estimated in terms of concentration which equals 3 standard deviations (on the blank) times the sensitivity (on the analyte). This fixes an analytical concentration limit for the isotopes, elements or species to be quantified. The amount DL's are estimated for the minimum volume that can be analyzed. From the theoretical side, a detection limit may be calculated

C.A. Degueldre, *The Analysis of Nuclear Materials and Their Environments*,
DOI 10.1007/978-3-319-58006-7_5

from the physical-chemical process and from the performance of the analytical unit, while it remains usually an experimental parameter.

For all analyses, sample volume, mass or amount, the flux of reagent, the size of the analyzed part of the sample and the acquisition time or time of analysis are key parameters linked to the detection limit. The nature and origin of the environmental sample may dictate the size of the sample. The size of the sample is also coupled with the analytical technique for which analysis time and detection limits are key parameters for its calculation. DL is a function of the number of sample atoms, volume of the sample, sub-sample excitation conditions, acquisition quality (detector), and interferences such as quenching or peak overlapping. It may however be desirable to split the investigated concentration range according to "major and minor" element concentration $>10^{-6}$ M and "trace" element concentration $<10^{-6}$ M (mol L^{-1}).

Information required—such as activity (chemical, radioisotopic), amount (mass, volume, ...), concentration (fraction, molarity, ... e.g. of dopant/contaminant) and structure (type, size...) in the studied phases—have to be determined at the nuclear (pm), atomic, molecular (nm), microscopic (μm), macroscopic structural (mm), bulk scale (cm), at the component or system scale (m), or at environmental or geographic scale (km) according to the requirements of the study. Identification concerns the actinides, fission products or activated products as isotopes or elements, but also their speciation that may not only be understood at the molecular scale but also in a broader sense such as at the environmental scale. The timescale passes from the fs accessible during FEL examination, e.g. of ultra-fast phenomena, through the ns, the ms, the ks, the Gs (31.5 Gs = 1000 a), the Ps with finally 31.5 Ps or 1 Ga the order of timescale of uranium-238 half-life or of the fossil natural geo-reactors. The explored energy range goes from the neV (Mössbauer spectroscopy), the meV (Infra-Red spectroscopy), the keV (e.g. gamma spectroscopy) to the GeV (e.g. muon tomography), for example.

Passive and active analytical methods have been reviewed with examples of their utilization in transmission, injection, diffusion or reflection modes. The sampling area, beam size and reagent quantities are either macroscopic, microscopic or nanoscopic in size, while spatial-temporal conditions make excitation versus detection direction through solid angle, with synchronous detection or with temporal delay.

The investigated analytical techniques have been classified according to their interaction (if any), between incident particles or injected reagents and the analyzed sample, and, for the signal detected or recorded. For passive techniques, excitations are absent and phonons, photons, leptons, neutrons or ions are detected or quantified for their energy, flux, chemical affinity, quantity or mass. For interactive techniques, irradiations or reagent additions are made with phonons, photons, leptons, neutrons or ions with a known energy, flux, chemical affinity or mass. The irradiation or injection is done locally while the reception may be carried out in a given space at a given angle from the stimuli direction or the incident beam, instantaneously or after a certain delay after irradiation.

The detection tools are spectroscopy (S), microscopy (M) or radiography (RAD) or tomography (TOM) instruments. The reaction takes place within or without a specific field such as electrical, magnetic, flow or mechanical acceleration. The detected signal may be the same in nature as the incident one, with the same energy (elastic interaction), or a signal with lower energy (inelastic interaction), with particles being again phonons, photons, leptons, neutrons or ions. In addition to these analytical tools or techniques, neutral species such as atoms or molecules may also be used to interrogate the material. They are treated as ions (from a mass point of view). The techniques are classified according to increasing energy of reagents or incident particles. The combination of all excitation or reagent addition, and reception or product detections make the analytic potential very rich to perform the identification of elements or isotopes, quantitative determination, and spatial speciation in a broad way.

There has been an optimization of technique and the discovery of new analytical tools over recent decades. Some of the techniques that are found today are obsolete while others reemerge due to new phase interest e.g. molten salts; some may be completed by combining the potential of one technique with another. There has been a constant challenge in pushing the use of the analytical techniques toward lower DL, better lateral and depth resolution, more extreme applications and more flexible uses.

Several of the advanced analytical techniques described in this work are very powerful methods for the characterization of nuclear materials or for determination of dopants, defects, fission products and/or actinides in nuclear materials or their environments. Better standards and reference materials are required as well as intercomparison exercises. For some techniques, standards are occasionally unavailable, for example: XRF (photonic), EMPA (electronic), LAICPMS (mixed), SIMS (ionic) and some surface analyses. Other techniques—such as AMS (ionic) and EELS (electronic) under vacuum and XAS (photonic) in solution—require internal calibrations; the sensitivity of the latter is expected to be improved in the future. However, the techniques presented are often complementary, and in many cases they provide unique information that cannot be obtained with alternative methods. Consequently, their continuous and expanded utilization is required for dopants or contaminants identification and speciation in nuclear materials and their environments. Development of analytical techniques such as Mössbauer absorption spectroscopy (MOAS) and electron paramagnetic resonance (EPR), which is more sensitive than nuclear magnetic resonance (NMR) for actinide identification and speciation (when possible), may be suggested in future research programs.

As far as the nuclear materials are concerned, studies must reflect the demanding environments (temperature, pressure and irradiation) under which they are used. These materials act as barriers and their properties are investigated with emphasis on mechanical performances, durability, plasticity and stability when damaged or loaded by dopants or contaminants. These materials range from fuels for thermal or fast reactors, to structural materials. Fuels are analyzed prior and after irradiation, after their reprocessing for recycling and later as waste forms. Macro-properties such as thermodynamic, thermophysical and mechanical as well as microstructural analysis of these materials have to be analyzed, for example comparing again properties before and after irradiation.

The following issues may be pointed out:

- Actinides are α emitters, and analysis by alpha spectroscopy (αS) or liquid scintillation counting (LSC) allows their detection down to ~10 μBq. αS requires more experimental efforts but permits higher resolution and more sensitive analysis (down to 10 μBq).
- Alternative to αS may be neutron activation analysis (NAA), for which limitations of low counting rates should be improved by activation. NAA has been found to be prone to interference, e.g. from uranium nuclides.
- Today, advanced inductively coupled plasma mass spectrometry (ICPMS) techniques are used instead of αS because ICPMS uses less material and is a less time-consuming technique. ICPMS may however require the use of a separation technique, which may be performed on-line.
- The following has been stated for contaminants e.g. actinides (An) analysis:

  ICPMS is recommended to be used for An with half-life >1000 years, and αS is recommended to be used for an with half-life <1000 years.

- Better standards, reference materials and benchmarks are required. For some techniques, standards are scarce, e.g. for XRF (photonic), EMPA (mixed), LAICPMS (mixed) and SIMS (ionic).
- RBS, AMS and EELS (under vacuum) and XAS (in solution), require internal calibrations, the sensitivity of the latter is expected to be improved in the future.

Advanced analytical techniques provide data on nuclear materials.

- Layers were characterized combining μXRF, μXRD, μXAFS, DRS, RBS …
- Layer species were determined by VIAXAFS and μXAFS, TEM/EELS …
- Defects and vacancies in Zircaloy have been identified by MuSR, XAFS, STXM, TEM…
- Fuel porosity and defects are characterized (by SANS, ToM, XAFS, NR …)
- Fuel may also be characterized by combining (μXRF, μXRD and μXAFS)
- The combination of these techniques provides complete pictures.

Combinations of techniques have also been discussed. For example, Seismic Reflection (SR), which cannot be used by it for identifying thorium or uranium, can be used in combination with other techniques as a prospecting tool. AFM morphological studies also provide useful information; however, they must be complemented with other technique results to provide the required identification result. Similarly, *Eh* electrode (EHE) measurements may contribute to the speciation of redox-sensitive actinides such as U, Np or Pu in waters. They are however generally completed by spectroscopic investigations. Chromatography, which is basically a separation technique, must be combined with detectors and has been also treated here.

As far as the environments of nuclear materials are concerned, one has to think about how the analyst and the environmental scientist would collaborate to produce data that can be used by modelers or authorities. The challenge is however to understand the behavior of actinide elements, fission products and other contaminants in the environment. Biogeochemical pathways have to be described, quantified and understood. The contaminated systems interact with the local environment that may modify actinide speciation by physical-chemical processes, which need to be studied. Simple processes such as actinide association with biofilm, bio-accumulation, specific co-precipitation, or sorption onto soil or rock require detailed and full quantification and speciation for understanding of process mechanisms. Transport of actinides, fission products and other contaminants in fluids such as air or water includes particulate or colloidal phases. These analyses must be integrated in the analytical strategy as specific species for modeling their biogeochemical affinity.

Data are provided by the analysts for the scientists and the modelers. The challenge is to understand, for example, the behaviour of actinide elements in the systems. In the environment, contaminant pathways have to be described. The contaminated systems interact with the local environment that may modify actinide speciation by physical-chemical processes. The analytical results must be integrated in the study for modeling their chemical affinity.

The challenge will be to find and develop direct analytical probes for full nuclear material characterization at very low defect/dopant/contaminant concentration to better characterize the damages/species/structures and to predict their behavior in homogeneous, heterogeneous or complex nuclear materials and their environments.

Index

C.A. Degueldre, *The Analysis of Nuclear Materials and Their Environments*,
DOI 10.1007/978-3-319-58006-7

9783319580043